T0240113

Springer Series in Chemical Physics

Volume 126

Series Editors

Jan Peter Toennies, Max Planck Institut für Dynamic und Selbstorganisation, Göttingen, Germany

Kaoru Yamanouchi, Department of Chemistry, University of Tokyo, Tokyo, Japan

The Springer Series in Chemical Physics consists of research monographs in basic and applied chemical physics and related analytical methods. The volumes of this series are written by leading researchers of their fields and communicate in a comprehensive way both the basics and cutting-edge new developments. This series aims to serve all research scientists, engineers and graduate students who seek up-to-date reference books.

M. A. Wahab

Symmetry Representations of Molecular Vibrations

 Springer

M. A. Wahab
Department of Physics
Jamia Millia Islamia
New Delhi, India

ISSN 0172-6218   ISSN 2364-9003 (electronic)
Springer Series in Chemical Physics
ISBN 978-981-19-2804-8   ISBN 978-981-19-2802-4   (eBook)
https://doi.org/10.1007/978-981-19-2802-4

This Springer imprint is published by the registered company Springer Nature Singapore Pte Ltd.
The registered company address is: 152 Beach Road, #21-01/04 Gateway East, Singapore 189721, Singapore

Preface

This book presents a comprehensive theoretical basis of symmetries, matrix representation of symmetries and the elements of group theory that are relevant to other symmetry elements/operations, crystallographic and molecular point groups, and the manner in which they are used to understand the reducible and irreducible representations of symmetry matrices and then able to derive the normal modes of vibration of different molecules by using suitable techniques independently with the help of the acquired knowledge. The determination of normal modes of vibration on a theoretical basis for a given molecule, in turn, not only provides an opportunity to learn the fundamental aspects of physics involved in the process but also allows us to compare and verify the same with the experimental results obtained from infrared (IR) and Raman techniques.

Research workers in the field of materials of physics, chemistry, biology, engineering, etc., invariably study their samples by using IR, Raman, Fourier transform infrared (FT-IR), and Fourier transform Raman (FT-Raman) techniques to confirm the presence of some specific functional groups by analyzing the types of vibrational modes contained in the resulting spectra. Getting these spectra recorded in any research laboratory in the field is an easy task. Also, because of the easily available correlation tables, the analysis of a given spectrum has now become just a routine matter. As a result of these easily available facilities, the researchers pay hardly any interest (in general) in understanding the actual theoretical details involved in the process which they must know even otherwise. Another important reason may be that one should have a comprehensive knowledge of symmetries, elements of group theory, matrix representation of symmetries, and a clear concept of reducible and irreducible representations of symmetry matrices to obtain the vibrational modes of a given molecule, which are by the way not very easy. I am able to say this because of the constant observation of the difficulties faced by the research scholars (including my own research students) of the university working in the field of material characterizations.

Keeping in view the difficult nature of the problem and the need of the students (research workers in general), I took up the challenge to write this book. I decided to help the graduate and, particularly, the research students who have been using the

IR and Raman-related characterizations to study their samples. I am hopeful that after going through this book, all those working in the field can easily understand the subject and be able to derive the normal modes of vibration for their molecular samples themselves. In the process, they will also learn various forms of symmetries and different ways of their representations. Keeping the students in view, the presentation has been kept systematic and simple. Further, to make the subject more understandable, a number of molecular cases have been taken as examples and have been solved, wherever felt necessary. Therefore, for those who wish to go no further, it is hoped that this book will provide them with a broad treatment, complete in itself, which explains the principles involved and adequately describes the present state of the subject concerned. However, for those who wish to go further, this book should be regarded as a foundation for further study.

This book is organized into five chapters. Chapter 1 deals with the molecular and crystal symmetries, similarities between them and their differences, matrix representations of symmetry operations, molecular and crystallographic point groups, determination of molecular point groups, and the point group notations for a comprehensive understanding of the subjects in the subsequent chapters. Chapter 2 deals with the elements of group theory, types of groups, point groups and classes, and construction of group multiplication tables for all 32 point groups in both international and Schoenflies notations. Chapter 3 deals with the great orthogonality theorem and its various properties, reducible and irreducible representations, construction of character tables, Mulliken symbols, and transformation properties. This chapter provides a comprehensive understanding of the character representation of symmetry elements/operations and the derivation of the character table for a given molecule belonging to a particular point group. Chapter 4 deals with the motions of diatomic and polyatomic molecules, the relationship between reducible and irreducible representations, the character of matrices of fundamental symmetry operations for nonlinear and linear molecules, determination of overall reducible representation of nonlinear molecules by following two different methods, representations of vibrational modes of nonlinear molecules by using two different methods, and determination of vibrational modes in some nonlinear and linear molecules. In this way, Chaps. 1–4 deal with the fundamental aspects that are needed to understand the subject and then able to derive the normal modes of vibration of a given molecule, theoretically. On the other hand, Chap. 5 (the final chapter) deals with the experimental aspects of IR and Raman-related techniques to obtain vibrational spectra of the given molecular sample. The addition of this chapter makes it possible to compare the experimental results with the theoretically calculated modes of vibration for a given molecule.

This book will serve as a reference book in the field mentioned above and will be useful for both the students as well as faculties. Further, this book can serve as a complementary or supplementary to a textbook for graduate students in the areas mentioned above. As mentioned above, this book will be extremely useful to all those who have been using the IR and Raman-related characterization techniques to study their samples. The most special feature of this book is that it contains the group multiplication tables for all 32 crystallographic point groups at one place, introduced for the first time in any literature, using both international and Schoenflies notations.

The salient features of the book are as follows:

1. It deals with the basic concepts of symmetries and elements of group theory required to understand the construction of a group multiplication table.
2. For the first time in the crystallographic history, the group multiplication tables of all 32 point groups in both international and Schoenflies notations are presented in this book at one place.
3. It explains the matrix and character representations of symmetries.
4. It provides a systematic procedure to construct the character tables needed for a molecule belonging to a particular point group.
5. It provides the procedure to derive the normal modes of vibration of both nonlinear and linear molecules in simplified ways.
6. It provides the means and ways to compare the theoretical results obtained in terms of the modes of vibration with the experimental results obtained from the spectra of IR and Raman-related techniques for different molecules.

Although proper care has been taken during the preparation of the manuscript, still some errors are expected to creep in. Any form of omissions, errors, or suggestions brought to the knowledge of the author will be appreciated and thankfully acknowledged.

I sincerely admit and acknowledge that I have learnt the art of systematic presentation of the research work or other related scientific matters from my supervisor, Prof. G. C. Trigunayat (late), Department of Physics, University of Delhi, India. I also sincerely acknowledge the work of my two sons Mr. Khurram Mujtaba Wahab and Mr. Shad Mustafa Wahab for making all the diagrams and formatting the manuscript. My special thanks to other members of my family for their continuous support and encouragement during the preparation of the manuscript.

I am indeed grateful to all the authors and publishers of books and journals mentioned in the bibliography for freely consulting them and even borrowing ideas during the preparation of this manuscript. I am also grateful to Springer Nature for the timely publication of this book.

New Delhi, India M. A. Wahab

The Electromagnetic Spectrum

Frequency (Hz)	Name of radiation	Photon energy (eV)	Wavelength (Å)
10^{22}			10^{-13}
10^{21}	Gamma rays	10^7	10^{-12}
10^{20}		10^6	10^{-11}
10^{19}		10^5	10^{-10}– $1(\text{Å})$
10^{18}	X-rays	10^4	10^{-9} – 1nm
10^{17}		10^3	10^{-8}
10^{16}	Ultraviolet	10^2	10^{-7}
10^{15}		10	10^{-6} – $1\mu\text{m}$
10^{14}	Visible light	1	10^{-5}
10^{13}	Infrared	10^{-1}	10^{-4}
10^{12}		10^{-2}	10^{-3}
10^{11}		10^{-3}	10^{-2} – 1cm
10^{10}	(UHF)	10^{-4}	10^{-1}
10^9	Shortwave	10^{-5}	1 m
10^8		10^{-6}	10
10^7		10^{-7}	10^2
1 MHz–10^6		10^{-8}	10^3 – 1km
10^5	(LF)	10^{-9}	10^4
10^4	Long wave	10^{-10}	10^5
1 kHz–10^3	(VLF)	10^{-11}	10^6

Contents

About the Author

M. A. Wahab is Former Professor and Head of the Department of Physics, Jamia Millia Islamia, New Delhi, India. He completed his Ph.D. (Physics) from the University of Delhi, India, and M.Sc. (Physics) from Aligarh Muslim University, India. Earlier, he served as a lecturer at the P. G. Department of Physics and Electronics, University of Jammu, Jammu and Kashmir, India, from 1981, and later, at the P. G. Department of Physics, Jamia Millia Islamia, from 1985. During these years, he taught electrodynamics, statistical mechanics, theory of relativity, advance solid state physics, crystallography, physics of materials, growth and imperfections of materials, and general solid state physics. He has authored 4 books: *Solid State Physics, Essentials of Crystallography*, and *Numerical Problems in Solid State Physics*, and *Numerical problems in Crystallography* (from Springer Nature). He has also published over 100 research papers in national and international journals of repute and supervised 15 Ph.D. theses during his career at Jamia Millia Islamia. Professor Wahab has published the discovery of hexagonal close packing (HCP) and rhombohedral close packing (RCP) as the two new space lattices, along with his son (Mr. Khurram Mujtaba Wahab), as their first joint paper after his retirement.

Chapter 1
Molecular and Crystal Symmetries

1.1 Introduction

In this chapter, the topics have been selected such that their knowledge will form the basis to understand the contents of the subsequent chapters. In the beginning of this chapter, we are going to study the fundamental aspects of molecular symmetries and the external symmetry elements/operations that are possible in crystalline solids.

A brief description of the types and properties of matrices has been made to develop the concept of matrix representation of symmetry operations in general. Molecular point groups including their classification, methods of determination, and comparison with the crystallographic point groups have been discussed. A new method of understanding the distribution of symmetry operations among the crystallographic point groups from the least symmetric system to the most symmetric system has been discussed.

At the end of this chapter, we are going to discuss two important point group notations. However, our emphasis will be on Schoenflies notation as it is widely used in the topics that are taken up in the subsequent chapters.

1.2 Symmetry Elements

A symmetry element is defined as a geometrical entity like a point, a line, or a plane in a body (object) about which an appropriate symmetry operation can be performed to transform it to self-coincidence.

A three-dimensional periodic pattern (a crystal) can have the following symmetry elements:

(i) **Translation Vector**: $T = n_1 \vec{a} + n_2 \vec{b} + n_3 \vec{c}$

where n_1, n_2, n_3 are integers and $\vec{a}, \vec{b}, \vec{c}$, are primitive lattice translations, also called the basis vectors (Fig. 1.1).

© The Author(s), under exclusive license to Springer Nature Singapore Pte Ltd. 2022
M. A. Wahab, *Symmetry Representations of Molecular Vibrations*, Springer Series
in Chemical Physics 126, https://doi.org/10.1007/978-981-19-2802-4_1

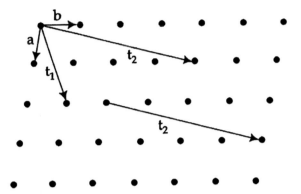

Fig. 1.1   Translation vectors, t_1 and t_2

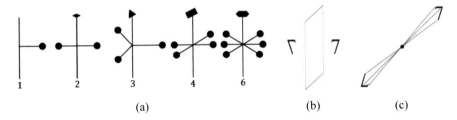

Fig. 1.2   a Five proper rotation axes **b** reflection **c** inversion of an object

(ii)   **Proper Rotation**: Through an angle, $\alpha = \frac{2\pi}{n}$

where $n = 1, 2, 3, 4,$ and 6. These rotation axes are termed monad, diad, triad, tetrad, and hexad, respectively (Fig. 1.2a).

(iii)   **Reflection**: Across a line (in 2D) or a plane (in 3D), m.

A mirror reflection (a line in 2D or a plane in 3D) has the property to transform a left-handed object into a right-handed object and vice-versa (Fig. 1.2b).

(iv)   **Inversion**: Through a point, $\bar{1}$.

An inversion is equivalent to a reflection through a point (instead of a line or a plane), called the inversion center or center of symmetry (Fig. 1.2c).

1.3   Symmetry Operations

A symmetry operation is an operation such as translation, rotation, reflection, inversion, or their compatible combinations that can be performed on a regular object (a crystal) to move it to self-coincidence.

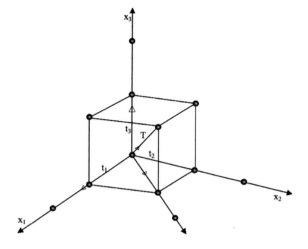

Fig. 1.3 Basis vectors and translation vector in a simple cubic crystal

1. **Translation Operation**: A translation operation performed on an atom in a 3-D system will displace it by unit translation t_1, t_2 or t_3 along x_1, x_2, or x_3 direction. For example, in a simple cubic crystal system (Fig. 1.3), a displacement T along the body diagonal is obtained by the translation operations t_1, t_2, and t_3 as

$$\vec{T} = \vec{t_1} + \vec{t_2} + \vec{t_3}$$

2. **Proper Rotation Operations**: A rotation operation has the property to rotate an object/crystal through an angle, $\alpha = \frac{2\pi}{n}$, and n such rotations will bring back the object/crystal into its initial position. For example, different numbers of rotation operations performed on a parallelogram and an equilateral triangle to bring them to a state of self coincidence are shown in Fig. 1.4.

In Sheonflies notation, the rotation operation is designated by the symbol C_n. Rotation operation has the following characteristic features.

(a) If a molecule/crystal has rotational symmetry C_n, then the rotation by $2\pi/n = 360°/n$ brings the object into an equivalent position.
(b) The value of n is the order of an n-fold rotation.
(c) If the molecule/crystal has more than one rotation axes, then the one with the highest value of n is the principal axis of rotation.
(d) Each n-fold rotation has $n-1$ operations associated with it (excluding $C_n^n = E$).
(e) There are certain important general relationships for C_n. They are as follows:

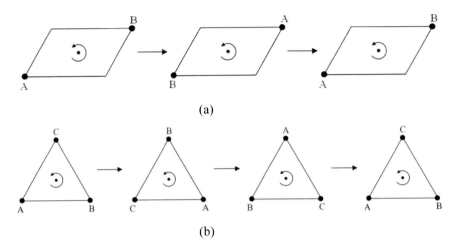

(a)

(b)

Fig. 1.4 a Two successive rotations of π bring the parallelogram to self-coincidence. **b** Three successive rotations of $\frac{2\pi}{3}$ bring the triangle to self-coincidence

$C_n{}^n = E$	–
$C_{2n}{}^n = C_2$	$(n = 2, 4, 6, 8\ldots\text{etc.})$
$C_n{}^m = C_{n/m}$	$(n/m = 2, 3, 4, 5\ldots\text{etc.})$
$C_n{}^{n+m} = C_n{}^m$	$(m < n)$

3. **Reflection/Mirror Operation**: A reflection operation has the property to change the character of the motif. It will change a right-handed object into a left-handed and vice–versa (Fig. 1.2b). Mirror coincident figures of this type are known as enantiomorphous pairs of objects. Mirror operation has the following other characteristic features:

(a) The mirror operation defines bilateral symmetry about its own plane (called mirror plane or reflection plane).

(b) For every point at a distance r along the normal to the mirror plane there exists an equivalent point at a distance −r.

(c) For a point (x, y, z), reflection across a mirror plane σ_{xz} takes the point into $(x, -y, z)$, similarly others.

(d) A mirror plane has only one operation associated with it, since $\sigma^2 = E$.

Different Orientations of Mirror Planes

Mirror planes can acquire three different orientations in crystals and molecular systems. They are horizontal, vertical, and dihedral and symbolically represented as σ_h, σ_v, and σ_d, respectively. They can be defined as follows:

(a) A horizontal mirror plane σ_h is defined as perpendicular to the principal axis of rotation.

(b) Both σ_v and σ_d, mirror planes are defined such that they contain a principal axis of rotation and are perpendicular to σ_h mirror plane.

(c) When both σ_v and σ_d mirror planes occur in the same system, they are differentiated by defining σ_v mirror plane to contain the greater number of atoms or to contain a principal axis of a reference Cartesian coordinate system (x or y axis).

(d) Any σ_d mirror planes typically will contain bond angle bisectors.

(e) Taking the example of a square planar MX_4 molecule, the five mirror plane symmetries are shown in the form of three classes as σ_h, $2\sigma_v$, and $2\sigma_d$ in Fig. 1.5.

4. **Inversion Operation**: Like reflection, an inversion operation also has the property to change the character of the motif. However, the basic difference in the two is that a reflection occurs through a plane of the mirror, while an inversion is equivalent to a reflection through a point (Fig. 1.2c). In fact, the center of inversion has the property of inverting all space (at present the number 7) through a point. Like a mirror, inversion operation has the following other characteristic features:

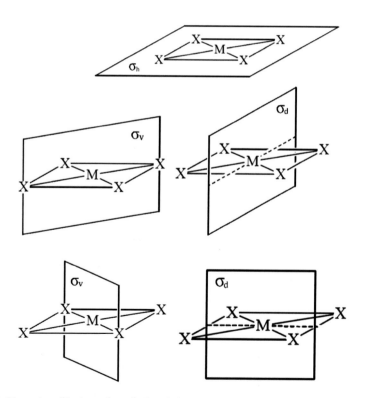

Fig. 1.5 Examples of horizontal, vertical, and dihedral mirror planes

(a) If there exists an inversion symmetry, then for every point (x, y, z) there exists an equivalent point $(-x, -y, -z)$.

(b) Atoms or molecules that have inversion symmetry are said to be centrosymmetric.

(c) Each inversion symmetry has only one operation associated with it because $i^2 = E$.

5. **Improper Rotation Operations**: We know that reflection and inversion operations produce enantiomorphous sets of objects in their own ways. However, it is possible to have some rotational operations relating to such enantiomorphous objects. These operations are called improper operations. Based on reflection and inversion operations, there exist two improper rotations; they are termed rotoreflection and rotoinversion operations, respectively.

Rotoreflection Operation

The term rotoreflection represents a combined operation of rotation and reflection (where the mirror plane is perpendicular to the axis of rotation). This combined operation repeats an initially (say) right-handed object into a left-handed and vice–versa. This produces a series of right-handed and left-handed objects such that the neighboring objects are always enantiomorphous, while the alternate objects are always identical. Therefore, corresponding to each of the five proper rotation axes, there exists a rotoreflection axis. They are symbolically represented as $\tilde{1}$ (read as 'one tilde'), $\tilde{2}$, $\tilde{3}$, etc. Three-dimensional views of $\tilde{1}$ and $\tilde{2}$ are shown in Fig. 1.6.

Rotoinversion Operation

Like rotoreflection, the term rotoinversion also represents a combined operation of rotation and inversion. Consequently, there exists a rotoinversion axis corresponding to each of the five rotation axis (Fig. 1.7). Rotoinversion axes are symbolically represented as $\bar{1}$ (read as 'one bar'), $\bar{2}$, $\bar{3}$, etc.

Figure 1.8 illustrates the plane projections of proper and improper rotation axes.

When comparing the five rotoreflection axes with five rotoinversion axes, they are found to be equivalent in pairs. That is,

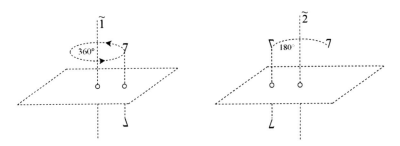

Fig. 1.6 Three-dimensional views of $\tilde{1}$ and $\tilde{2}$

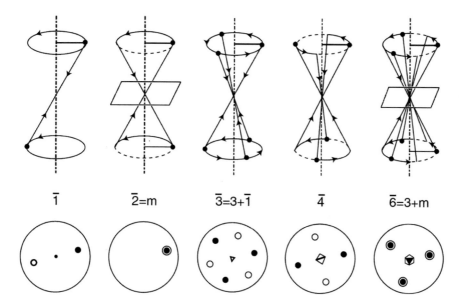

Fig. 1.7 Five rotoinversion axes $\bar{n}$

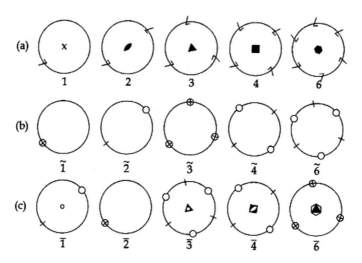

Fig. 1.8 Showing projections of **a** proper rotation axes, **b** rotoreflection axes, and **c** rotoinversion axes

$$\tilde{1} = \frac{1}{m} = m = \bar{2},\ \tilde{2} = \bar{1},\ \tilde{3} = \frac{3}{m} = \bar{6},\ \tilde{4} = \bar{4}\ and\ \tilde{6} = \bar{3}$$

On the basis of such observations, the following empirical rules can be formulated which in turn can be used for conversion from one system to another, i.e., from rotoreflection to rotoinversion and vice–versa. They are as follows:

$$\tilde{n}_{odd} = \overline{2n} = \frac{n}{m} = n\sigma_h \text{ (mirror perpendicular to the axis)}$$

$$\widetilde{2n} = \tilde{n}_{odd} = ni$$

$$and \ \widetilde{4n} = \overline{4n}$$

where $n = 1, 2, 3, 4,$ and 6, m is a mirror plane, and i is the center of inversion, respectively. The international/conventional symbol used to represent these axes is provided below in Table 1.1.

Simple relationships between a symmetry element and its corresponding symmetry operation are provided in Table 1.2.

Example 1: Determine the Schoenflies equivalents of the following rotoinversion symmetry axes: $\overline{2}, \overline{3}, \overline{4},$ and $\overline{6}$.

Solution: *Given*: The rotoinversion symmetry axes are $\overline{2}, \overline{3}, \overline{4},$ and $\overline{6}$. Let us determine their equivalent in Schoenflies notation, one by one.

(a)   The symbol $\overline{2}$ indicates that the crystal has rotoinversion axis of order 2. Therefore,

$$\overline{2} = C_2^1 . i = C_2^1 . C_2^1 . \sigma_h = E \ \sigma_h = \sigma \text{ (mirror plane)}$$

Table 1.1 Conventional symbol for improper axes

Rotoinversion axes	Rotoreflection axes	Conventional symbol	International symbol
$\overline{1}$	$\tilde{2}$	Center of symmetry	$\overline{1}$
$\overline{2}$	$\tilde{1}$	Mirror plane	m
$\overline{3}$	$\tilde{6}$	threefold rotoinversion	$\overline{3}$
$\overline{4}$	$\tilde{4}$	Fourfold rotoinversion	$\overline{4}$
$\overline{6}$	$\tilde{3}$	Sixfold rotoinversion	$\overline{6}$

Table 1.2 Relationships between symmetry elements and symmetry operations

Symmetry element	Symmetry operation
Proper rotation axis C_n (n)	One or more rotations through an angle $\alpha = \frac{2\pi}{n}$ about the axis
Plane of symmetry σ (m)	One or more reflections in the plane
Improper rotation axis S_n ($\tilde{n}$)	Rotation followed by reflection in a plane perpendicular to the rotation axis
Center of symmetry i ($\overline{1}$)	Inversion of space through the center of symmetry
Identity element E (1)	This operation leaves the object unchanged

where $C_2^1 . C_2^1 = E$ is the identity element.

(b) The symbol $\bar{3}$ indicates that the crystal has rotoinversion axis of order 3. Therefore,

$$\bar{3} = C_3^1 . i = C_3^1 . C_2^1 . \sigma_h = C_6^2 . C_6^3 . \sigma_h = C_6^5 . \sigma_h = S_6^5$$

This is equivalent to the rotoreflection axis of order 6.

(c) The symbol $\bar{4}$ indicates that the crystal has rotoinversion axis of order 4. Therefore,

$$\bar{4} = C_4^1 . i = C_4^1 . C_2^1 . \sigma_h = C_4^1 . C_4^2 . \sigma_h = C_4^3 . \sigma_h = S_4^3$$

This is equivalent to the rotoreflection axis of order 4.

(d) The symbol $\bar{6}$ indicates that the crystal has rotoinversion axis of order 6. Therefore,

$$\bar{6} = C_6^1 . i = C_6^1 . C_2^1 . \sigma_h = C_6^1 . C_6^3 . \sigma_h = C_3^2 . \sigma_h$$
$$= C_3^3 . C_3^2 . \sigma_h = C_3^5 . \sigma_h = S_3^5$$

This is equivalent to the rotoreflection axis of order 3.

1.4 Molecular Symmetries

A molecule is defined as the smallest portion of a substance capable to exist independently and retains the properties of the original substance. Further, a molecule may have a few atoms (called a small molecule) or a large number of atoms (called a large molecule).

It is true that a vast majority of molecules when considered in their entirety do not possess any symmetry. However, many molecules (particularly the smaller ones) do possess real symmetries. Depending on the number of symmetry elements/symmetry operations present in molecules, they can be classified as low symmetry molecules or high symmetry molecules. Unlike crystallographic symmetries (which we know are limited) the molecular symmetries are unlimited. Table 1.3 provides a list of symmetry properties to distinguish the molecular symmetries from the crystallographic symmetries.

Table 1.3 Distinction between molecular and crystallographic symmetries

S. no	Molecules	Crystals
1	C_n, axis of order n (where $n = 1, 2,..., \infty$) is present	C_n, axis of order n (where $n = 1, 2, 3, 4$ and 6) is present
2	Total number of point groups is not limited	Total number of point groups is limited to 32
3	Roto-reflection operation is considered	Roto-inversion operation is considered
4	Translational symmetry is absent or Space group is absent	Translational symmetries (such as screw axes, glide planes) are assumed to be present or Space group is assumed to be present
5	Schoenflies notation is used	Hermann−Mauguin (International) notation is (generally) used

1.5   Elements of Matrices

Matrix

A matrix is an array of numbers or symbols for numbers. The order or dimension of a matrix is defined by its number of rows and columns. Thus a matrix which has m rows and n columns has the dimension $m \times n$ (pronounced as m by n). For example, a rectangular matrix of order $m \times n$ is defined as follows:

$$A = \begin{pmatrix} a_{11} & a_{12} & \cdots & a_{1n} \\ a_{21} & a_{22} & \cdots & a_{2n} \\ \cdots & \cdots & \cdots & \cdots \\ a_{m1} & a_{m2} & \cdots & a_{mn} \end{pmatrix} \tag{1.1}$$

where in each element a_{ij}, i denotes the rows (horizontal sets) and j denotes the columns (vertical sets). A special case of this is a square matrix where the number of rows is equal to the number of columns. We will consider only the square matrix in our further discussion.

Diagonal Matrix

A square matrix is called the diagonal matrix if the elements of the matrix

$$a_{ij} \neq 0 \; for \; i = j$$

$$a_{ij} = 0 \; for \; i \neq j$$

For example, a diagonal matrix of order n is

$$A = \begin{pmatrix} a_{11} & \cdots & \cdots & \cdots \\ \cdots & a_{22} & \cdots & \cdots \\ \cdots & \cdots & \cdots & \cdots \\ \cdots & \cdots & \cdots & a_{nn} \end{pmatrix} \qquad (1.2)$$

Unit/Identity Matrix

The matrix A is called a unit matrix if the elements

$$a_{ij} = 0 \; for \; i \neq j$$

$$\ldots = 1 \; for \; i = j$$

Thus a unit/identity matrix of the order 3 is

$$I = \begin{pmatrix} 1 & 0 & 0 \\ 0 & 1 & 0 \\ 0 & 0 & 1 \end{pmatrix} \qquad (1.3)$$

Unitary Matrix

A unitary matrix is a matrix A such that $A^{-1} = A^{1}$, whose elements are real and the value of its determinant is ± 1.

Singular Matrix

A matrix is a singular matrix whose value for its determinant is zero. For example, the matrix

$$A = \begin{pmatrix} 2 & 4 \\ 4 & 8 \end{pmatrix} \qquad (1.4)$$

is a singular matrix of order 2.

Adjoint of a Square Matrix

Adjoint of a square matrix is the transpose of the matrix whose elements are the cofactors of elements of the given matrix, i.e.,

$$Adj(A) = C_{ji} \qquad (1.5)$$

$$\text{where the cofactor } C_{ij} = (-1)^{i+j} M_{ij}(A) \qquad (1.6)$$

In Eq. 1.6, M_{ij} are called the minors of matrix A.

Inverse of a Matrix

The inverse of a matrix is conceptually the same as the reciprocal of a number, such that if we multiply the given matrix by its inverse, the result will be an identity matrix. Therefore, if a given square matrix has a nonzero determinant, then that matrix will have a 'unique inverse'. The inverse of a matrix A is generally denoted as A^{-1} and not by $1/A$.

Therefore, for every non-singular square matrix A, there exists an inverse matrix A^{-1}, such that $A\,A^{-1} = I$, where I is a unit or identity matrix. From Eq. 1.1, we can consider a 3×3 square matrix given by

$$A = \begin{pmatrix} a_{11} & a_{12} & a_{13} \\ a_{21} & a_{22} & a_{23} \\ a_{31} & a_{32} & a_{33} \end{pmatrix}$$

Let the given square matrix be a non-singular matrix of order 3×3. This implies that $\Delta = |A| \neq 0$, which means A^{-1} exists. Its inverse can be determined by using the formula,

$$A^{-1} = \frac{Adj(A)}{|A|} \tag{1.7}$$

For this purpose, we should adopt the following procedure:

1. Check the singularity of the given square matrix.
2. Calculate the cofactors using Eq. 1.6.
3. Determine the adjoint of the resulting matrix.
4. Divide the adjoint by the determinant of the original matrix.

Let A_{ij} be the cofactor of A in $|A|$, then using Eq. 1.6, we can easily find

$$A_{11} = (-1)^{1+1} \begin{vmatrix} a_{22} & a_{23} \\ a_{32} & a_{33} \end{vmatrix}, \ A_{12} = (-1)^{1+2} \begin{vmatrix} a_{21} & a_{23} \\ a_{31} & a_{33} \end{vmatrix},$$

$$A_{13} = (-1)^{1+3} \begin{vmatrix} a_{21} & a_{22} \\ a_{31} & a_{32} \end{vmatrix}$$

$$A_{21} = (-1)^{2+1} \begin{vmatrix} a_{12} & a_{13} \\ a_{32} & a_{33} \end{vmatrix}, \ A_{22} = (-1)^{2+2} \begin{vmatrix} a_{11} & a_{13} \\ a_{31} & a_{33} \end{vmatrix},$$

$$A_{23} = (-1)^{2+3} \begin{vmatrix} a_{11} & a_{12} \\ a_{31} & a_{32} \end{vmatrix}$$

$$A_{31} = (-1)^{3+1} \begin{vmatrix} a_{12} & a_{13} \\ a_{22} & a_{23} \end{vmatrix}, \ A_{32} = (-1)^{3+2} \begin{vmatrix} a_{11} & a_{13} \\ a_{21} & a_{23} \end{vmatrix},$$

$$A_{33} = (-1)^{3+3} \begin{vmatrix} a_{11} & a_{12} \\ a_{21} & a_{22} \end{vmatrix}$$

The matrix of the cofactors A_{ij} is given by

$$A_{ij} = \begin{pmatrix} A_{11} & A_{12} & A_{13} \\ A_{21} & A_{22} & A_{23} \\ A_{31} & A_{32} & A_{33} \end{pmatrix}$$

Transpose of the matrix A_{ij} is Adj(A)

$$Adj(A) = C_{ji} = \begin{pmatrix} A_{11} & A_{21} & A_{31} \\ A_{12} & A_{22} & A_{32} \\ A_{13} & A_{23} & A_{33} \end{pmatrix}$$

Hence, the inverse is obtained as

$$A^{-1} = \frac{Adj(A)}{|A|} = \frac{1}{|A|} \begin{pmatrix} A_{11} & A_{21} & A_{31} \\ A_{12} & A_{22} & A_{32} \\ A_{13} & A_{23} & A_{33} \end{pmatrix}$$

Some Important Points on Inverse of a Matrix

The following points are helpful to understand more clearly the idea of the inverse of a matrix.

1. If the inverse of a square matrix exists, it is unique.
2. If A and B are two invertible matrices of the same order, then

$$(AB)^{-1} = B^{-1}A^{-1}$$

3. The inverse of a square matrix A exists only if its determinant is a nonzero value, $|A| \neq 0$.
4. If a square matrix A has an inverse A^{-1}, then

$$AA^{-1} = A^{-1}A = I \tag{1.8}$$

Example 2: Find the inverse of a 2 × 2 square matrix $A = \begin{pmatrix} 3 & 4 \\ 1 & 2 \end{pmatrix}$.

Solution: *Given*: the matrix $A = \begin{pmatrix} 3 & 4 \\ 1 & 2 \end{pmatrix}$. Following the steps mentioned above, let us determine the inverse of the given matrix.

First of let us find the determinant of the given matrix

$$|A| = \begin{vmatrix} 3 & 4 \\ 1 & 2 \end{vmatrix} = 6 - 4 = 2.$$

$\Rightarrow A^{-1}$ exists.

Now, we determine the cofactors of all elements,

$$A_{11} = (-1)^{1+1}.2 = 2, A_{12} = (-1)^{1+2}.1 = -1,$$
$$A_{21} = (-1)^{2+1}.4 = -4, A_{22} = (-1)^{2+2}.3 = 3.$$

Next, to find the adjoint matrix of A using Eq. 1.5, we obtain

$$\text{Adj}(A) = \begin{pmatrix} A_{11} & A_{21} \\ A_{12} & A_{22} \end{pmatrix} = \begin{pmatrix} 2 & -4 \\ -1 & 3 \end{pmatrix}$$

Finally, the inverse of A is

$$A^{-1} = \frac{\text{Adj}(A)}{|A|} = \frac{1}{2} \begin{pmatrix} 2 & -4 \\ -1 & 3 \end{pmatrix} = \begin{pmatrix} 1 & -2 \\ -1/2 & -3/2 \end{pmatrix}$$

Making use of Eq. 1.8, let us check the correctness of the above result. Therefore, we write

$$\begin{pmatrix} 3 & 4 \\ 1 & 2 \end{pmatrix} \times \frac{1}{2} \begin{pmatrix} 2 & -4 \\ -1 & 3 \end{pmatrix} = \frac{1}{2} \begin{pmatrix} 2 & -4 \\ -1 & 3 \end{pmatrix} \times \begin{pmatrix} 3 & 4 \\ 1 & 2 \end{pmatrix} = \frac{1}{2} \begin{pmatrix} 2 & 0 \\ 0 & 2 \end{pmatrix} = \begin{pmatrix} 1 & 0 \\ 0 & 1 \end{pmatrix}$$

The product is an identity matrix, which confirms that the above result is correct.

Example 3: Find the inverse of the matrix $A = \begin{pmatrix} 0 & 1 & 1 \\ 1 & 0 & 1 \\ 1 & 1 & 0 \end{pmatrix}$ and verify that $A^{-1}A$

$= AA^{-1} = I.$

Solution: *Given*: the matrix $A = \begin{pmatrix} 0 & 1 & 1 \\ 1 & 0 & 1 \\ 1 & 1 & 0 \end{pmatrix}$.

The determinant of the given matrix is

$$|A| = \begin{vmatrix} 0 & 1 & 1 \\ 1 & 0 & 1 \\ 1 & 1 & 0 \end{vmatrix} = 0(0-1) - 1(0-1) + 1(1-0) = 1 + 1 = 2$$

$\Rightarrow A^{-1}$ exists.

Now, let A_{ij} be the cofactors of the matrix elements A, then from Eq. 1.6, we have

$$A_{11} = (-1)^{1+1} \begin{vmatrix} 0 & 1 \\ 1 & 0 \end{vmatrix} = -1, A_{12} = (-1)^{1+2} \begin{vmatrix} 1 & 1 \\ 1 & 0 \end{vmatrix} = 1,$$

$$A_{13} = (-1)^{1+3} \begin{vmatrix} 1 & 0 \\ 1 & 1 \end{vmatrix} = 1$$

$$A_{21} = (-1)^{2+1} \begin{vmatrix} 1 & 1 \\ 1 & 0 \end{vmatrix} = 1, \ A_{22} = (-1)^{2+2} \begin{vmatrix} 0 & 1 \\ 1 & 0 \end{vmatrix} = -1,$$

$$A_{23} = (-1)^{2+3} \begin{vmatrix} 0 & 1 \\ 1 & 1 \end{vmatrix} = 1$$

$$A_{31} = (-1)^{3+1} \begin{vmatrix} 1 & 1 \\ 0 & 1 \end{vmatrix} = 1, \ A_{32} = (-1)^{3+2} \begin{vmatrix} 0 & 1 \\ 1 & 1 \end{vmatrix} = 1,$$

$$A_{33} = (-1)^{3+3} \begin{vmatrix} 0 & 1 \\ 1 & 0 \end{vmatrix} = -1$$

So that the matrix of the cofactors is $A_{ij} = \begin{pmatrix} -1 & 1 & 1 \\ 1 & -1 & 1 \\ 1 & 1 & -1 \end{pmatrix}$

$$\Rightarrow \text{Adj}(A) = C_{ji} = A_{ji} = \begin{pmatrix} -1 & 1 & 1 \\ 1 & -1 & 1 \\ 1 & 1 & -1 \end{pmatrix}$$

Hence, $A^{-1} = \frac{\text{Adj}(A)}{|A|} = \frac{1}{2} \begin{pmatrix} -1 & 1 & 1 \\ 1 & -1 & 1 \\ 1 & 1 & -1 \end{pmatrix}$

Now, $A^{-1}A = \frac{1}{2} \begin{pmatrix} -1 & 1 & 1 \\ 1 & -1 & 1 \\ 1 & 1 & -1 \end{pmatrix} \begin{pmatrix} 0 & 1 & 1 \\ 1 & 0 & 1 \\ 1 & 1 & 0 \end{pmatrix} = \begin{pmatrix} 1 & 0 & 0 \\ 0 & 1 & 0 \\ 0 & 0 & 1 \end{pmatrix} = I$

Similarly, $AA^{-1} = \begin{pmatrix} 0 & 1 & 1 \\ 1 & 0 & 1 \\ 1 & 1 & 0 \end{pmatrix} \begin{pmatrix} -\frac{1}{2} & \frac{1}{2} & \frac{1}{2} \\ \frac{1}{2} & -\frac{1}{2} & \frac{1}{2} \\ \frac{1}{2} & \frac{1}{2} & -\frac{1}{2} \end{pmatrix} = \begin{pmatrix} 1 & 0 & 0 \\ 0 & 1 & 0 \\ 0 & 0 & 1 \end{pmatrix} = \text{I}.$

$\Rightarrow A^{-1}A = AA^{-1} = I.$

Example 4: General form of a rotation matrix through an angle α about any principal axis (say x_3-axis) is given by

$$A = A_{ij} = \begin{pmatrix} \cos \alpha & -\sin \alpha & 0 \\ \sin \alpha & \cos \alpha & 0 \\ 0 & 0 & 1 \end{pmatrix}$$

Find its inverse and show that $A^{-1}A = AA^{-1} = I.$

Solution: *Given*: The matrix $A = \begin{pmatrix} \cos \alpha & -\sin \alpha & 0 \\ \sin \alpha & \cos \alpha & 0 \\ 0 & 0 & 1 \end{pmatrix}$

$$|A| = \begin{vmatrix} \cos \alpha & -\sin \alpha & 0 \\ \sin \alpha & \cos \alpha & 0 \\ 0 & 0 & 1 \end{vmatrix} = \cos^2 \alpha + \sin^2 \alpha + 0 = 1$$

$\Rightarrow$ Inverse of the matrix A exists.

Let A_{ij} be the cofactors of a_{ij} in $|A|$, then from Eq. 1.6, we have

$$
\begin{array}{lll}
A_{11} = \cos\alpha & A_{12} = -\sin\alpha & A_{13} = 0 \\
A_{21} = \sin\alpha & A_{22} = \cos\alpha & A_{23} = 0 \\
A_{31} = 0, & A_{32} = 0, & A_{33} = 1
\end{array}
$$

The matrix of the cofactors A_{ij} is given by

$$
A_{ij} = \begin{pmatrix} \cos\alpha & -\sin\alpha & 0 \\ \sin\alpha & \cos\alpha & 0 \\ 0 & 0 & 1 \end{pmatrix}
$$

$$
\Rightarrow \mathrm{Adj}(A) = \begin{pmatrix} \cos\alpha & \sin\alpha & 0 \\ -\sin\alpha & \cos\alpha & 0 \\ 0 & 0 & 1 \end{pmatrix}
$$

Hence, $A^{-1} = \frac{\mathrm{Adj}(A)}{|A|} = \frac{1}{1}\begin{pmatrix} \cos\alpha & \sin\alpha & 0 \\ -\sin\alpha & \cos\alpha & 0 \\ 0 & 0 & 1 \end{pmatrix} = \begin{pmatrix} \cos\alpha & \sin\alpha & 0 \\ -\sin\alpha & \cos\alpha & 0 \\ 0 & 0 & 1 \end{pmatrix}$

$Now,\ A^{-1}A = \begin{pmatrix} \cos\alpha & \sin\alpha & 0 \\ -\sin\alpha & \cos\alpha & 0 \\ 0 & 0 & 1 \end{pmatrix}\begin{pmatrix} \cos\alpha & \sin\alpha & 0 \\ -\sin\alpha & \cos\alpha & 0 \\ 0 & 0 & 1 \end{pmatrix}$

$$
= \begin{pmatrix} \cos^2\alpha + \sin^2\alpha & -\cos\alpha\sin\alpha + \cos\alpha\sin\alpha & 0 \\ -\sin\alpha\cos\alpha + \sin\alpha\cos\alpha & \sin^2\alpha + \cos^2\alpha & 0 \\ 0 & 0 & 1 \end{pmatrix}
$$

$$
= \begin{pmatrix} 1 & 0 & 0 \\ 0 & 1 & 0 \\ 0 & 0 & 1 \end{pmatrix} = I
$$

$Similarly,\ AA^{-1} = \begin{pmatrix} \cos\alpha & -\sin\alpha & 0 \\ \sin\alpha & \cos\alpha & 0 \\ 0 & 0 & 1 \end{pmatrix}\begin{pmatrix} \cos\alpha & \sin\alpha & 0 \\ -\sin\alpha & \cos\alpha & 0 \\ 0 & 0 & 1 \end{pmatrix}$

$$
= \begin{pmatrix} \cos^2\alpha + \sin^2\alpha & \cos\alpha\sin\alpha - \cos\alpha\sin\alpha & 0 \\ \sin\alpha\cos\alpha - \sin\alpha\cos\alpha & \sin^2\alpha + \cos^2\alpha & 0 \\ 0 & 0 & 1 \end{pmatrix}
$$

$$
= \begin{pmatrix} 1 & 0 & 0 \\ 0 & 1 & 0 \\ 0 & 0 & 1 \end{pmatrix} = I
$$

$$\Rightarrow A^{-1}A = AA^{-1} = I.$$

Skew-symmetric/Anti-symmetric Matrix

A matrix is called a skew-symmetric or anti-symmetric matrix if and only if $a_{ij} = -a_{ji}$ for all pairs of (i, j). Accordingly, the matrix

$$A = \begin{pmatrix} 0 & -1 \\ 1 & 0 \end{pmatrix} \tag{1.9}$$

is an example of a skew-symmetric matrix of order 2.

In order to be anti-symmetric or skew-symmetric, a matrix must necessarily be a square matrix. Further, the diagonal elements of a skew-symmetric matrix must be zero since $a_{ij} = -a_{ji}$ if and only if $a_{ij} = 0$.

Transpose of a Matrix

The transpose of a matrix A is obtained by interchanging its rows and columns and is represented as A^T. Therefore, if $A = ((a_{ij}))$, then $A^T = ((a_{ji}))$. As a result, the transpose of a column matrix (vector) is a row matrix (vector). If $a_{ij} = a_{ji}$ for all pairs (i, j) for matrix A, then A is a symmetric matrix and $A^T = A$. However, if $a_{ij} = -a_{ji}$ for all pairs of (i, j) for matrix A, then A is a skew-symmetric matrix and $A^T = -A$.

Orthogonal Matrix

A square matrix is called an orthogonal matrix if $AA^T = A^TA = I$, where I is the identity matrix and A^T is the transpose matrix of A. For an orthogonal matrix, one can immediately verify that $A^T = A^{-1}$. Since $AA^T = I$, it follows that:

$$\sum_{k=1}^{n} A_{ik} A_{jk} = \delta_{ij}$$

$$= 1, \; if \; i = j$$
$$= 0, \; if \; i \neq j \tag{1.10}$$

This can be stated as:

(i) The sum of squares of the elements in any row (or any column) is equal to 1, as provided in Table 1.4.

Table 1.4 Sum of the squares of elements in a row/column

Sum of the squares in a row	Sum of the squares in a column
$A_{11}^2 + A_{12}^2 + A_{13}^2 = 1$	$A_{11}^2 + A_{21}^2 + A_{31}^2 = 1$
$A_{21}^2 + A_{22}^2 + A_{23}^2 = 1$	$A_{12}^2 + A_{22}^2 + A_{32}^2 = 1$
$A_{31}^2 + A_{32}^2 + A_{33}^2 = 1$	$A_{13}^2 + A_{23}^2 + A_{33}^2 = 1$

Table 1.5 Sum of products of elements from different rows/columns

Sum of the products from different rows	Sum of the products from different columns
$A_{11}A_{21} + A_{12}A_{22} + A_{13}A_{23} = 0$	$A_{11}A_{12} + A_{21}A_{22} + A_{31}A_{32} = 0$
$A_{11}A_{31} + A_{12}A_{32} + A_{13}A_{33} = 0$	$A_{11}A_{13} + A_{21}A_{23} + A_{31}A_{33} = 0$
$A_{21}A_{31} + A_{22}A_{32} + A_{23}A_{33} = 0$	$A_{12}A_{13} + A_{22}A_{23} + A_{32}A_{33} = 0$
$A_{31}A_{11} + A_{32}A_{12} + A_{33}A_{13} = 0$	$A_{13}A_{11} + A_{23}A_{21} + A_{33}A_{31} = 0$
$A_{21}A_{11} + A_{22}A_{12} + A_{23}A_{13} = 0$	$A_{12}A_{11} + A_{22}A_{21} + A_{32}A_{31} = 0$

(ii)   The sum of products of the elements from one row (or one column) and the corresponding element of another row (or another column) is equal to 0, as provided in Table 1.5.

Example 5: Find out the most general form of an orthogonal matrix of order 2.
Solution: Let us start with an arbitrary matrix of order 2, which can be written as

$$A = \begin{pmatrix} a & b \\ c & d \end{pmatrix}$$

where $a, b, c,$ and d are any scalars, real, or complex. If matrix A is to be an orthogonal matrix, then its elements must satisfy the following orthogonal conditions, i.e.,

$$a^2 + b^2 = 1 \qquad (x)$$
$$c^2 + d^2 = 1 \qquad (y)$$
$$and \quad ac + bd = 0 \qquad (z)$$

The general solution of Eq. (x) is $a = \cos\theta$ and $b = \sin\theta$, where θ is real or complex. Similarly, the solution of Eq. (y) can be $c = \cos\phi$ and $d = \sin\phi$, where ϕ is a scalar. In order to satisfy Eq. (z), we can find that θ and ϕ must be related through

$$\cos\theta \cos\phi + \sin\theta \sin\phi = 0$$

$$or \; \cos(\theta - \phi) = 0$$

$$\Rightarrow \theta - \phi = \pm\frac{\pi}{2}$$

Therefore, the most general form of the orthogonal matrix of order 2 becomes

$$A = \begin{pmatrix} \cos\theta & \sin\theta \\ \mp\sin\theta & \pm\cos\theta \end{pmatrix}$$

Choosing upper signs, we get $\Delta = +1$. On the other hand, the lower signs will give $\Delta = -1$. Thus, the most general form is

$$A = \begin{pmatrix} \cos\theta & \sin\theta \\ -\sin\theta & \cos\theta \end{pmatrix}$$

Example 6: Show that matrix A corresponding to a 6-fold rotation is an orthogonal matrix. Obtain its inverse. The matrix A is as follows:

$$A = \begin{pmatrix} \frac{1}{2} & -\frac{\sqrt{3}}{2} & 0 \\ \frac{\sqrt{3}}{2} & \frac{1}{2} & 0 \\ 0 & 0 & 1 \end{pmatrix}$$

Solution: Let matrix A be an orthogonal matrix. For this,
$A^{-1} = A^T$ and $AA^{-1} = AA^T = E$

$$So\,that,\ A^T = \begin{pmatrix} \frac{1}{2} & \frac{\sqrt{3}}{2} & 0 \\ -\frac{\sqrt{3}}{2} & \frac{1}{2} & 0 \\ 0 & 0 & 1 \end{pmatrix}$$

$$and,\ AA^T = \begin{pmatrix} \frac{1}{2} & -\frac{\sqrt{3}}{2} & 0 \\ \frac{\sqrt{3}}{2} & \frac{1}{2} & 0 \\ 0 & 0 & 1 \end{pmatrix} \begin{pmatrix} \frac{1}{2} & \frac{\sqrt{3}}{2} & 0 \\ -\frac{\sqrt{3}}{2} & \frac{1}{2} & 0 \\ 0 & 0 & 1 \end{pmatrix}$$

$$= \begin{pmatrix} \frac{1}{4}+\frac{3}{4} & \frac{\sqrt{3}}{4}-\frac{\sqrt{3}}{4} & 0 \\ \frac{\sqrt{3}}{4}-\frac{\sqrt{3}}{4} & \frac{3}{4}+\frac{1}{4} & 0 \\ 0 & 0 & 1 \end{pmatrix} = \begin{pmatrix} 1 & 0 & 0 \\ 0 & 1 & 0 \\ 0 & 0 & 1 \end{pmatrix} = E$$

$\Rightarrow$ The given matrix A is an orthogonal matrix

Trace/Character of a Matrix

The trace/character of a matrix is the sum of the diagonal elements of a square matrix and is given by

$$T_r A = \sum_{k=0}^{n} A_{ii} \tag{1.11}$$

The trace/character of various matrices corresponding to different symmetry elements/operations is provided in Table 1.6.

Table 1.6 Symmetry Operations and Character of Matrix

Symmetry operation	Trace/character of matrix
Identity	3
Proper rotation	$2\cos\theta^a + 1$
Inversion	-3
Improper rotation	$2\cos\theta - 1$
Reflection	1

$^a\theta$ refers to the angle of rotation about an axis

Property of Trace/Character

A similarity transformation leaves the trace/character of the matrix unchanged (invariant).

Proof Let us start with a diagonal element of the given unitary matrix

$$A'_{ii} = \sum_{k=1}^{n}\sum_{l=1}^{n} T_{il}^{-1} A_{lk} T_{ki}$$

$$so\ that\ T_r(A') = \sum_{i=1}^{n}\sum_{k=1}^{n}\sum_{l=1}^{n} T_{il}^{-1} A_{lk} T_{ki} \tag{1.12}$$

Since the summation contains only finite terms, therefore we can rearrange the order of summation, i.e.,

$$T_r(A') = \sum_{k=1}^{n}\sum_{l=1}^{n} A_{lk} \sum_{l=1}^{n} T_{il}^{-1} T_{ki}$$

$$but\ \sum_{i=1}^{n} T_{il}^{-1} T_{ki} = \delta_{lk}$$

$$Therefore,\ T_r(A') = \sum_{k=1}^{n} A_{kk} = T_r(A) \tag{1.13}$$

This proves that the similarity transformation leaves the trace/character of the matrix unchanged (invariant).

Example 7: If the two matrices, respectively, are

$$A = \begin{pmatrix} 1 & 2 & 1 \\ 2 & 2 & 1 \\ 1 & 0 & 2 \end{pmatrix}\ and\ B = \begin{pmatrix} 2 & 1 & 2 \\ 1 & 2 & 1 \\ 1 & -1 & 2 \end{pmatrix}$$

Show that the character remains unchanged irrespective of the order in which the matrices are multiplied.

Solution: *Given:* Matrices, $A = \begin{pmatrix} 1 & 2 & 1 \\ 2 & 2 & 1 \\ 1 & 0 & 2 \end{pmatrix}$ and $B = \begin{pmatrix} 2 & 1 & 2 \\ 1 & 2 & 1 \\ 1 & -1 & 2 \end{pmatrix}$, character of

AB and $BA = ?$

Let us consider the two products, i.e.,

$$AB = \begin{pmatrix} 1 & 2 & 1 \\ 2 & 2 & 1 \\ 1 & 0 & 2 \end{pmatrix} \begin{pmatrix} 2 & 1 & 2 \\ 1 & 2 & 1 \\ 1 & -1 & 2 \end{pmatrix} = \begin{pmatrix} 5 & 4 & 6 \\ 7 & 5 & 8 \\ 4 & -1 & 6 \end{pmatrix}$$

$$\Rightarrow \chi(AB) = 5 + 5 + 6 = 16$$

$$BA = \begin{pmatrix} 2 & 1 & 2 \\ 1 & 2 & 1 \\ 1 & -1 & 2 \end{pmatrix} \begin{pmatrix} 1 & 2 & 1 \\ 2 & 2 & 1 \\ 1 & 0 & 2 \end{pmatrix} = \begin{pmatrix} 6 & 6 & 7 \\ 6 & 6 & 5 \\ 1 & 0 & 4 \end{pmatrix}$$

$$\Rightarrow \chi(BA) = 6 + 6 + 4 = 16$$

$$\Rightarrow \chi(AB) = \chi(BA)$$

Conjugate of a Matrix

The conjugate of a matrix is defined as the matrix obtained from a given matrix A by replacing its elements with the corresponding conjugate complex numbers. It is denoted by $\overline{A}$. For example, if the given matrix is

$$A = \begin{pmatrix} 1 & i \\ -i & 1 \end{pmatrix}, \quad \text{then } \overline{A} = \begin{pmatrix} 1 & -i \\ i & 1 \end{pmatrix}$$

Characters of Conjugate Matrices

An important property of a square matrix is its character. This is simply the sum of its diagonal elements, usually denoted by a symbol χ (chi). Thus, the character of the matrix A is defined in a slightly different form (from Eq. 1.11 given above) as

$$\chi(A) = \sum_{i=1}^{n} a_{ii}$$

Let us discuss two important cases related to the behavior of the character.

Case I: If the matrices $C = AB$ and $D = BA$, then the characters of the matrices C and D are equal.

Proof As defined above, the character of the matrix C can be written as

$$\chi(C) = \sum_{i=1}^{n} a_{ii} = \sum_{i=1}^{n} \sum_{j=1}^{n} a_{ij} b_{ji}$$

Similarly, the character of the matrix D can be written as

$$\chi(D) = \sum_{i=1}^{n} d_{jj} = \sum_{j=1}^{n} \sum_{i=1}^{n} b_{ji} a_{ij} = \sum_{i=1}^{n} \sum_{j=1}^{n} a_{ij} b_{ji} = \chi(C)$$

Case II: Conjugate matrices have identical characters.

Proof We know that conjugate elements of a group are related by a similarity transformation. In the same way, conjugate matrices will also be related. Therefore, if the matrices R and P are conjugate, then there is some other matrix Q such that R $= Q^{-1}PQ$.

Since associative law holds for matrix multiplication, this can be proved as

$$\chi \ of \ R = \chi \ of \ Q^{-1}PQ = \chi \ of \left(Q^{-1}P\right)Q$$

$$= \chi \ of \ Q\left(Q^{-1}P\right) = \chi \ of \left(QQ^{-1}\right)P$$

$$= \chi \ of \ P$$

1.6 Matrix Representation of Symmetry Operations

We know that the crystallographic coordinate axes x, y, and z are generally used for the unambiguous description of planes and directions in crystals. However, the description of the physical properties of crystals as well as the analytical representations of their point symmetry groups are based on orthogonal axes say x_1, x_2, and x_3. The standard rules for orienting these axes with respect to the crystallographic axes are given in Table 1.7. Both the coordinate systems are always chosen right-handed such that if we look at them from the end of say x_3 (or z) axis, a rotation from x_1 to x_2 (or x to y) occurs in the counter-clockwise direction.

Further, we know that any symmetry operation of a crystal class (a point group) transforms the old set of axes x_1, x_2, x_3 into a new set of axes x'_1, x'_2, x'_3. The angular relationships between the two sets of axes in terms of their direction cosines are provided in Table 1.8, while they have been shown diagrammatically in Fig. 1.9. Also coordinates in the two sets of axes are conveniently described by the following linear equations:

Table 1.7 Standard rules for axes orientation

Crystal system	Orthogonal axes		
	X_3	X_2	X_1
Triclinic	[001]	In the plane $\perp$ to [001]	
Monoclinic	[001]	[010]	In the plane (010)
Orthorhombic	[001]	[010]	[100]
Tetragonal	[001]	[010]	[100]
Trigonal and Hexagonal	[0001]	$[01\bar{1}0] = [120]$	$[\bar{2}1\bar{1}0] = [100]$
Cubic	[001]	[010]	[100]

Table 1.8 Relationships between two sets of axes

Axes	Old →	x_1	x_2	x_3
	New ↓			
	x_1'	C_{11}	C_{12}	C_{13}
	x_2'	C_{21}	C_{22}	C_{23}
	x_3'	C_{31}	C_{32}	C_{33}

Fig. 1.9 Directions cosines

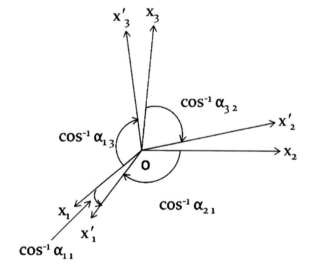

$$x_1' = C_{11}x_1 + C_{12}x_2 + C_{13}x_3$$
$$x_2' = C_{21}x_1 + C_{22}x_2 + C_{23}x_3 \qquad (1.14)$$
$$x_3' = C_{31}x_1 + C_{32}x_2 + C_{33}x_3$$

The set of equations can be written in the matrix form as

$$\begin{pmatrix} x_1' \\ x_2' \\ x_3' \end{pmatrix} = \begin{pmatrix} C_{11} & C_{12} & C_{13} \\ C_{21} & C_{22} & C_{23} \\ C_{31} & C_{32} & C_{33} \end{pmatrix} \begin{pmatrix} x_1 \\ x_2 \\ x_3 \end{pmatrix} \tag{1.15}$$

Equation 1.15 in turn can be written in short as

$$x_i' = \sum_{j=1}^{3} C_{ij} x_j \, (i = 1, \, 2, \, 3) \tag{1.16}$$

where the nine C_{ij} coefficients are not independent of one another and can be written as

$$C_{ij} = \left\{ \begin{array}{l} 1 \ for \ i = j \\ 0 \ for \ i \neq j \end{array} \right\} \tag{1.17}$$

The first script in the symbol C_{ij} refers to the new axes and the second to the old ones. The matrix elements C_{ij} may be conveniently recognized as the direction cosines between new and old axes in the right-handed coordinate system taken in the order old $\rightarrow$ new as shown in Fig. 1.9. Accordingly, the elements of the matrix C_{ij} in Eq. 1.15 can be written in terms of direction cosines as follows:

$$C_{ij} = \begin{pmatrix} \cos(x_1 x_1') & \cos(x_1 x_2') & \cos(x_1 x_3') \\ \cos(x_2 x_1') & \cos(x_2 x_2') & \cos(x_2 x_3') \\ \cos(x_3 x_1') & \cos(x_3 x_2') & \cos(x_3 x_3') \end{pmatrix} \tag{1.18}$$

This is the basis to understand the matrix representation of any given symmetry operation. For example, Eq. 1.18 can be used to obtain the matrix for a rotation operation through an angle α (Fig. 1.10) about any principal axis (say x_3 axis) as

$$C_{ij} = \begin{pmatrix} \cos \alpha & \cos(90 + \alpha) & \cos 90 \\ \cos(90 - \alpha) & \cos \alpha & \cos 90 \\ \cos 90 & \cos 90 & \cos 0 \end{pmatrix}$$

This on simplification gives us

$$C_{ij} = \begin{pmatrix} \cos \alpha & -\sin \alpha & 0 \\ \sin \alpha & \cos \alpha & 0 \\ 0 & 0 & 1 \end{pmatrix} \tag{1.19}$$

where $\alpha = \frac{2\pi}{n}$, $(n = 1, 2, 3, 4, \text{ and } 6)$.

The matrices of some fundamental symmetry operations are termed as the generating elements. They are ten in number including five proper rotation axes, i.e., 1, 2, 3,

Fig. 1.10 Rotation of axes
about x_3 (x_3') axis

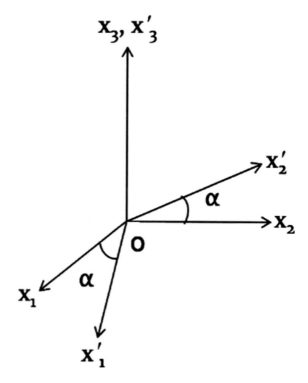

4, and 6; one mirror symmetry, m, and four rotoinversion axes, i.e., $\bar{1}$, $\bar{3}$, $\bar{4}$, and $\bar{6}$.
Let us first obtain the matrices of these generating elements one by one for two
systems of axes, frequently used in dealing with crystallographic problems.

(a) **Orthogonal Axes**
(i) **1 $(\bar{1})$-*fold operation***

1-fold operation is equivalent to a no rotation or a rotation of 360° around any
direction in a crystal. This operation does not bring about any change in the axes.
This gives us $x_1' = x_1, x_2' = x_2$, and $x_3' = x_3$ (Fig. 1.11). The matrix corresponding to
this operation is obtained either from Eq. 1.18 by substituting the values of direction
cosines or directly from Eq. 1.19 with $\alpha = 0$. This is known as the identity matrix
and is represented as

$$1 \equiv \begin{pmatrix} 1\,0\,0 \\ 0\,1\,0 \\ 0\,0\,1 \end{pmatrix} \qquad (1.20)$$

On the other hand, $\bar{1}$-operation inverts the whole space through a point, called the
center of symmetry.

Fig. 1.11 Two sets of axes

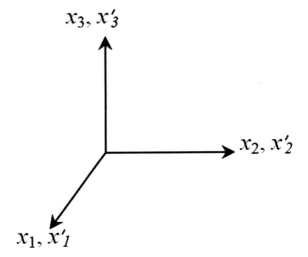

This operation gives us $x_1' = -x_1, x_2' = -x_2$, and $x_3' = -x_3$ as shown in Fig. 1.12. The corresponding matrix is represented as

$$\bar{1} = \begin{pmatrix} -1 & 0 & 0 \\ 0 & -1 & 0 \\ 0 & 0 & -1 \end{pmatrix} \qquad (1.21)$$

This matrix can also be obtained by simply changing the sign of the digits of the matrix in Eq. 1.20.

Fig. 1.12 Transformation of axes by inversion operation

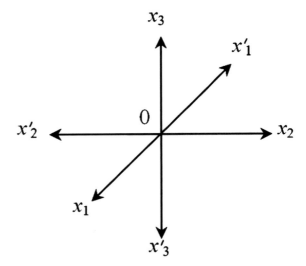

Fig. 1.13 Transformation of
axes by a two-fold symmetry
operation parallel to x_3

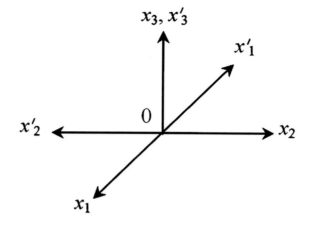

(ii) $2(\bar{2})$-fold operation

In general, 2 and $(\bar{2})$-fold operations are common to all principal crystallographic
directions except [111].

A 2-fold proper rotation along x_3 (or z) axis is shown in Fig. 1.13. This suggests
that us $x_1' = -x_1$, $x_2' = -x_2$, and $x_3' = x_3$. In Miller index notation, this operation is
denoted as 2[001]. The corresponding matrix can be obtained either from Eq. 1.18
or from Eq. 1.20 directly by changing the sign of the first two axes. This operation
is represented as

$$2[001] \equiv \begin{pmatrix} -1 & 0 & 0 \\ 0 & -1 & 0 \\ 0 & 0 & 1 \end{pmatrix} \tag{1.22}$$

On the other hand, a $\bar{2}$-operation is an improper rotation equivalent to a mirror
plane, i.e., $\bar{2} \equiv m$. A mirror plane perpendicular to the x_3 (z) axis is shown in
Fig. 1.14. This suggests that $x_1' = x_1$, $x_2' = x_2$, and $x_3' = -x_3$.

The corresponding matrix is represented as

$$\bar{2}[001] \equiv m[001] \equiv \begin{pmatrix} 1 & 0 & 0 \\ 0 & 1 & 0 \\ 0 & 0 & -1 \end{pmatrix} \tag{1.23}$$

Similarly, the operations 2[010] and m[010] gives us $x_1' = -x_1$, $x_2' = x_2$, $x_3' = -x_3$ and $x_1' = x_1$, $x_2' = -x_2$, $x_3' = x_3$, respectively. The corresponding matrices are

$$2[010] \equiv \begin{pmatrix} -1 & 0 & 0 \\ 0 & 1 & 0 \\ 0 & 0 & -1 \end{pmatrix} \; and \; m[010] \equiv \begin{pmatrix} 1 & 0 & 0 \\ 0 & -1 & 0 \\ 0 & 0 & 1 \end{pmatrix} \tag{1.24}$$

Fig. 1.14 Transformation of
axes by a mirror plane
perpendicular to x_3

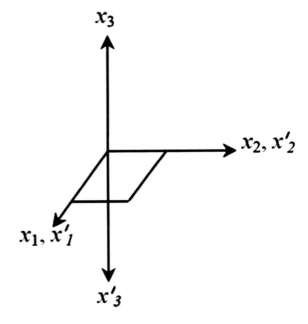

Fig. 1.15 Crystallographic
axes chosen for an
equilateral triangle

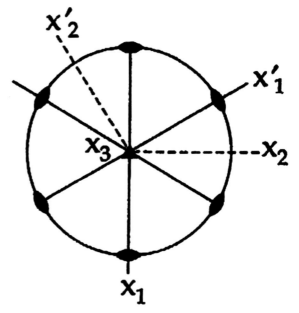

(iii)   3 and 6-fold operations

A 3-fold operation coupled with the orthogonal axes x_1, x_2, x_3 is shown in Fig. 1.15. A proper rotation about x_3-axis transforms x_1 into x_1' and x_2 into x_2' each of which is thrown 120° away from its initial position. Substituting the values α (=120°) in Eq. 1.19, the corresponding matrix can be obtained as

$$3[001] = \begin{pmatrix} \cos 120° & -\sin 120° & 0 \\ \sin 120° & \cos 120° & 0 \\ 0 & 0 & 1 \end{pmatrix} = \begin{pmatrix} -\frac{1}{2} & -\frac{\sqrt{3}}{2} & 0 \\ \frac{\sqrt{3}}{2} & -\frac{1}{2} & 0 \\ 0 & 0 & 1 \end{pmatrix} \qquad (1.25)$$

In a similar manner, the matrix corresponding to a 6-fold rotation is obtained as

$$6[001] = \begin{pmatrix} \cos 60° & -\sin 60° & 0 \\ \sin 60° & \cos 60° & 0 \\ 0 & 0 & 1 \end{pmatrix} = \begin{pmatrix} \frac{1}{2} & -\frac{\sqrt{3}}{2} & 0 \\ \frac{\sqrt{3}}{2} & \frac{1}{2} & 0 \\ 0 & 0 & 1 \end{pmatrix} \qquad (1.26)$$

On the other hand, a proper 3-fold rotation along the body diagonal of a cube, i.e., $3[111]$ transforms x_1 into x_1' and x_2 into x_2' and x_3 into x_3', each of these is thrown by 90° away from its original position. This operation gives us $x_1' = x_2$, $x_2' = x_3$, and $x_3' = x_1$ as shown in Fig. 1.16. The corresponding matrix can be obtained by substituting the values of direction cosines in Eq. 1.18, we have

$$3[111] = \begin{pmatrix} \cos 90° & \cos 90° & \cos 0° \\ \cos 0° & \cos 90° & \cos 90° \\ \cos 90° & \cos 0° & \cos 90° \end{pmatrix} = \begin{pmatrix} 0 & 0 & 1 \\ 1 & 0 & 0 \\ 0 & 1 & 0 \end{pmatrix} \qquad (1.27)$$

Fig. 1.16 Rotational operation along 3[111] direction

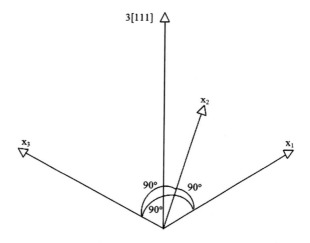

(iv)   **4 ($\overline{4}$)-fold operation**

A 4 ($\overline{4}$)-fold operation is possible along the three crystallographic directions [100], [010], and [001]. These directions are equivalent to orthogonal axes. A 4-fold rotation along [001] transforms the axis x_1 into x_1' and x_2 into x_2', each of which is thrown 90° away from its initial position. This gives us $x_1' = x_2$, $x_2' = -x_1$, and $x_3' = x_3$. The corresponding matrix can be represented as

$$4[001] = \begin{pmatrix} 0 & -1 & 0 \\ 1 & 0 & 0 \\ 0 & 0 & 1 \end{pmatrix} \tag{1.28}$$

This matrix can also be obtained from Eq. 1.19 by substituting $\alpha = 90°$. On the other hand, the matrix corresponding to $\overline{4}[001]$ symmetry operation can be obtained from Eq. 1.28 by changing the sign of the digit 1, i.e.,

$$\overline{4}[001] = \begin{pmatrix} 0 & 1 & 0 \\ -1 & 0 & 0 \\ 0 & 0 & -1 \end{pmatrix} \tag{1.29}$$

The matrices of generating elements corresponding to orthogonal axes are provided in Table 1.9.

(b)   Crystallographic Axes

The orthogonal axes and the crystallographic axes differ for some crystal systems. Let us consider the trigonal/hexagonal crystal system and examine 2-, 3-, and 6-fold operations under crystallographic axes (Fig. 1.17) separately.

(i)   **2-fold operation**

2-fold operations are possible along the three crystallographic axes [100], [010], and [001] as shown in Fig. 1.17. A rotation of 180° (π) along [100] gives us $a' = a$, $b' = -a-b$, and $c' = -c$. Therefore, the position vector of a point at (x, y, z).

$$r = x_1 a + x_2 b + x_3 c$$

changes after 180° rotation to

$$\begin{aligned} r' &= x_1 a' + x_2 b' + x_3 c' \\ &= x_1 a + x_2(-a - b) + x_3(-c) \\ &= (x_1 - x_2)a + (-x_2 b) + (-x_3)c \end{aligned}$$

The corresponding matrix can be represented as

Table 1.9   Generating elements and their matrices (orthogonal axes)

Identity, $I = \begin{pmatrix} 1 & 0 & 0 \\ 0 & 1 & 0 \\ 0 & 0 & 1 \end{pmatrix}$	Inversion, $\bar{1} = \begin{pmatrix} -1 & 0 & 0 \\ 0 & -1 & 0 \\ 0 & 0 & -1 \end{pmatrix}$
$2[001] = \begin{pmatrix} -1 & 0 & 0 \\ 0 & -1 & 0 \\ 0 & 0 & 1 \end{pmatrix}$	$\bar{2}[001] = m[001] = \begin{pmatrix} 1 & 0 & 0 \\ 0 & 1 & 0 \\ 0 & 0 & -1 \end{pmatrix}$
$2^*[010] = \begin{pmatrix} -1 & 0 & 0 \\ 0 & 1 & 0 \\ 0 & 0 & -1 \end{pmatrix}$	$\bar{2}^*[010] = m[010] = \begin{pmatrix} 1 & 0 & 0 \\ 0 & -1 & 0 \\ 0 & 0 & 1 \end{pmatrix}$
$3[001] = \begin{pmatrix} -\frac{1}{2} & -\frac{\sqrt{3}}{2} & 0 \\ \frac{\sqrt{3}}{2} & -\frac{1}{2} & 0 \\ 0 & 0 & 1 \end{pmatrix}, \bar{3}[001] = \begin{pmatrix} \frac{1}{2} & \frac{\sqrt{3}}{2} & 0 \\ -\frac{\sqrt{3}}{2} & \frac{1}{2} & 0 \\ 0 & 0 & -1 \end{pmatrix}$	$3^*[111] = \begin{pmatrix} 0 & 0 & 1 \\ 1 & 0 & 0 \\ 0 & 1 & 0 \end{pmatrix}$
$4[001] = \begin{pmatrix} 0 & -1 & 0 \\ 1 & 0 & 0 \\ 0 & 0 & 1 \end{pmatrix}$	$\bar{4}[001] = \begin{pmatrix} 0 & 1 & 0 \\ -1 & 0 & 0 \\ 0 & 0 & -1 \end{pmatrix}$
$6[001] = \begin{pmatrix} \frac{1}{2} & -\frac{\sqrt{3}}{2} & 0 \\ \frac{\sqrt{3}}{2} & \frac{1}{2} & 0 \\ 0 & 0 & 1 \end{pmatrix}$	$\bar{6}[001] = \begin{pmatrix} -\frac{1}{2} & \frac{\sqrt{3}}{2} & 0 \\ -\frac{\sqrt{3}}{2} & -\frac{1}{2} & 0 \\ 0 & 0 & -1 \end{pmatrix}$

* Supplementary generating elements

$$2[100] = \begin{pmatrix} 1 & -1 & 0 \\ 0 & -1 & 0 \\ 0 & 0 & -1 \end{pmatrix} \qquad (1.30)$$

Similarly, we can obtain other matrices. They are given as

$$2[010] = \begin{pmatrix} -1 & 0 & 0 \\ -1 & 1 & 0 \\ 0 & 0 & -1 \end{pmatrix} \qquad (1.31)$$

$$and\ 2[001] = \begin{pmatrix} -1 & 0 & 0 \\ 0 & -1 & 0 \\ 0 & 0 & 1 \end{pmatrix} \qquad (1.32)$$

Fig. 1.17 The hexagonal
system of axes

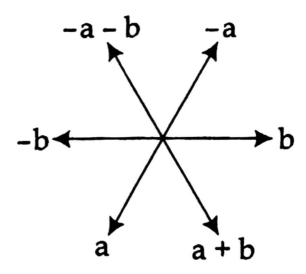

(ii)   3- *and* 6-*fold operations*

In a hexagonal system of axes (Fig. 1.17), a 3-fold axis is possible only along [001]
axis. A rotation of $120°\left(\frac{2\pi}{3}\right)$ along [001] direction gives us $a' = b, b' = -a - b$,
and $c' = c$. Therefore, the change in the position vector is given by

$$r' = x_1 a' + x_2 b' + x_3 c'$$

$$= x_1 b + x_2(-a - b) + x_3(-c)$$

$$= -x_2 a + (x_1 - x_2)b + (x_3)c$$

and the corresponding matrix can be represented as

$$3[001] = \begin{pmatrix} 0 & -1 & 0 \\ 1 & -1 & 0 \\ 0 & 0 & 1 \end{pmatrix} \tag{1.33}$$

Like a 3-fold axis, a 6-fold is also possible only along [001] axis. Proceeding in
a similar manner, a 6-fold rotation, i.e., a rotation of $60°\left(\frac{\pi}{3}\right)$ along [001], will lead
us to obtain $a' = a + b, b' = -a$, and $c' = c$.

Therefore, the change in the position vector is given by

$$r' = x_1 a' + x_2 b' + x_3 c'$$

$$= x_1(a + b) + x_2(-a) + x_3(c)$$

$$= (x_1 - x_2)a + (x_1)b + x_3 c$$

and the corresponding matrix can be represented as

$$6[001] = \begin{pmatrix} 1 & -1 & 0 \\ 1 & 0 & 0 \\ 0 & 0 & 1 \end{pmatrix} \tag{1.34}$$

Matrices corresponding to rotoinversion axes can be obtained by simply interchanging the sign of the digits appearing in proper rotation matrices. Table 1.10 provides the matrices of generating elements corresponding to crystallographic axes.

With the help of the generating elements provided in Tables 1.9 and 1.10 and taking into account the group conditions, the matrix representation of 32-point groups can be obtained.

Example 8: Making use of the general form of rotation matrix obtained in Eq. 1.19, determine the representative matrices corresponding to five rotoreflection axes, S_1, S_2, S_3, S_4, and S_6.

Solution: *Given*: Five rotoreflecion axes: S_1, S_2, S_3, S_4, and S_6. Determine their matrices.

We know that the rotoreflection operation is a combination of rotation and reflection (perpendicular to the principal c-axis) operations. Therefore, to find the five rotoreflection axes, we take the product of σ_h with C_2, we have

$$S_n = \sigma_h C_n = \begin{pmatrix} 1 & 0 & 0 \\ 0 & 1 & 0 \\ 0 & 0 & -1 \end{pmatrix} \begin{pmatrix} \cos\theta & -\sin\theta & 0 \\ \sin\theta & \cos\theta & 0 \\ 0 & 0 & 1 \end{pmatrix}$$

$$= \begin{pmatrix} \cos\theta & -\sin\theta & 0 \\ \sin\theta & \cos\theta & 0 \\ 0 & 0 & -1 \end{pmatrix}$$

Now, considering different cases, we have.

(i) For $n = 1$, $\theta = 0°$ or $360°$,

$\Rightarrow \cos\theta = 1$, $\sin\theta = 0$, and hence,

Table 1.10 Generating elements and their matrices (crystallographic axes)

Identity, $I = \begin{pmatrix} 1 & 0 & 0 \\ 0 & 1 & 0 \\ 0 & 0 & 1 \end{pmatrix}$	Inversion, $\bar{1} = \begin{pmatrix} -1 & 0 & 0 \\ 0 & -1 & 0 \\ 0 & 0 & -1 \end{pmatrix}$
$2[001] = \begin{pmatrix} -1 & 0 & 0 \\ 0 & -1 & 0 \\ 0 & 0 & 1 \end{pmatrix}$	$\bar{2}[001] = m[001] = \begin{pmatrix} 1 & 0 & 0 \\ 0 & 1 & 0 \\ 0 & 0 & -1 \end{pmatrix}$
$3[001] = \begin{pmatrix} 0 & -1 & 0 \\ 1 & -1 & 0 \\ 0 & 0 & 1 \end{pmatrix}, \bar{3}[001] = \begin{pmatrix} 0 & 1 & 0 \\ -1 & 1 & 0 \\ 0 & 0 & -1 \end{pmatrix}$	$3^*[111] = \begin{pmatrix} 0 & 0 & 1 \\ 1 & 0 & 0 \\ 0 & 1 & 0 \end{pmatrix}$
$4[001] = \begin{pmatrix} 0 & -1 & 0 \\ 1 & 0 & 0 \\ 0 & 0 & 1 \end{pmatrix}$	$\bar{4}[001] = \begin{pmatrix} 0 & 1 & 0 \\ -1 & 0 & 0 \\ 0 & 0 & -1 \end{pmatrix}$
$H \rightarrow 2^*[010] = \begin{pmatrix} -1 & 0 & 0 \\ -1 & 1 & 0 \\ 0 & 0 & -1 \end{pmatrix}$	$H \rightarrow \bar{2}^*[010] = m[010] = \begin{pmatrix} 1 & 0 & 0 \\ 1 & -1 & 0 \\ 0 & 0 & 1 \end{pmatrix}$
$6[001] = \begin{pmatrix} 1 & -1 & 0 \\ 1 & 0 & 0 \\ 0 & 0 & 1 \end{pmatrix}$	$\bar{6}[001] = \begin{pmatrix} -1 & 1 & 0 \\ -1 & 0 & 0 \\ 0 & 0 & -1 \end{pmatrix}$

* Supplementary generating elements

$$S_1 = \begin{pmatrix} 1 & 0 & 0 \\ 0 & 1 & 0 \\ 0 & 0 & -1 \end{pmatrix} = \sigma_h$$

(ii)   For $n = 2$, $\theta = 180°$,

⇒ $\cos\theta = -1$, $\sin\theta = 0$, and hence,

$$S_2 = \begin{pmatrix} -1 & 0 & 0 \\ 0 & -1 & 0 \\ 0 & 0 & -1 \end{pmatrix} = i$$

(iii)   For $n = 3$, $\theta = 120°$,

$$\Rightarrow \cos\theta = -\frac{1}{2}, \ \sin\theta = \frac{\sqrt{3}}{2}$$

$$S_3 = \begin{pmatrix} -\frac{1}{2} & -\frac{\sqrt{3}}{2} & 0 \\ \frac{\sqrt{3}}{2} & -\frac{1}{2} & 0 \\ 0 & 0 & -1 \end{pmatrix}$$

(iv) For $n = 4$, $\theta = 90°$,

$\Rightarrow \cos\theta = 0$, $\sin\theta = 1$, and hence,

$$S_4 = \begin{pmatrix} 0 & -1 & 0 \\ 1 & 0 & 0 \\ 0 & 0 & -1 \end{pmatrix}$$

(v) For $n = 6$, $\theta = 60°$,

$$\Rightarrow \cos\theta = \frac{1}{2}, \ \sin\theta = \frac{\sqrt{3}}{2}$$

$$S_6 = \begin{pmatrix} \frac{1}{2} & -\frac{\sqrt{3}}{2} & 0 \\ \frac{\sqrt{3}}{2} & \frac{1}{2} & 0 \\ 0 & 0 & -1 \end{pmatrix}$$

From the above discussion, we can infer some general characteristic features of S_n operations. They are as follows:

(a) S_n exists if the operations C_n followed by σ_h (or vice versa) bring the object to an equivalent position.

(b) If both C_n and σ_h exist, then S_n also exists.

For example: S_4 is collinear with C_4 in a square planar system such as MX$_4$ molecule.

(c) Neither C_n nor σ_h needs to exist for S_n to exist.

For example: S_4 is collinear with C_2 in a tetrahedral system such as a tetrahedral molecule MX$_4$ inscribed in a cube.

(d) About the same axis, two successive S_4 operations are identical to a single C_2 operation, i.e., $S_4{}^2 = C_2$ (Fig. 1.18).

(e) Therefore, for the tetrahedral MX$_4$ molecule there are only two possible operations (i.e., S_4 and $S_4{}^3$) about this axis in this class.

Also, there are some important equivalences of successive S_n operations. They are as follows:

(a) If n is even, $S_n{}^n = E$

Fig. 1.18 Showing S_4
collinear with C_2 in
tetrahedral molecule

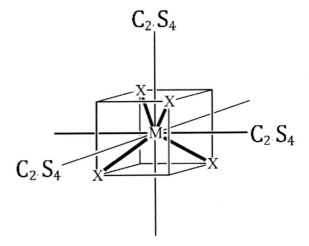

(b)   If n is odd, $S_n{}^n = \sigma$ and $S_n{}^{2n} = E$
(c)   If m is even, $S_n{}^m = C_n{}^m$ when $m < n$ and $S_n = C_n{}^{m-n}$ when $m > n$
(d)   If S_n with even n exists, then $C_{n/2}$ also exists
(e)   If S_n with odd n exists, then both C_n and σ perpendicular to C_n also exist.

1.7   Molecular Point Groups

As we know that real molecules can have different geometries or shapes (such
as linear, planar, tetrahedral, pyramidal, and octahedral), hence they will have
different symmetries and different point groups. Further, unlike crystallographic
point groups, molecular point groups are unlimited. The molecular point groups
can be conveniently divided into the following four categories:

(1)   Non-rotational point groups.
(2)   Mono-axial point groups (where $n = 2, 3, \ldots \infty$).
(3)   Dihedral point groups (where $n = 2, 3, \ldots \infty$).
(4)   Cubic point groups.

Non-rotational Point Groups

Non-rotational point groups do not possess any axial symmetry and are the lowest
symmetry point groups. C_1 is the point group of asymmetric molecules. The point
group C_s describes the symmetry of bilateral objects with only E and σ_h symmetries.
The independent occurrence of the point group C_i is not common in molecules, since
most of the centrosymmetric molecules tend to have other symmetries as well.

Mono-axial Point Groups

The simplest family under this category is C_n, from which one can generate $(n-1)$ other members as C_n, C_n^2, ..., $C_n^n = E$. These groups are cyclic groups.

The point group family C_{nv} contains n vertical mirrors intersecting each other along the C_n principal axis. The point groups C_{2v}, C_{3v}, etc. of the C_{nv} family are the same as crystallographic point groups. On the other hand, the point group $C_{\infty v}$ has an infinity fold (C_∞) principal axis. It is an important member of the C_{nv} family because all non-centrosymmetric linear molecules belong to this point group.

The point group family C_{nh} can be generated by adding a horizontal plane mirror, which is perpendicular to the C_n principal axis. The point groups C_{2h}, C_{3h}, etc. of the C_{nh} family are the same as crystallographic point groups.

The point group family S_{2n} can be generated by carrying out $2n$ successive $2n$-fold improper rotation operations about a single axis. The last operation in the series is equivalent to an identity element, i.e., $S_{2n}^{2n} = E$.

Dihedral Point Groups

The dihedral point groups are those point groups which have n 2-fold (C_2) axes (called dihedral axes) perpendicular to n-fold (C_n) principal axis. There are three families of dihedral point groups. They are D_n, D_{nd}, and D_{nh}.

The D_n point groups can be generated by adding n dihedral (C_2) axes to C_n principal axis. Here, unlike C_n groups, D_n groups are not cyclic.

In a similar manner, the D_{nd} family of point groups can be generated by adding n dihedral axes to the C_{nv} family of groups. In these point groups, the combination of rotation and vertical mirror (reflection) operations generates a series of S_{2n} operations about an axis collinear with the principal axis.

Similar to the above cases, the D_{nh} family of point groups can be generated by adding n dihedral axes to the C_{nh} family of point groups. The D_{nh} point groups include n-fold improper axes when $n > 2$ and are centrosymmetric when n is even. The point group $D_{\infty h}$ has an infinity fold (C_∞) principal axis. It is an important member of D_{nh} family because all linear centrosymmetric molecules belong to this point group.

Cubic Point Groups

The cubic point groups are associated with highly symmetrical polyhedra originating from a cube. Such molecules that are frequently encountered belong to three cubic point groups: T_d, O_h, and I_h.

Tetrahedral molecules (such as SiF_4) belong to the point group T_d. It has 24 symmetry elements. Similarly, an octahedral molecule (such as SF_6) belongs to the point group O_h. It has 48 symmetry elements, while both the regular icosahedral and dodecahedral molecules (such as $B_{12} H_{12}^{2-}$ and $B_{12} Cl_{12}^{2-}$) belong to the point group I_h. It has 120 symmetry elements. Table 1.11 provides a list of chemically important point groups and their principal symmetry operations. The centrosymmetric linear, non-centrosymmetric linear, and cubic groups together are called 'special groups'. It

Table 1.11 Molecular point groups and their principal operations

Symbols	Symmetry operations
	Nonrotational groups
C_1	E (asymmetric)
C_S	E, σ_h
C_i	E, i
	Mono-axial groups ($n = 2, 3, \ldots, \infty$)
C_n	E, C_n,..., C_n^{n-1}
C_{nv}	E, C_n,..., C_n^{n-1}, $n\sigma_v$ ($n/2\ \sigma_v$ and $n/2\ \sigma_d$, if n is even)
C_{nh}	E, C_n,..., C_n^{n-1}, σ_h
S_{2n}	E, S_{2n},..., S_{2n}^{2n-1}
$C_{\infty v}$	$C_{\infty v}\ \infty\sigma_v$ (non-centrosymmetric linear)
	Dihedral groups ($n = 2, 3, \ldots, \infty$)
D_n	E, C_n,..., C_n^{n-1}, nC_2 ($\perp C_n$)
D_{nd}	E, C_n,..., C_n^{n-1}, S_{2n},..., S_{2n}^{2n-1}, nC_2 ($\perp C_n$), $n\sigma_d$
D_{nh}	E, C_n,..., C_n^{n-1}, nC_2 ($\perp C_n$), σ_h, $n\sigma_v$
$D_{\infty h}$	E, C_∞, S_∞, ∞C_2 ($\perp C_\infty$), σ_h, $\infty\sigma_v$, i (Centrosymmetric linear)
	Cubic groups
T_d	E, $4C_3$, $4C_3^2$, $3C_2$, $3S_4$, $3S_4^3$, $6\sigma_d$ (tetrahedron)
O_h	E, $4C_3$, $4C_3^2$, $6C_2$, $3C_4$, $3C_4^3$, $3C_2(= C_4^2)$, i, $3S_4$, $3S_4^3$, $4S_6$, $4S_6^5$, $3\sigma_h$, $6\sigma_d$ (octahedron)
I_h	E, $6C_5$, $6C_5^2$, $6C_5^3$, $6C_5^4$, $10C_3$, $10C_3^2$, $15C_2$, i, $6S_{10}$, $6S_{10}^3$, $6S_{10}^7$, $6S_{10}^9$, $10S_6$, $10S_6^5$, 15σ (icosahedron, dodecahedron)

is because of the fact that these point groups can be readily assigned to the concerned molecule.

1.8 Determination of Molecular Point Groups

The identification of the molecular point group of a given molecule is an important and necessary step before we apply group theory for any analytical purpose. Since the number of symmetry elements/operations in some molecules is very large (may approach even infinity), identifying the corresponding point groups by finding all symmetry operations is practically not possible. However, fortunately it is not necessary. In such cases, the best option is to find out the essential symmetries of the molecules that can classify their point group uniquely. This is extremely useful as we know that many symmetry operations are related to each other (e.g., C_3 and C_3^2 represent different forms of the same axis, S_3 axis requires the presence of C_3 axis and σ_h, and so on) in some way or the other.

There are two fundamental ways to determine the point group of a given molecule:

(i)   By using the flow-chart method, and
(ii)   By using geometry and formula of the molecule

Flow Chart Method

The success of this method to determine the point group depends on one's ability to find the key (essential) symmetry elements that are present in the given molecule. This is a skill which can only be developed by continued practice and a clear understanding of the exact relationships among the symmetry elements/operations.

"Special Groups"

A flow chart can be prepared on the basis of a question asked about the presence (yes)/absence (no) of a particular symmetry element/operation (point group) in the given molecule each time in a sequential manner until a correct point group is finally determined. The sequence of the flow chart is shown in Fig. 1.19. This begins by asking a question about the presence/absence of any of the point groups in the given molecule from among the "special groups".

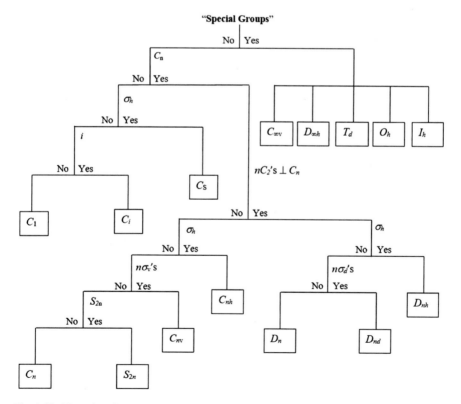

Fig. 1.19 Flow chart for systematically determining the point group of a molecule

Geometry and Formula Method

As the name suggests, in this method the point group of the given molecule is determined on the basis of the information obtained from its geometry and chemical
formula. For example, HCl and FBeCl are non − centrosymmetric linear molecules,
while CO_2 and $BeCl_2$ are centrosymmetric linear molecules, hence their point groups
are $C_{\infty v}$ and $D_{\infty h}$, respectively. Similarly, CH_4 and SO_4^{2-} are tetrahedral, while SF_6
is octahedral in shape, hence their point groups are T_d and O_h, respectively. Table
1.12 provides a list of point groups based on their geometry and chemical formula.

1.9 Crystallographic Point Groups

It is possible to find an individual or a collection of symmetry operations to form
a crystallographic group such that it obeys the group conditions completely. When
these groups satisfy all the properties of a mathematical group they are called point
groups. These are so named because there exists at least one point in space that
remains invariant (unchanged) during symmetry operation(s) of a point group.

On a similar line, a crystal system can have one or more point groups depending on
the shape of its primitive unit cell. This has the ability to display a range of symmetry
elements. The minimum symmetry which is inherently related to the unit cell is
known as essential symmetry. Similarly, the maximum symmetry is its characteristic
symmetry and is also known as holohedry. Therefore, the possible symmetry elements
of a given crystal system will lie in the range.

$$\text{Essential symmetry} \leq \text{intermediate symmetry} \leq \text{holosymmetry}$$

Thus in order to derive the possible point groups for a given crystal system,
one can start with the essential (inherent) symmetry element and then go on
adding a symmetry element which is compatible with the unit cell shape until the
maximum symmetry (holosymmetry) corresponding to the given crystal system is
obtained. Here it is important to mention that in a recent study on symmetry, Wahab
(2020) found that the mirror is the only fundamental symmetry in crystalline solids
because all other symmetries, such as rotation, inversion, rotoreflection, rotoinversion, and translational periodicity, can be easily derived from suitable combinations
of mirrors. Further, on the basis of the newly developed mirror combination scheme,
he suggested some inevitable changes in the allocation of point group symmetries in
low symmetry crystal systems, such as triclinic, monoclinic, and orthorhombic. Point
groups and their corresponding symmetry elements/operations for various crystal
systems after taking into account the required changes are provided in Table 1.13.

Table 1.12 Geometry, chemical formula, and point groups of some selected molecules

Chemical formula	Molecular geometry	Point group (symmetry)
Formaldehyde		$C_{2v}, E, C_2, \sigma_{xz}, \sigma_{yz}$
Hypochlorous acid		C_s, E, σ
Dinitrogen-difluoride		$C_{2h}, E, C_2, i, \sigma_h$
Pyridine		$C_{2v}, E, C_2, \sigma_{xz}, \sigma_{yz}$
Tetrahedral SiF_4		$T_d, E, 8C_3, 3C_2, 6\sigma_d, 6S_4$
Ammonia		$C_{3v}, E, C_3, C_3^2, \sigma_v, \sigma_v', \sigma_v''$
Cubane C_8H_8		$T, T_d, T_h, O, O_h, E, 8C_3, 3C_2,$ $\ldots\ldots$
Carbon-dioxide acetylene		$C_{\infty v}, E, C_\infty, \sigma_v, D_{\infty h}, E, C_\infty,$ S_∞

1.10 Point Group Notations

We know that there are two types of improper rotation axes. They are rotoinversion axes and rotoreflection axes. Based on these, two different types of point group notations have been developed. They are as follows:

Table 1.13 Point groups and symmetry elements/operations

S. no	Crystal system	Point groups		Symmetry elements/operations	Order of the group
		International notations	Schoenflies notations		
1	Triclinic	1	C_1	E	1
		m	C_h (C_S)	E σ_h	2
		$\bar{1}$	C_i (S_2)	E i	2
2	Monoclinic	2	C_2	E C_2	2
		mm2	C_{2v}	E C_2 σ_v' σ_v''	4
		($\bar{1}$)	(S_2)	(E i) Not counted in terms of number (see Wahab 2020)	(2)
3	Orthorhombic	222	D_2 (V)	E C_2 C_2' C_2''	4
		2/m	C_{2h}	E C_2 i σ_h	4
		2/mmm	D_{2h} (V_h)	E C_2 C_2' C_2'' i σ_h σ_v' σ_v''	8
4	Rhombehedron	3	C_3	E $2C_3$	3
		$\bar{3}$	S_6 (C_{3i})	E $2C_3$ i $2S_6$	6
		32	D_3	E $2C_3$ $3C_2$	6
		3 m	C_{3v}	E $2C_3$ $3\sigma_v$	6
		$\bar{3}$ m	D_{3d}	E $2C_3$ $3C_2$ i $2S_6$ $3\sigma_v$	12
5	Trigonal/hcp	3	C_3	E $2C_3$	3
		$\bar{3}$	S_6 (C_{3i})	E $2C_3$ i $2S_6$	6
		32	D_3	E $2C_3$ $3C_2$	6
		3 m	C_{3v}	E $2C_3$ $3\sigma_v$	6
		$\bar{3}$ m	D_{3d}	E $2C_3$ $3C_2$ i $2S_6$ $3\sigma_v$	12
		$\bar{6}$	C_{3h}	E $2C_3$ σ_h $2S_3$	6
		$\bar{6}$m2	D_{3h}	E $2C_3$ $3C_2$ σ_h $2S_3$ $3\sigma_v$	12
6	Tetragonal	4	C_4	E $2C_4$ C_2	4
		$\bar{4}$	S_4	E $2S_4$ C_2	4
		4/m	C_{4h}	E $2C_4$ C_2 i $2S_4$ σ_h	8
		422	D_4	E $2C_4$ C_2 2 C_2' 2 C_2''	8
		4 mm	C_{4v}	E $2C_4$ C_2 $2\sigma_v$ $2\sigma_d$	8
		$\bar{4}$ 2m	D_{2d} (V_d)	E C_2 2 C_2' $2\sigma_d$ 2 S_4	8
		4/mmm	D_{4h}	E $2C_4$ $C_2$2C_2' 2 C_2'' i $2S_4$ σ_h $2\sigma_v$ $2\sigma_d$	16
7	Hexagonal	6	C_6	E $2C_6$ $2C_3$ C_2	6
		6/m	C_{6h}	E $2C_6$ $2C_3$ C_2 i $2S_3$ $2S_6$ σ_h	12
		622	D_6	E $2C_6$ $2C_3$ C_2 3 C_2' 3 C_2''	12
		6 mm	C_{6v}	E $2C_6$ $2C_3$ C_2 $3\sigma_v$ $3\sigma_d$	12
		6/mmm	D_{6h}	E $2C_6$ $2C_3$ C_2 3 C_2' 3 C_2'' i $2S_3$ $2S_6$ σ_h $3\sigma_v$ $3\sigma_d$	24

(continued)

Table 1.13 (continued)

S. no	Crystal system	Point groups		Symmetry elements/operations	Order of the group
		International notations	Schoenflies notations		
8	Cubic	23	T	E $8C_3$ $3C_2$	12
		m3	T_h	E $8C_3$ $3C_2$ i $8S_6$ $3\sigma_h$	24
		432	O	E $8C_3$ $3C_2$ 6 C_2' $6C_4$	24
		$\bar{4}$ 3 m	T_d	E $8C_3$ $3C_2$ $6\sigma_d$ $6S_4$	24
		m3m	O_h	E $8C_3$ $3C_2$ 6 C_2' $6C_4$ i $8S_6$ $3\sigma_h$ $6\sigma_d$ $6S_4$	48

(i)   The Hermann-Mauguin (also known as 'International') notation is based on rotoinversion axes, and

(ii)   The Schoenflies notation is based on rotoreflection axes.

Both the point group notations are important on different basis. For example, the solid-state physicists and the crystallographers prefer to use the International notation because.

(i)   It specifies the direction of symmetry axes more clearly, and

(ii)   Visualization of translation periodicity is easier in this.

On the other hand, the chemists and spectroscopists prefer to use the Schoenflies notation as they find it more systematic to visualize the molecular symmetries.

Let us briefly describe the general representation of symmetry symbols used in the two-point group notations. The symbols used in the cubic case are somewhat specialized.

Hermann-Mauguin (International) Notation

1.   Each component of a point group is supposed to refer a direction of its own.

2. n groups ($n = 1, 2, 3, 4$ and 6) are the cyclic groups, where $n = 1$ is the identity element E. They are known as principal axes.

3. m group (m = mirror plane) is a symbol of mirror plane. The position of m in different point group symbols indicates the directions of the normal to the mirror plane.

4. $\bar{n}$ groups ($\bar{n} = \bar{1}, \bar{2}, \bar{3}, \bar{4}$ and $\bar{6}$) are the rotoinversion axes (rotation followed by inversion) axes, where $\bar{1}$ = center of symmetry, $\bar{2} = m$, etc.

5. $\frac{n}{m}$ groups ('n over m') are the n-fold axes with mirror plane normal to them. Here, $\frac{1}{m} = m = \bar{2}$ and $\frac{3}{m} = \bar{6}$, etc.

6. nm (or nmm) groups are the symbols of the system of one or more mirror planes containing the principal axis (e.g., $1m, 2mm, 3m, 4mm, 6mm$). Here, $1m = m$ and $2mm$ or $mm2$ is sometimes written as mm only.

7. $\frac{n}{m}\frac{2}{m}\frac{2}{m}$ groups ($n = 4, 6$) denote n-fold axis with a mirror plane normal to it and two mirror planes parallel to it. In short, this symbol can be written as $\frac{n}{mmm}$.

8. $n2$ (or $n22$) groups denote n-fold as principal axis. The second digit refers to an axis normal to the first and the third digit refers to the diagonal 2-fold axis.

9.   The cubic system poses a special problem because it has symmetry axes which are neither parallel nor perpendicular to the crystallographic axes. These are 3-fold axes along <111> directions present in every cubic point group. Thus, the presence of a number 3 in the second position is an indication of the cubic system. The first component of a cubic point group symbol refers to the cube axes and the third component refers to the face diagonal of the cube.
10. In the International notation, the point groups are merely a brief list of symmetry elements associated with the three crystallographic directions.

Schoenflies Notation

1. C_n groups ($C =$ cyclic): A C_n group is a simple nth order cyclic group of rotation about a single n-fold axis. Its elements are $C_n, C_n{}^2, ..., C_n{}^n = E$. For example, the symmetry elements of C_3 group are C_3, C_3^2 and $C_3^3 = E$. They also form an ablian group.
2.   Schoenflies represented the rotoreflection (rotation followed by reflection) axes $\tilde{1}, \tilde{2}, \tilde{3}, \tilde{4}, \tilde{6}$ by S_1, S_2, S_3, S_4 and S_6. Here $S_1 = C_s$ and $S_2 = C_i$ (centre of inversion). When n is an odd number (e.g., $n = 3$), S_n and C_{nh} are equivalent, but the C_{nh} notation is preferred. The S_6 group is sometime called C_{3i} because its symmetry elements include both, a C_3 axis and a centre of inversion. The S_4 group is a new class whose Hermann-Mauguin equivalent is $\bar{4}$.
3. C_{nh} group ($h =$ horizontal): A C_{nh} group is obtained by placing a mirror plane perpendicular to a C_n axis. For $n = 1$, the symmetry elements are σ and E the group is called C_s. $C_{2h} = C_2 \times C_i$, i.e., the symmetry elements of C_{2h} are C_2, σ_h, i and E. They also form an abelian group.
4. C_{nv} groups ($v =$ vertical): A C_{nv} group has a C_n rotation axis and $n\sigma_v$ mirror planes containing the rotation axis. For $n = 1$, there is only one mirror plane, and $C_{1v} = C_{1h}, = C_s$. A C_{nv} group is of order $2n$.

The above-mentioned four groups, i.e., C_n, C_{nh}, C_{nv}, and S_n deal with the monoaxial rotational system only, while the remaining groups to be mentioned below, deal with the polyaxial rotational systems.

5. D_n group ($D =$ dihedral): A D_n group is obtained by adding a 2-fold axis perpendicular to the principal C_n axis. In analogy to C_{nv}, a D_n group is of order $2n$.
6. D_{nh} group: A D_{nh} group is obtained by adding a mirror plane containing all of the 2-fold axes to D_n. This automatically generates n vertical mirror planes. This group can also be obtained as a product of D_n and C_i. That is

$$D_{nh} = D_n \times C_i \ for \ n \ even$$
$$= D_n \times C_s \ for \ n \ odd$$

This group is of order $4n$.

7. D_{nd} group (d = diagonal): A D_{nd} (n = 2, 3) group is obtained by adding n vertical mirror planes (σ_d) that bisect the angles between the 2-fold axes in D_n. This group is of order $4n$.

The above preceding cyclic and dihedral groups are illustrated by particular examples in Fig. 1.20. The remaining point groups that are mentioned below are often referred to collectively as the cubic groups because the symmetry axis and planes occurring in them can be selected from those of a cube. They are illustrated collectively in Fig. 1.21 and point group wise in Fig. 1.22, respectively.

8. T group (T = tetrahedral): It consists of all the axes of symmetry of a regular tetrahedron. The 2-fold axes pass through the centre of the opposite edges of the cube, whereas the 3-fold axes are formed by the body diagonals of the cube. It is of order 12.

9. T_d group is the complete symmetry group of the regular tetrahedron. It is obtained from T by adding mirror planes, each of which contains one 2-fold

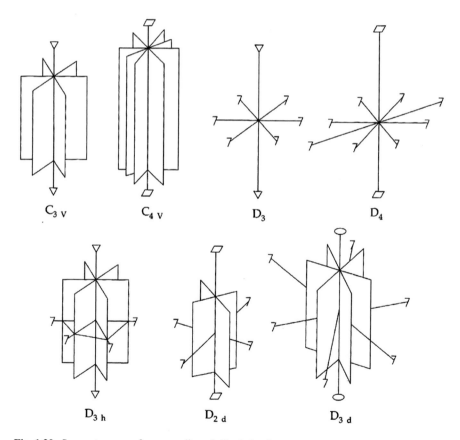

Fig. 1.20 Symmetry axes of some cyclic and dihedral point groups

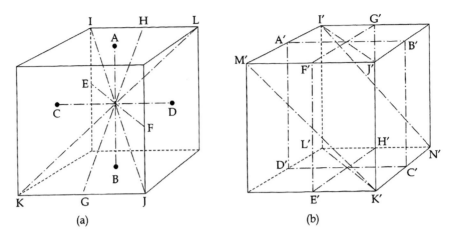

Fig. 1.21 Symmetry axes and planes of a cube

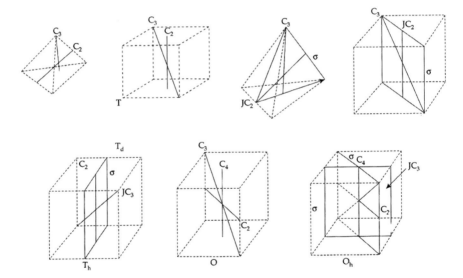

Fig. 1.22 Symmetry planes and axes for the cubic point groups. F or T and T_d, the regular tetrahedral are also shown

   axis and one 3-fold axis. Each of these planes contains two diagonally opposite cube edges and the two face diagonals connecting them. It is of order 24.

10. T_h group is also obtained from T by adding a center of symmetry (C_i). That is $T_h = T \times C_i$. This produces three mirror planes, which divide the cube into the usual octants and turn a proper C_3 axis into an improper C_3 axis. It is of order 24.

Table 1.14 Symmetry Elements, symmetry operations and their symbols in two notations

Symmetry		Symbols	
Element	Operation	Schoenflies	Hermann-Mauguin (international)
Rotation axis	Counter-clockwise rotation of $\frac{360°}{n}$ about an axis	$C_1(E), C_2, C_3, C_4, C_6$	1, 2, 3, 4, 6
Mirror plane Center of Inversion (center of symmetry)	Reflection through a plane All points are inverted through the center of symmetry	σ i	m $\bar{1}$
Rotoreflection axis	Rotation of $\frac{360°}{n}$ followed by reflection in a plane $\perp$ to the axis	$S_1, S_2, S_3, S_4, S_6(C_{3i})$	
Rotoinversion axis	Rotation of $\frac{360°}{n}$ followed by inversion through a point on the axis		$\bar{1}, \bar{2}, \bar{3}, \bar{4}, \bar{6}$

11. O group (O = octahedral) is the set of proper rotations that carries a regular octahedron into itself. It contains the symmetry axes, C_2, C_3, and C_4 of a cube. It is of order 24.

12. O_h group is the complete symmetry group of a cube. O_h is obtained from O by adding a centre of symmetry (C_i). That is, $O_h = O \times C_i$. It is of order 48. Orders of all point groups are listed in Table 1.11.

Symmetry elements, symmetry operations and their symbols in two crystallographic notations are provided in Table 1.14.

1.11 Summary

1. A three-dimensional periodic pattern (a crystal) can have the following symmetry elements:

(a) Translation Vector: $T = n_1 \vec{a} + n_2 \vec{b} + n_3 \vec{c}$

where n_1, n_2, n_3 are integers and $\vec{a}, \vec{b}, \vec{c}$, are primitive lattice translations, also called the basis vectors (Fig. 1.1).

(b) Proper Rotation (through an angle): $\alpha = \frac{2\pi}{n}$

where $n = 1, 2, 3, 4$ and 6. These rotation axes are termed as monad, diad, triad, tetrad, and hexad, respectively (Fig. 1.2a).

(c) Reflection (a mirror reflection): m (a line in 2D or a plane in 3D) has the property to transform a left-handed object into a right-handed object and vice-versa (Fig. 1.2b)

(d) Inversion (through a point): $\bar{1}$ an inversion is equivalent to a reflection through a point (instead of a line or a plane), called inversion center or center of symmetry (Fig. 1.2c).

2. A 3-D symmetry element/operation can be represented by a matrix of order 3 × 3. However, the matrix elements of different symmetry elements or symmetry operations are different.

3. Two different kinds of axes, i.e., orthogonal axes and crystallographic axes are used to obtain the matrix elements of different symmetry elements or symmetry operations.

4. Ten symmetry elements/operations of lower order (in each category) are called generating elements.

5. With the help of the generating elements provided in Tables 1.9 and 1.10 and taking into account the group conditions, the matrix representation of 32-point groups can be obtained.

6. Crystals are found to have a limited number of point groups, i.e., 32, while molecules can have theoretically infinite point groups.

7. Based on rotoreflection axes and rotoinversion axes, two different types of point group notations have been developed, one each by Schoenflies and Hermann-Mauguin (also known as International notation).

Chapter 2
Elements of Group Theory and Multiplication Tables

2.1 Introduction

In the last chapter, we studied in detail about the symmetry elements/operations, their matrix representations, molecular and crystallographic point groups, and the two point group notations.

In this chapter, we are going to study about the elements of group theory that are needed to understand the molecular and crystallographic point groups. For the purpose, we shall briefly discuss various types of groups and their orders, concepts of subgroups and super groups, ways of determining the symmetry elements/operations in a point group to form separate classes, and the classification of point groups on a different basis. However, our main objective in this chapter will be to construct the group multiplication tables for all 32 crystallographic point groups using both International and Schoenflies notations.

2.2 Elements of Group Theory

Although we do not necessarily require group theoretical concepts for the derivation of 32 point groups exhibited by crystals, it is useful to appreciate the intimate connections found between the symmetry operations of a point group and the mathematical group. Therefore, we shall discuss certain elementary aspects of the group theory in this section.

Group Concepts

A group in a mathematical sense is a set of abstract numbers { 1, 2, 3…} or rational, integer, or real numbers {a, b, c…}, while in chemistry/crystallography, it is the symmetry elements, for all an operation is defined such that a third element is associated with any ordered pair "multiplication". This operation must satisfy the four group conditions (also called group axioms) given below. From a group, here we actually

© The Author(s), under exclusive license to Springer Nature Singapore Pte Ltd. 2022   49
M. A. Wahab, *Symmetry Representations of Molecular Vibrations*, Springer Series
in Chemical Physics 126, https://doi.org/10.1007/978-981-19-2802-4_2

mean a discrete group (applicable to point group symmetries of molecules/crystalline solids).

Group Axioms

1. **Closure**

A product of two symmetry elements A and B in a group is equivalent to a symmetry element C, also an element of the same group, such that

$$AB = C \tag{2.1}$$

where the product AB means the operation B followed by the operation A. In general, the product $AB \neq BA$. However, if $AB = BA$, the group is said to be commutative or abelian.

2. **Associativity**

Every element of the group obeys the associative law of combination. If A, B, and C are the elements of the group, then

$$(AB)C = A(BC) \tag{2.2}$$

3. **Identity Element**

Every group contains one element called the identity element E, such that

$$AE = EA = A \tag{2.3}$$

4. **Inverse Element**

Every element in the group has its inverse also in the group, such that

$$AA^{-1} = A^{-1}A = E \tag{2.4}$$

Simple Example of a Group

For a simple example of a group, let us consider the permutation group of three elements, P(3). Below we provide the list of $3! = 6$ possible permutations that can be carried out, where the top row denotes the initial arrangement of the three numbers shown in Fig. 2.1 and the bottom row denotes the final arrangement after each permutation.

$$E = \begin{pmatrix} 1\ 2\ 3 \\ 1\ 2\ 3 \end{pmatrix} \quad A = \begin{pmatrix} 1\ 2\ 3 \\ 2\ 1\ 3 \end{pmatrix} \quad B = \begin{pmatrix} 1\ 2\ 3 \\ 1\ 3\ 2 \end{pmatrix}$$

$$C = \begin{pmatrix} 1\ 2\ 3 \\ 3\ 2\ 1 \end{pmatrix} \quad D = \begin{pmatrix} 1\ 2\ 3 \\ 3\ 1\ 2 \end{pmatrix} \quad F = \begin{pmatrix} 1\ 2\ 3 \\ 2\ 3\ 1 \end{pmatrix}$$

Fig. 2.1 C_{3v} (3m) symmetry operations

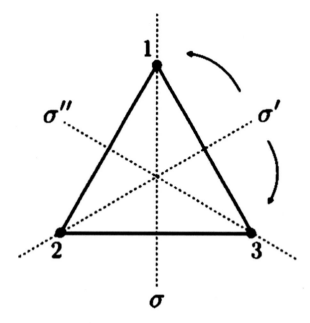

This group is found to be identical to the symmetry operations on an equilateral triangle shown in Fig. 2.1, where the six elements on the top row correspond to three points (represented by 1, 2, and 3) of the triangle in the initial state and the bottom row as the effect of the six distinct symmetry operations that can be performed on these three points. We call each symmetry operation an element of the group. They are the identity E, counter clockwise and clockwise 120° rotations (one each), and three reflections σ, σ', and σ''. They belong to the point group C_{3v}, and the corresponding group multiplications are provided in Table 19 (Appendix).

The equilateral triangle has the following six symmetries:
(i) Identity E, (ii) rotation C_3 by $2\pi/3$, (iii) rotation C_3^2 by $-2\pi/3$, (iv) reflection σ, (v) reflection σ', and (vi) reflection σ''. Based on the symmetry operations, one can deduce the corresponding six matrices that satisfy the group multiplication Table 19, and the symmetry matrices are as follows:

$$E = \begin{pmatrix} 1 & 0 \\ 0 & 1 \end{pmatrix}, C_3 = \frac{1}{2}\begin{pmatrix} -1 & -\sqrt{3} \\ \sqrt{3} & -1 \end{pmatrix}, C_3^2 = \frac{1}{2}\begin{pmatrix} -1 & \sqrt{3} \\ -\sqrt{3} & -1 \end{pmatrix}$$

$$\sigma = \begin{pmatrix} -1 & 0 \\ 0 & 1 \end{pmatrix}, \sigma' = \frac{1}{2}\begin{pmatrix} 1 & \sqrt{3} \\ \sqrt{3} & -1 \end{pmatrix}, \sigma'' = \frac{1}{2}\begin{pmatrix} 1 & -\sqrt{3} \\ -\sqrt{3} & -1 \end{pmatrix}$$

These matrices constitute a matrix representation of the group that is isomorphic to P(3) and to the symmetry operations on an equilateral triangle.

Order of the Group

In general, the order of the group is the number of non-equivalent symmetry elements in the group. For example, the point group C_{2h} (2/m) has four non-equivalent symmetry elements E (1), C_2 (2), σ_h (m), and S_2 ($\bar{1}$). Hence, the order of this point group is 4.

Example 2.1 A pyridine molecule has the following symmetry elements E, C_{2x}, C_{2y}, and C_{2z}. Show that they constitute a group. Determine its order and the point group.

Solution: *Given*: Four symmetry elements of a pyridine molecule are as follows: E, C_{2x}, C_{2y}, and C_{2z} (refer to Table 1.12); order of group = ? point group = ?
   Let us check that they follow the group conditions.

(i)   To check the closure property, let us take their products

$$C_{2x}C_{2y} = C_{2z}$$
$$C_{2y}C_{2z} = C_{2x}$$
$$C_{2z}C_{2x} = C_{2y}$$

       These products can be verified from Fig. 2.2 and also from the group multiplication Table 7 (Appendix).
(ii)   To cheek the associative property, let us consider the triple product

Fig. 2.2 Symmetry operations of point group D_2 (222)

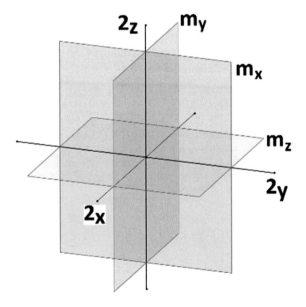

$$C_{2x}(C_{2y}C_{2z}) = (C_{2x}C_{2y})C_{2z}$$
$$\text{LHS} = C_{2x}(C_{2y}C_{2z}) = C_{2x}C_{2x} = C_{2x}^2 = E$$
$$\text{RHS} = (C_{2x}C_{2y})C_{2z} = C_{2z}C_{2z} = C_{2z}^2 = E$$
$$\Rightarrow \text{LHS} = \text{RHS}$$

(iii) The given symmetry elements contain one identity element, E which leaves the other members of the group unchanged, i.e.,

$$EC_{2x} = C_{2x}E = C_{2x}$$
$$EC_{2y} = C_{2y}E = C_{2y}$$
$$EC_{2z} = C_{2z}E = C_{2z}$$

(iv) To check the inverse, we observe that the inverse of E is E itself, while that of

$$C_{2x} \text{ is } C_{2x}^{-1} = C_{2x}^{-1}C_{2x}^2 = C_{2x}$$
$$C_{2y} \text{ is } C_{2y}^{-1} = C_{2y}^{-1}C_{2y}^2 = C_{2y}$$
$$C_{2z} \text{ is } C_{2z}^{-1} = C_{2z}^{-1}C_{2z}^2 = C_{2z}$$
$$\Rightarrow \text{ Every symmetry element has its own inverse.}$$

Since four symmetry elements of the pyridine molecule are independent and satisfy all the group conditions, hence they form a group of order 4. Further, from Fig. 2.2, we observe that there are three mutually perpendicular mirror planes providing three mutually perpendicular twofold axes, one each along the axis of their intersection; therefore, the point group is D_2 (222).

Example 2.2 A dinitrogen-difluoride molecule has the following symmetry elements E, C_2, σ_h, and i. Show that they constitute a group. Determine its order and its point group.

Solution: *Given:* Four symmetry elements of dinitrogen-difluoride molecule are as follows: E, C_2, σ_h, and i (refer to Table 1.12); order of group = ? point group = ?
Let us check that they follow the group conditions.

(i) To check the closure property, let us take their products

$$C_2\sigma_h = i$$
$$C_2 i = \sigma_h$$
$$\sigma_h i = C_2$$

These products can be verified from Fig. 1.13 (where x_3 is twofold) and also from the group multiplication Table 6 (Appendix).

(ii) To cheek the associative property, let us consider the triple product

$$C_2(\sigma_h i) = (C_2\sigma_h)i$$
$$LHS = C_2(\sigma_h i) = C_2 C_2 = C_2^2 = E$$
$$RHS = (C_2\sigma_h)i = i.i = E$$
$$\Rightarrow LHS = RHS$$

(iii)   The given symmetry elements contain one identity element, E which leaves the other members of the group unchanged, i.e.,

$$EC_2 = C_2 E = C_2$$
$$E\sigma_h = \sigma_h E = \sigma_h$$
$$Ei = iE = i$$

(iv)   To check the inverse, we observe that the inverse of E is E itself, while that of

$$C_2 \text{ is } C_2^{-1} = C_2^{-1}C_2^2 = C_2$$
$$\sigma_h \text{ is } \sigma_h^{-1} = \sigma_h^{-1}\sigma_h^2 = \sigma_h$$
$$\text{and i is } i^{-1} = i^{-1}i^2 = i$$
$$\Rightarrow \text{ Every symmetry element has its own inverse.}$$

Since four symmetry elements of the dinitrogen-difluoride molecule are independent and satisfy all the group conditions, hence they form a group of order 4. Further, there is one horizontal mirror plane (σ_h) perpendicular to twofold rotation axis; therefore, the point group is C_{2h} (2/m).

Cyclic Group

A group is said to be cyclic if all elements of the group can be generated by one element, such as A, A^2, A^3, ... A^n (=E). The element A is called the generator of the group, where n refers to the total number of elements in the group and is called the order of the group. Cyclic groups are abelian but the converse is not true. The point groups C_2 (2), C_3 (3), C_4 (4), and C_6 (6) are some examples of cyclic groups.

Example 2.3 Show that the symmetry elements E, C_4, $C_4^2 = (C_2)$, and C_4^3 constitute a cyclic group. What is the point group?

Solution: *Given*: Four symmetry elements are as follows: E, C_4, $C_4^2 = (C_2)$, and C_4^3; point group = ?

From the definition of a cyclic group, we know that only one element can generate all other elements of the group. Here, it is C_4 which can generate all other members. For example,

$$C_4 C_4 = C_4^2 = C_2, C_4 C_2 = C_4 C_4^2 = C_4^3, C_4 C_4^3 = C_4^4 = E$$

Now, let us check that they follow the group conditions.

(i)   To check the closure property, let us take their products, i.e.,

Fig. 2.3 Symmetry operations of point group C_4 (4)

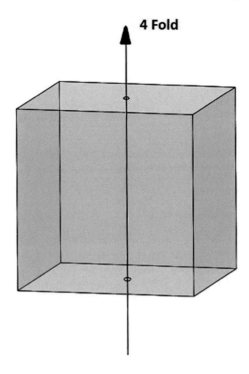

4 Fold

$$C_4C_2 = C_4^3$$
$$C_2C_4^3 = C_4$$
$$C_4C_4 = C_2$$

These operations can be verified from Fig. 2.3 and also from group multiplication Table 9 (Appendix).

(ii)   To check the associative property, let us consider the triple product, i.e.,

$$C_4(C_4^2 C_4^3) = (C_4 C_4^2)C_4^3$$
$$\text{LHS} = C_4(C_4^2 C_4^3) = C_4 C_4 = C_4^2 = C_2$$
$$\text{RHS} = (C_4 C_4^2)C_4^3 = C_4^3 C_4^3 = C_4^2 = C_2$$
$$\Rightarrow \text{LHS} = \text{RHS}$$

(iii)   The given symmetry elements contain one identity element E, which leaves the other members of the group unchanged, i.e.,

$$EC_4 = C_4 E = C_4$$
$$EC_4^2 = C_4^2 E = C_4^2$$
$$EC_4^3 = C_4^3 E = C_4^3$$

(iv)   Every given symmetry element is found to have its own inverse, i.e.,

$$C_4 \text{ is } C_4^{-1} = C_4^{-1}C_4^4 = C_4^3$$
$$C_2 \text{ is } C_2^{-1} = C_2^{-1}C_2^2 = C_2$$
$$C_4^3 \text{ is } C_4^{-3} = C_4^{-3}C_4^4 = C_4$$
$$\Rightarrow \text{ Every symmetry element has its own inverse.}$$

Since the four given symmetry elements are independent and satisfy all the group conditions, therefore they form a group of order 4. Since $C_4^4 = E$, the point group, C_4 (4), is a cyclic group.

Abelian Group

A group is said to be abelian if all its elements commute with one another. Further, the two elements A and B are said to commute with one another if $AB = BA$. In abelian groups, each element is in a class by itself, since

$$XAX^{-1} = AXX^{-1} = AE = A$$

Example 2.4 Do the point groups C_1 (1), C_2 (2), C_3 (3), C_4 (4), and C_6 (6) under proper rotation are cyclic groups? Taking at least one case as an example, show that they are abelian also.

Solution: *Given*: Point groups of proper rotation are as follows: C_1 (1), C_2 (2), C_3 (3), C_4 (4), and C_6 (6), show they are cyclic, abelian also.

Let us consider C_6 point group as a test case. This is the last member of proper rotation which generates the other five members as follows:

$$C_6 \cdot C_6 = C_6^2 = (C_3), C_6^3 = (C_2), C_6^4 = \left(C_3^2\right), C_6^5 \text{ and } C_6^6 = E$$

Since the element C_6 generates the other five members of the group, therefore it is cyclic.

Now to show the C_6 group is abelian, it should commute with all other elements of the group. That is C_6 must follow the condition $AB = BA$ with other members of the group element, and let us check the following:

$$EC_6 = C_6E = C_6$$
$$C_6C_6 = C_6^2 = C_3$$
$$C_6^2C_6 = C_6C_6^2 = C_6^3 = C_2$$
$$C_6^3C_6 = C_6C_6^3 = C_6^4 = C_3^2$$
$$C_6^4C_6 = C_6C_6^4 = C_6^5$$
$$C_6^5C_6 = C_6C_6^5 = E$$

This exercise shows that C_6 commutes with all other members of the group, hence this is abelian also. In a similar manner, other cyclic point groups can be shown to be abelian also.

Example 2.5 Show that the point group C_{3v} (3m) whose symmetries are E, C_3, C_3^2, σ_x, σ_y, and σ_{xy}, belong to non-abelian group.

Solution: *Given*: For point group C_{3v} (3m), symmetries are as follows: E, C_3, C_3^2, σ_x, σ_y, and σ_{xy}; group = ?

We know that a group is said to be abelian if all its elements commute with one another. From the group multiplication Table 19 (Appendix), we find

$$C_3\sigma_x = \sigma_{xy} \text{ and } \sigma_x C_3 = \sigma_y \quad \Rightarrow C_3\sigma_x \neq \sigma_x C_3$$

Similarly,

$$C_3\sigma_y = \sigma_x \text{ and } \sigma_y C_3 = \sigma_{xy} \quad \Rightarrow C_3\sigma_y \neq \sigma_y C_3$$

$$C_3\sigma_{xy} = \sigma_y \text{ and } \sigma_{xy}C_3 = \sigma_x \quad \Rightarrow C_3\sigma_{xy} \neq \sigma_{xy} C_3$$

Carrying out a similar exercise with C_3^2, we find

$$C_3^2\sigma_x = \sigma_y \text{ and } \sigma_x C_3^2 = \sigma_{xy} \quad \Rightarrow C_3^2\sigma_x \neq \sigma_x C_3^2$$

Similarly,

$$C_3^2\sigma_y = \sigma_{xy} \text{ and } \sigma_y C_3^2 = \sigma_x \quad \Rightarrow C_3^2\sigma_y \neq \sigma_y C_3^2$$

$$C_3^2\sigma_{xy} = \sigma_x \text{ and } \sigma_{xy} C_3^2 = \sigma_y \quad \Rightarrow C_3^2\sigma_{xy} \neq \sigma_{xy} C_3^2$$

Since we find that C_3 (or C_3^2) and σ_v do not commute with each other, therefore the point group C_{3v} (3m) is non-abelian.

Isomorphic Group

Two or more groups are said to be isomorphic if they obey the same group multiplication table. This means that there is a one-to-one correspondence between elements A, B, ... of one group and those A', B', ... of the other, such that AB = C implies A'B' = C' and vice versa.

Example 2.6 Show that three point groups of order 2 and a mathematical group containing the elements 1, −1 are isomorphic.

Solution: *Given*: There are three point groups of order 2, viz. C_2, C_s, and C_i; one mathematical group containing the elements 1, −1.

Let us write their group multiplication tables and check the one-to-one correspondence between their elements. Group multiplication tables of C_2, C_s, and C_i and the elements 1, -1 are as follows:

C_2	E	C_2
E	E	C_2
C_2	C_2	E

C_s	E	σ_h
E	E	σ_h
σ_h	σ_h	E

C_i	E	i
E	E	i
i	i	E

$\overline{1}$	1	-1
1	1	-1
-1	-1	1

Here, we observe that $1 \leftrightarrow E$, $C_2 \leftrightarrow \sigma_h$, $\sigma_h \leftrightarrow i$, $i \leftrightarrow -1$, and $-1 \leftrightarrow C_2$ are related.

Elements of any one group show one-to-one correspondence with elements of any other group. Hence, they are isomorphic.

Example 2.7 Prove the following: (a) A group formed by the elements 1, i, –1, –i is only a mathematical group (MG) and does not have a point group, (b) the group formed by the elements 1, i, –1, –i is abelian, and (c) the above abelian mathematical group (MG) resembles with the point group C_4.

Solution: *Given*: One mathematical group (MG) contains elements 1, i, –1, –i as members. Show that they form a group, the group is abelian and resembles C_4.

(a)   Let us check the group conditions:

$$AB = C; 1 \times i = i, i \times -1 = -i, \text{etc.}$$
$$(AB)C = A(BC); (1 \times i) \times -1 = 1(i \times -1) = -i, \text{etc.}$$
$$AE = EA = A; 1 \times i = i \times 1 = i, \text{etc.}$$
$$\text{There exist an additive inverse; } 1 + (-1) = 0, \text{etc.}$$

Since the group elements follow the group conditions, therefore they form a group. However, these numbers are not associated with the symmetry operation of any molecular or crystallographic system.

(b)   Let us check the commutative aspect:

The combination of any two elements is also an element of the group. That is their products irrespective of their order of combination are all equal and are also an element of the group.

$$1 \times 1 = 1$$
$$1 \times i = i = i \times 1$$
$$1 \times -i = -i = -i \times 1$$
$$-1 \times 1 = -1 = 1 \times -1$$

The multiplications are commutative; hence, the group is abelian.

(iii)   Let us check the resemblance with the point group C_4:

Comparing the elements of the two groups, we observe that they follow a one-to-one criterion, i.e.,

C_4	C_4^2	C_4^3	$C_4^4 (=E)$
i	i^2	i^3	i^4
i	-1	$-i$	1

$\Rightarrow$ The two groups resemble with each other and are isomorphic.

Finite Group

A group containing a finite number of elements is called a finite group. For example, crystallographic point groups and space groups are finite groups.

Generators of a Finite Group

It is possible to generate all elements of a group by starting from a certain set of elements (at the most three) and taking their power and products. However, it is to be noted that the definition of a generator is not always unique. For example, in the point group (D_2) 222, the possible generator is 2[100] and 2[010] or 2[100] and 2[001] or 2[010] and 2[001].

Subgroups and Super Groups

A set of symmetry elements is said to be a subgroup of a bigger group (considered as a subgroup of the group) if the set itself forms a group and satisfies the group conditions. In general, every group has two trivial subgroups, the identity element and the group itself. However, in the simplest term, it can be said that the addition of symmetry elements to a point group produces super groups while the suppression of the symmetry elements from the point group produces subgroups. For example, we know that the point group E (1) is the least symmetric and is the subgroup of all other 31 point groups. On the other hand, the point groups D_{6h} (6/mmm) and O_h (m3m) can have no super group because no symmetry elements could be added to them to obtain any new point group. Subgroups, super groups, and order of the point groups are illustrated in Fig. 2.4.

A group is called a proper group if there are symmetry elements of the super group not contained in the subgroup. For example, the set of point groups E (1), C_2 (2), σ_h (m), and S_2 (i) is a proper subgroup of the point group C_{2h} (2/m).

Classes

There is a procedure following which the elements of a group can be separated into smaller sets, and such sets are called classes. Two elements A and B in a group will belong to the same class if there is an element X within the group such that

$$X^{-1}A\,X = B \qquad\qquad (2.5)$$

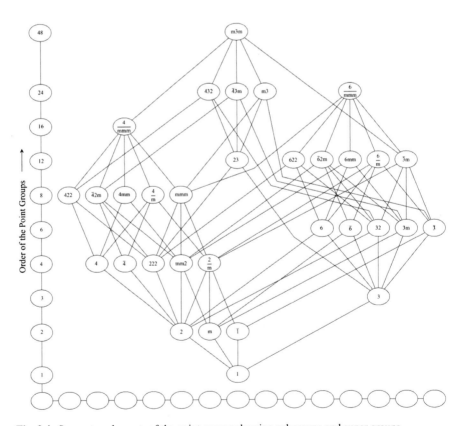

Fig. 2.4 Symmetry elements of the point groups showing subgroups and super groups

where X^{-1} is the inverse of X. From Eq. 2.5, we can say that B is a similarity transform of A by X, or that A and B are conjugate to one another. By making use of the similarity transform of one element by other elements, one can determine whether a set of elements form classes or not. Conjugate has the following properties:

(i)   Every element of a group is conjugate with itself, i.e.,

$$X^{-1}AX = A$$

(ii)   If A is conjugate with B, then B is conjugate with A.

$$A = X^{-1}BX$$
$$B = Y^{-1}AY$$

where Y is another element of the group.

(iii)   If A is conjugate with B and C, then B and C are conjugate with each other.

A complete set of elements that are conjugate with one another is called a class of the group.

The order of a class (c) of the group must be an integral factor of the order of the group (g), i.e.,

$$g = mc \qquad\qquad (2.6)$$

The similarity transformation method is sometimes too elaborate to find the classes particularly in high symmetry systems. Therefore, an alternative method may be used to find the classes of symmetry elements/operations in a group. This method consists of the following steps. They can be considered as rules.

1.   If symmetry element/operation commutes with all others (elements/operations), then it is in a separate class. For example, the elements E (1), C_i $(\bar{1})$, σ_h (or m_h), and C_2 (2) belong to separate classes.
2.   A rotation operation (proper or improper) and its inverse belong to the same class if there are n vertical mirrors or n perpendicular C_2 (2)-axes.
3.   Two rotations (proper or improper) about different axes belong to the same class if there is a third operation that interchanges points on these two axes.
4.   Two reflections through two different mirrors belong to the same class if there is a third operation that interchanges points on the two mirror planes.
5.   In addition to the above rules, according to the general relationship of group theory, the number of classes is equal to the number of irreducible representations of the group.

Example 2.8 H_2O (water) molecule has the symmetry elements E, C_2, σ_{xz}, and σ_{yz}. They belong to the point group C_{2v} (mm2). Show that they are members of different classes.

Solution: *Given*: Symmetry elements of H_2O molecule are as follows: E, C_2, σ_{xz}, and σ_{yz}, point group C_{2v} (mm2), classes = ?

From the above discussion, we know that the symmetry elements E, C_2, σ_{xz}, and σ_{yz} are inverses of their own. Now, applying similarity transform on the symmetry element C_2 and referring to the group multiplication table of C_{2v} (mm2), we get

$$EC_2E = EC_2 = C_2$$
$$C_2C_2C_2 = C_2E = C_2$$
$$\sigma_{xz}C_2\sigma_{xz} = \sigma_{xz}\sigma_{yz} = C_2$$
$$\sigma_{yz}C_2\sigma_{yz} = \sigma_{yz}\sigma_{xz} = C_2$$

$\Rightarrow$ All similarity transforms generated the same symmetry element C_2, and hence, it forms a class of its own.

Similarly, applying similarity transforms on other symmetry operations, we obtain

$$E\sigma_{xz}E = E\sigma_{xz} = \sigma_{xz}$$
$$C_2\sigma_{xz}C_2 = C_2\sigma_{yz} = \sigma_{xz}$$
$$\sigma_{xz}\sigma_{xz}\sigma_{xz} = \sigma_{xz}E = \sigma_{xz}$$
$$\sigma_{yz}\sigma_{xz}\sigma_{yz} = \sigma_{yz}C_2 = \sigma_{xz}$$

and

$$E\sigma_{yz}E = E\sigma_{yz} = \sigma_{yz}$$
$$C_2\sigma_{yz}C_2 = C_2\sigma_{xz} = \sigma_{yz}$$
$$\sigma_{xz}\sigma_{yz}\sigma_{xz} = \sigma_{xz}C_2 = \sigma_{yz}$$
$$\sigma_{yz}\sigma_{yz}\sigma_{yz} = \sigma_{yz}E = \sigma_{yz}$$

Like C_2, we find σ_{xz} and σ_{yz} operations also form their own separate classes. Hence, all the four members of the point group C_{2v} (mm2) belong to separate classes.

Example 2.9 Show that the symmetry elements C_3 and C_3^2 belong to different classes when they are members of the point group C_3 (3).

Solution: *Given*: Symmetry elements: C_3 and C_3^2, point group C_3 (3), class = ?

We know that the point group C_3 (3) has three members in it. They are E, C_3, and C_3^2, where the symmetry element E is the inverse of itself, while C_3 and C_3^2 are the inverses of each other. Now to check their classes, let us apply similarity transforms on the given symmetry element, and by referring to the group multiplication table of the point group C_3 (3), we get

$EEE = EE = E$	$EC_3E = EC_3 = C_3$
$C_3EC_3^2 = C_3C_3^2 = E$	$C_3C_3C_3^2 = C_3E = C_3$
$C_3^2EC_3 = C_3^2C_3 = E$	$C_3^2C_3C_3 = C_3^2C_3 = C_3$
$EC_3^2E = EC_3^2 = C_3^2$	
$C_3C_3^2C_3^2 = C_3C_3 = C_3^2$	
$C_3^2C_3^2C_3 = C_3^2E = C_3^2$	

From this exercise, we observe that different similarity transforms give different symmetry operations/elements. This implies that each symmetry element forms its own class. Since point group C_3 is cyclic and hence abelian. Accordingly, the symmetry elements C_3 and C_3^2 belong to different classes when they are members of the point group C_3.

Example 2.10 A complex ion tris(ethylenediamine)cobalt(III) $\left[Co(en)_3\right]^{3+}$ possesses idealized D_3 (32) point group symmetry. Determine the number of classes in it.

Solution: *Given*: For point group D_3 (32), symmetry elements are as follows: E, C_3, C_3^2, C_{2x}, C_{2y}, and C_{2xy}. The number of classes = ?

First of all, checking the inverses of the symmetry elements, we find that E is the inverse of itself, C_3 and C_3^2 are the inverses of each other, and C_{2x}, C_{2y}, and C_{2xy} are inverses of their own. Further, the identity element E belongs to a class of its own. Now, applying similarity transforms on C_3 and C_3^2, we obtain the following:

$$EC_3E = EC_3 = C_3$$
$$C_3C_3C_3^2 = C_3E = C_3$$
$$C_3^2C_3C_3 = C_3^2C_3^2 = C_3$$
$$C_{2x}C_3C_{2x} = C_{2x}C_{2xy} = C_3^2$$
$$C_{2y}C_3C_{2y} = C_{2y}C_{2x} = C_3^2$$
$$C_{2xy}C_3C_{2xy} = C_{2xy}C_{2y} = C_3^2$$

Similarly,

$$EC_3^2E = EC_3^2 = C_3^2$$
$$C_3C_3^2C_3^2 = C_3C_3 = C_3^2$$
$$C_3^2C_3^2C_3 = C_3^2E = C_3^2$$
$$C_{2x}C_3^2C_{2x} = C_{2x}C_{2y} = C_3$$
$$C_{2y}C_3^2C_{2y} = C_{2y}C_{2xy} = C_3$$
$$C_{2xy}C_3^2C_{2xy} = C_{2xy}C_{2x} = C_3$$

We observe that they have generated either C_3 or C_3^2 which means that the two symmetry operations are the members of the same class (also according to rule 2, mentioned above). Further, applying similarity transform on C_{2x}, we obtain

$$EC_{2x}E = EC_{2x} = C_{2x}$$
$$C_3C_{2x}C_3^2 = C_3C_{2xy} = C_{2y}$$
$$C_3^2C_{2x}C_3 = C_3^2C_{2y} = C_{2xy}$$
$$C_{2x}C_{2x}C_{2x} = C_{2x}E = C_{2x}$$
$$C_{2y}C_{2x}C_{2y} = C_{2y}C_3 = C_{2xy}$$
$$C_{2xy}C_{2x}C_{2xy} = C_{2xy}C_3^2 = C_{2y}$$

This generates C_{2x}, C_{2y}, and C_{2xy}. Similar results can be obtained when similarity transforms are taken on C_{2y} and C_{2xy}. The results imply that the operations C_{2x}, C_{2y}, and C_{2xy} belong to the same class (also according to rule 3, mentioned above). Therefore, the six symmetry elements E, C_3, C_3^2, C_{2x}, C_{2y}, and C_{2xy} belong to three different classes, and they are represented as E, $2C_3$, and $3C_2$.

Example 2.11 The point group D_4 (422) possesses the symmetry elements E, C_4, C_2, C_4^3, C_{2x}, C_{2y}, $C_2(1)$, and $C_2(2)$. Determine the number of classes in it.

Solution: *Given*: For point group D_4 (422), symmetry elements are as follows: E, C_4, C_4^2, C_4^3, C_{2x}, C_{2y}, $C_2(1)$, and $C_2(2)$, No. of classes = ?

We know that E is the inverse of itself. Similarly, C_2 is the inverse of itself. C_4 and C_4^3 are the inverses of each other. All other C_2 operations are inverses of their own.

Since the number of symmetry elements is large, let us use the rules instead of similarity transforms to obtain the number of classes. Thus, using the rules, we obtain the following:

1. According to rule 1, the identity element E forms a class of its own.
2. According to rule 2, C_2 belongs to a separate class.
3. Also, according to rule 2, both C_4 and its inverse C_4^3 belong to one class (due to the presence of a vertical C_2-axis).
4. According to rule 4, the first two twofold symmetries C_{2x} and C_{2y} belong to one class while the other two twofold symmetries $C_2(1)$ and $C_2(2)$ together belong to another class.

Hence, in all, there are five separate classes. Therefore, eight symmetry elements of the point group D_4 (422) belong to five classes. They are E, C_2, $2C_4$, $2C_{2x}$, and $2C_2(1)$.

Example 2.12 The point group D_{2d} ($\overline{4}2m$) possesses the symmetry elements E, C_2, S_4, S_4^3, C_{2x}, C_{2y}, σ_1, and σ_2. Determine the number of classes in it.

Solution: *Given*: For point group D_{2d} ($\overline{4}2m$), symmetry elements are as follows: E, C_2, S_4, S_4^3, C_{2x}, C_{2y}, σ_1, and σ_2, No. of classes = ?

We know that E and C_2 are inverses of their own. Similarly, S_4 and S_4^3 are the inverses of each other. C_{2x}, C_{2y}, σ_1, and σ_2 are inverses of their own.

Let us obtain the number of classes using the above-mentioned rules.

1. According to rule 1, the identity element E forms a class of its own.
2. According to rule 2, C_2 belongs to a separate class.

3. According to rule 2, S_4 and S_4^3 belong to one class (due to the vertical C_2-axis).
4. According to rule 3, C_{2x} and C_{2y} belong to another separate class (due to mirror plane σ_1 or σ_2 which interchanges points on the two axes).
5. According to rule 4, σ_1 and σ_2 belong to a separate class (due to the S_4-axis which interchanges points on the two mirrors).

Hence, in all, there are five separate classes. Therefore, eight symmetry elements of the point group D_{2d} ($\overline{4}2m$) belong to five classes. They are E, C_2, $2S_4$, $2C_{2x}$, and $2\sigma_1$.

2.3 Classifications of Crystallographic Point Groups

Based on the criterion, whether two symmetry elements/operations commute or not (or whether the operation AB equals BA or not), the crystallographic point groups can be classified as abelian or non-abelian, respectively. It is interesting to see that the numbers of abelian and non-abelian point groups in 3D are equal. However, 1D and 2D situations are relatively different. In 1D, there are two point groups E (1) and σ (m), both are cyclic, hence are abelian. On the other hand, in 2D, seven point groups out of ten are abelian and one out of seven is non-cyclic. It is interesting to see that in 3D, the abelian and non-abelian point groups are equal, i.e., they are 16 each. The classification of 2D and 3D point groups is provided in Tables 2.1 and 2.2, respectively.

In terms of percentage, 1D, 2D, and 3D crystals show 100, 70, and 50% abelian character, i.e., there is a decrease of abelian character by 50% in moving from 1 to 3D. Cyclic point groups are also found to decrease in the order 100%, 60%, and 31%, respectively, with the increase in crystal dimensions. They are shown in Fig. 2.5.

Table 2.1 Abelian and non-abelian points groups (2D)

Abelian		Order of the group/classes	Non-abelian		Order of the group	Classes
International notation	Schoenflies notation		International notation	Schoenflies notation		
1	C_1	1	3m	C_{3v}	6	3
2	C_2	2	4mm	C_{4v}	8	5
3	C_3	3	6mm	C_{6v}	12	6
4	C_4	4	–	–	–	–
6	C_6	6	–	–	–	–
m	$C_h(C_s)$	2	–	–	–	–
mm2	C_{2v}	4	–	–	–	–

Table 2.2 Abelian and non-abelian points groups (3D)

Abelian		Order of the group/classes	Non-abelian		Order of the group	Classes
International notation	Schoenflies notation		International notation	Schoenflies notation		
1	E	1	32	D_3	6	3
$\bar{1}$	$C_i(S_2)$	2	3m	C_{3v}		
2	C_2		$\bar{3}$m	D_{3d}	12	6
m	$C_h(C_s)$		422	D_4	8	5
mm2	C_{2v}	4	4mm	C_{4v}		
222	D_2 (V)		$\bar{4}$2m	$D_{2d}(V_d)$		
2/m	C_{2h}		4/mmm	D_{4h}	16	10
2/mmm	D_{2h}	8	622	D_6	12	6
3	C_3	3	6mm	C_{6v}		
$\bar{3}$	$S_6(C_{3i})$	6	$\bar{6}$m2	D_{3h}		
4	C_4	4	6/mmm	D_{6h}	24	12
$\bar{4}$	S_4		23	T	12	4
4/m	C_{4h}	8	m3	T_h	24	8
6	C_6	6	432	O		5
$\bar{6}$	C_{3h}		$\bar{4}$3m	T_d		
6/m	C_{6h}	12	m3m	O_h	48	10

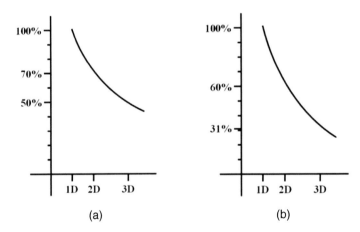

Fig. 2.5 a Variation of abelian character with crystal dimension. **b** Variation of cyclic character with crystal dimension

2.4 Construction of Group Multiplication Tables of 32 Point Groups

In a group, the product of two (or more) symmetry elements/operations can be shown to be equivalent to a single symmetry element/operation. The resultant of the products to a single operation can be completely summarized in a table called the group multiplication table. Such a table fully defines the multiplicative operations of symmetry elements. This implies that all symmetry operations and their products form a small, self-contained, closed set. For a given point group, a multiplication table can be constructed according to the following procedure.

1. In a product $AB = C$, B is the first operation (taken as the operations in the first row on the top of the table) and A is the second operation (taken as the operations in the first column on the extreme left of the table). For the sake of clarity, both the first row and the first column of all the group multiplication tables have been shaded.
2. The product elements/operations should follow all the rules of the group.
3. The first member of both operations (first and second operations) is an identity element/operation; therefore, the resultant of the products in the first row and the first column of the table is just the duplicates of the list of operations in the header row and extreme left column (Table 2.3).
4. Each row/column shows every element/operation once and only once. In other words, each row/column contains each element of the group once.
5. Point group elements/operations in the first row and first column of the table are set in a manner such that the product element on the diagonal of the group multiplication table is an identity element.

Based on the above-mentioned steps, we can construct a group multiplication table by taking the product of two cyclic groups, A (whose elements are E, a, a^2.... a^{n-1}) and B (whose elements are E, b, b^2....b^{n-1}) such that $ab = ba$ as shown in Table 2.3. The elements of the resultant group are E, a, b, ab (=ba), a^2 b, etc. It is a super group of both A and B.

Now, by making use of the procedure mentioned above, we can easily construct the group multiplication tables for all 32 point groups, while only some of them are found in the literature. Therefore, for the benefit of readers, we provide them all at

Table 2.3 Group multiplication table of (AB)

Cyclic group	First Operation			
Second Operation	E	a	b	c(= ab)
E	E	a	b	c
a	a	E	c	b
b	b	c	E	a
c	c	b	a	E

one place for the first time. The group multiplication tables for all point groups are constructed in pairs; one is based on H-M (International) and the other on Shoenflies notations, respectively. They are provided separately in the Appendix at the end of this chapter with the titles as Table 1(a) using International notation, Table 1(b) using Shoenflies notation, and so on. Here, it is important to note from Table 2.2 that the order of the following four point group D_{6h} (6/mmm) of hexagonal crystal system and T_h (m3), O (432), and T_d ($\overline{4}$3m) of cubic crystal system is 24 and that of the point group O_h (m3m) of cubic crystal system is 48, respectively. Accordingly, the sizes of their group multiplication tables are large, each of which cannot be accommodated in one page. Hence, the tables corresponding to the first four point groups (of order 24) have been conveniently divided into two parts in the form 2(12 × 24), where each part contains the elements of first and second operations as shown in Fig. 2.6. Similarly, the biggest table corresponding to the last point group (of order 48) is divided into eight parts in the form 8(12 × 24), where each part contains the elements of the first and second operations.

Points Groups: C_1 (1), $C_i(\overline{1})$, $C_s(\sigma_h = m_h)$, and C_2 (2)

The first point group C_1 (1) is a group of the lowest order, one. It consists of only one symmetry element/operation, called the identity element/operation, E (1), which brings no change in the position of the object as illustrated in Fig. 1.10. This is the least symmetric among 32 crystallographic point groups and common to all others. The corresponding group multiplications are provided in Table 1.

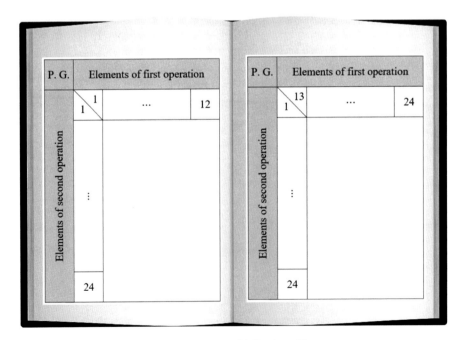

Fig. 2.6 Model for representing bigger group multiplication tables

The point group C_i ($\bar{1}$) is a cyclic group of order 2. It consists of inversion center (through the origin) and identity E (1) symmetry elements/operations. An inversion operation changes the sign of all three axes as illustrated in Fig. 1.11. The corresponding group multiplications are provided in Table 2.

The point group C_s or C_{1h} (m_h) is a cyclic group of order 2. It consists of reflection σ_h (m_h) and identity E (1) symmetry elements/operations. The reflection operation perpendicular to an axis (say z-axis) changes its handedness (Fig. 1.13). The corresponding group multiplications are provided in Table 3.

The point group C_2 (2) is a cyclic group of order 2. It consists of C_2 (twofold rotation) and identity E (1) symmetry elements/operations. The C_2 (2) operation along a particular axis (say z-axis) changes the sign of the other two axes shown in Fig. 1.12. However, according to the newly developed mirror combination scheme, when two mirrors intersect each other at 90° (right angle), a twofold axis is generated along the axis of their intersection as shown in Fig. 2.7. The corresponding group multiplications are provided in Table 4.

Point Groups: C_{2v} (mm2), C_{2h} (2/m), D_2 (222), and D_{2h} (mmm)

The point group C_{2v} (mm2) is an abelian group of order 4. It consists of two mirror planes (σ_{xz} and σ_{yz}) perpendicular to each other, a twofold rotation axis passing through the intersection of the two mirrors (Fig. 2.7), and an identity E (1) symmetry elements/operations, respectively. The corresponding group multiplications are provided in Table 5.

Fig. 2.7 Two mutually perpendicular mirrors and twofold axes

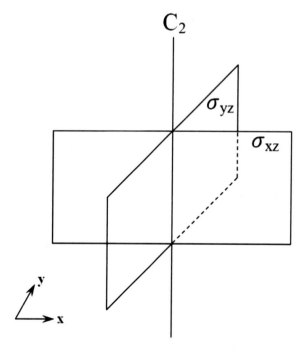

The point group C_{2h} (2/m) is an abelian group of order 4. This point group symmetry can be created only when a third mirror is introduced such that it is perpendicular to the two mirror combination producing mm2 symmetry as shown in Fig. 2.1. In fact, it is a combination of three mirrors and they are perpendicular to each other. The point group C_{2h} (2/m) consists of a C_2 (twofold rotation) axis, a C_s (a reflection through the plane perpendicular to the z-axis), a C_i (an inversion through the center, taken as origin), and an identity E (1) symmetry elements/operations, respectively. The corresponding group multiplications are provided in Table 6.

The point group D_2 (222) is an abelian group of order 4. The combination of three mutually perpendicular mirrors will produce three twofold axes, one each along their respective intersections as shown in Fig. 2.1. Therefore, the D_2 (222) point group consists of three mutually perpendicular twofold axes and an identity E (1) symmetry elements/operations. The corresponding group multiplications are provided in Table 7.

The point group D_{2h} (mmm) is an abelian group of order 8. It consists of three mutually perpendicular mirror planes (m_x, m_y, and m_z), three mutually perpendicular C_2 (twofold rotation) axes, an inversion through the center, taken as origin (Fig. 2.1), and identity E (1) symmetry elements/operations, respectively. The corresponding group multiplications are provided in Table 8.

Point Groups: C_4 (4), $S_4(\bar{4})$, C_{4h} (4/m), D_4 (422), C_{4v} (4mm), $D_{2d}(\bar{4}2m)$, and D_{4h} (4/mmm)

The point groups C_4 (4) and S_4 ($\bar{4}$) are the member of a cyclic group of order 4. They are shown in Fig. 2.8. They consist of one C_4 or S_4 (i.e., 4 or $\bar{4}$-fold rotation) along the z-axis and its powers. The corresponding group multiplications are provided in Tables 9 and 10, respectively.

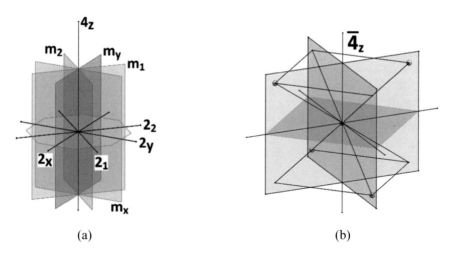

(a)   (b)

Fig. 2.8 An illustration of the groups (**a**) $C_4(4)$, D4(422), C_{4h} (4/m), C_{4v} (4mm), and D_{4h} (4/mmm); (**b**) $S_4(\bar{4})$ and $D_{2d}(\bar{4}2m)$ symmetry

The point group C_{4h} (4/m) is an abelian group of order 8. Its illustration in the mirror combination scheme is shown in Fig. 2.8a. It consists of the members of both C_4 (4) and S_4 ($\bar{4}$) cyclic groups (when counting the common members only once), a reflection σ_h through a plane perpendicular to the axis of rotation (z-axis), and a center of inversion, respectively. The corresponding group multiplications are provided in Table 11.

The point group D_4 (422) is a group of order 8. It consists of the members of C_4 (4) cyclic group and four twofold axes perpendicular to the fourfold axis [two each along (x and y) and (2_1 and 2_2) with inter-axial angle of 45°] as clearly shown in Fig. 2.8a. The corresponding group multiplications are provided in Table 12.

The point group C_{4v} (4mm) is a group of order 8. It consists of the members of C_4 cyclic group and four vertical mirrors [two each along (x and y) and (2_1 and 2_2) with inter-planar angle of 45°] whose intersections coincide with the fourfold rotation along the z-axis as shown in Fig. 2.8a. The corresponding group multiplications are provided in Table 13.

The point group D_{2d} ($\bar{4}$2m) is a group of order 8. It consists of the members of S_4 ($\bar{4}$) cyclic group, two vertical mirrors whose intersections coincide with the S_4-axis and two twofold axes along x- and y-directions as shown in Fig. 2.8b. The corresponding group multiplications are provided in Table 15.

The point group D_{4h} (4/mmm) is a group of order 16. Its extended form (4/m 2/m 2/m) is clearly seen to be exhibited in Fig. 2.8a. It consists of all the eight members of D_4 (422) group and other members can be obtained by using the product $D_4 \times C_i$. The corresponding group multiplications are provided in Table 14.

Point Groups: C_3 (3) and C_{3i} ($\bar{3}$)

The point group C_3 (3) is a cyclic group of order 3. It consists of one C_3 (threefold) rotation (where inter-planar angle is 60° as shown in Fig. 2.9a) along the z-axis and its powers. The corresponding group multiplications are provided in Table 16.

The point group C_{3i} or S_6 ($\bar{3}$) is a cyclic group of order 6. It consists of six members and three members of C_3 (3) point group, and the other three members can be obtained by using the product $C_3 \times C_i$. The corresponding group multiplications are provided in Table 17.

Point Groups: D_3 (32), C_{3v} (3m), and D_{3d} ($\bar{3}$m)

The point group D_3 (32) is a group of order 6. It consists of a threefold axis and three twofold axes making 120° angles with one another in a plane perpendicular to the threefold axis (Fig. 2.10a). The corresponding group multiplications are provided in Table 18.

The point group C_{3v} (3m) is a group of order 6. It consists of all the three members of C_3 (3) point group and the other three from vertical mirrors making 120° angles with one another, the intersections of which coincide with the threefold axis (Fig. 2.10a). The corresponding group multiplications are provided in Table 19.

The point group D_{3d} ($\bar{3}$m) is a group of order 12. It consists of all the six members of C_{3v} (3m) point group and the other six members can be obtained by using the product $C_{3v} \times C_i$. The corresponding group multiplications are provided in Table 20.

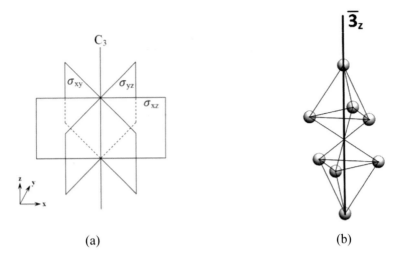

Fig. 2.9 a Symmetry operations of point group C_3 (3) and **b** symmetry operations of point group C_{3i} ($\bar{3}$)

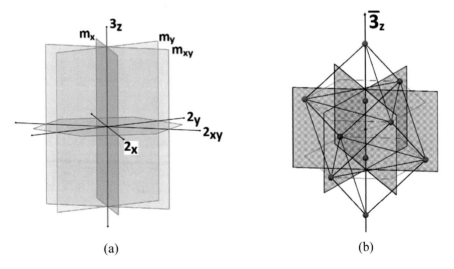

Fig. 2.10 a Symmetry operations of point group D_3 (32) and C_{3v} (3m) and **b** symmetry operations of point group D_{3d} ($\bar{3}$m)

Point Groups: C_{3h} ($\bar{6}$) and D_{3h} ($\bar{6}$2m)

The point group C_{3h} or S_3 ($\bar{6}$) is a cyclic group of order 6 (Fig. 2.11a). It consists of a S_3 ($\bar{6}$-fold rotation) rotoinversion axis and its powers. The corresponding group multiplications are provided in Table 21.

The point group D_{3h} ($\bar{6}$2m) is a group of order 12. It consists of the members of S_6 ($\bar{3}$) cyclic group, three vertical mirrors whose intersections coincide with the S_6 ($\bar{3}$)

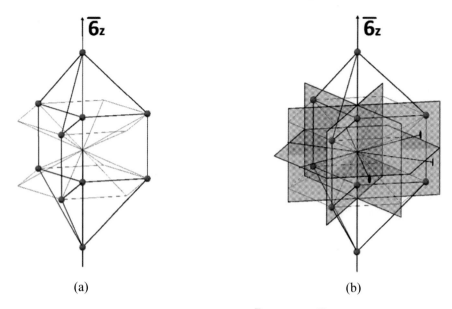

Fig. 2.11 Symmetry operations of point group **a** C$_{3h}$ ($\bar{6}$) and **b** D$_{3h}$ ($\bar{6}$2m)

rotoreflection axis and three twofold axes along x-, y-, and xy-directions as shown in Fig. 2.11b. The corresponding group multiplications are provided in Table 22.

Point Groups: C$_6$ (6), C$_{6h}$ (6/m), D$_6$ (622), C$_{6v}$ (6mm), and D$_{6h}$ (6/mmm)

The point group C$_6$ (6) is a cyclic group of order 6. It consists of one C$_6$ (sixfold rotation) axis and its powers. The corresponding group multiplications are provided in Table 23 (Fig. 2.12).

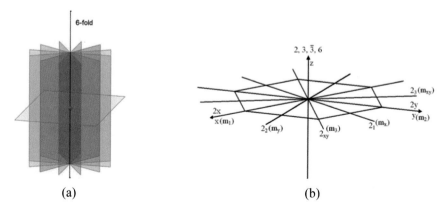

Fig. 2.12 Symmetry operations of point group C$_6$ (6), C$_{6h}$ (6/m), D$_6$ (622), C$_{6v}$ (6mm), and D$_{6h}$ (6/mmm); **a** mirror combinations and **b** line diagram

The point group C_{6h} (6/m) is an abelian group of order 12. It consists of all the six members of C_6 (6) point group and the other six members can be obtained by using the product $C_6 \times C_i$. The corresponding group multiplications are provided in Table 24.

The point group D_6 (622) is a group of order 12. It consists of a sixfold axis and six twofold axes making 60° angles with one another in a plane perpendicular to the sixfold axis. The corresponding group multiplications are provided in Table 25.

The point group C_{6v} (6mm) is a group of order 12. It consists of all the six members of C_6 (6) group and six vertical mirrors making a 60° angle with one another, the intersection of which coincides with the sixfold axis. The corresponding group multiplications are provided in Table 26.

The point group D_{6h} (6/mmm) is a group of order 24. It consists of all the twelve members of D_6 (622) point group and the other twelve members can be obtained by using the product $D_6 \times C_i$. The corresponding group multiplications are provided in Table 27. It is the first big table which is divided into two parts for its representation in the form of a model shown in Fig. 2.6.

Point Groups: T (23), T$_h$ (m3), T$_d$ ($\overline{4}$3m), O (432), and O$_h$ (m3m)

The point group T (23) of order 12 is the first member of the cubic crystal system. It consists of eight threefold proper rotations (4 each of first and second order) about four body diagonals and three twofold proper rotations along the cube axes, respectively, as shown in Fig. 2.13. The corresponding group multiplications are provided in Table 28.

The point group T_h (m3) is a group of order 24. It consists of all the twelve members of T (23) group, six S_4 ($\overline{4}$) improper rotations about the cube axes, and six diagonal mirror planes (σ_d) passing through the center of the cube and containing six edges of the tetrahedron. The corresponding group multiplications are provided in two parts in Table 29.

The point group T_d ($\overline{4}$3m) is a group of order 24. It consists of all twelve members of T (23) group, six S_4 ($\overline{4}$) improper rotations about the cube axes, and six dihedral mirror (σ_d) planes passing through the center of the cube and containing six edges of

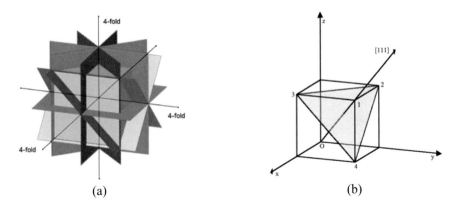

(a)   (b)

Fig. 2.13 **a** Showing cubic symmetries and **b** showing tetrahedral axes

the tetrahedron. The corresponding group multiplications are provided in two parts in Table 30.

The point group O (432) is a group of order 24. It consists of all the twelve members of T (23) group, six fourfold proper rotations about the cube axes, and six twofold rotations about the six axes extending from the center of the cube to the midpoints of the cube faces. The corresponding group multiplications are provided in two parts in Table 31.

The point group O_h (m3m) is of the highest order 48 and is the last member of the cubic crystal system. It consists of all 24 members of O (432) group and the other 24 members are obtained from the product $O \times C_i$. This is the most symmetric among the 32 crystallographic point groups. The corresponding group multiplications are provided in eight parts in Table 32.

2.5 Summary

1. A set of symmetry elements/operations forms a group if and only if the following four group conditions (also called as group axioms) are satisfied.

    (i)   Closure property
    (ii)   Associative property
    (iii) Property of Identity
    (iv)   Property of Inverse

2. A group may be of different types: cyclic, abelian, isomorphic, finite, super groups, and subgroups.

3. In a group, the product of two (or more) symmetry elements/operations can be shown to be equivalent to a single symmetry element/operation. The resultant of the products to a single operation can be completely summarized in a table called the group multiplication table. For a given point group, a multiplication table can be constructed according to the following procedure.

    (i)   In a product AB = C, B is the first operation (taken as the operations in the first row on the top of the table) and A is the second operation (taken as the operations in the first column on the extreme left of the group multiplication table).
    (ii)   The product elements/operations should follow all the rules of the group.
    (iii) The first member of both operations (first operation and second operation) is an identity element/operation; therefore, the resultant of the products in the first row and the first column of the table just duplicates the list of operations in the header row and extreme left column.
    (iv)   Each row/column shows every element/operation once and only once. In other words, each row/column contains each element of the group once.
    (v)   Point group elements/operations in the first row and first column of the table are set in a manner such that the product element on the diagonal of the group multiplication table is an identity element.

4. Based on the procedure mentioned above, group multiplication tables of all 32 crystallographic point groups have been constructed for the first time using the two point group notations and provided in the Appendix at the end of this chapter.

Appendix

Group Multiplication Tables of 32 Point Groups

Table 1(a)

1	1
1	1

Table 1(b)

E	E
E	E

Table 2(a)

$\bar{1}$	1	$\bar{1}$
1	1	$\bar{1}$
$\bar{1}$	$\bar{1}$	1

Table 2(b)

C_i	E	i
E	E	i
i	i	E

Table 3(a)

m	1	m
1	1	m
m	m	1

Table 3(b)

C_s	E	σ_h
E	E	σ_h
σ_h	σ_h	E

Table 4(a)

2	1	2
1	1	2
2	2	1

Table 4(b)

C_2	E	C_2
E	E	C_2
C_2	C_2	E

Table 5(a)

mm2	1	2	m_1	m_2
1	1	2	m_1	m_2
2	2	1	m_2	m_1
m_1	m_1	m_2	1	2
m_2	m_2	m_1	2	1

Table 5(b)

C_{2v}	E	C_2	σ_{xz}	σ_{yz}
E	E	C_2	σ_{xz}	σ_{yz}
C_2	C_2	E	σ_{yz}	σ_{xz}
σ_{xz}	σ_{xz}	σ_{yz}	E	C_2
σ_{yz}	σ_{yz}	σ_{xz}	C_2	E

Table 6(a)

2/m	1	2	m	$\bar{1}$
1	1	2	m	$\bar{1}$
2	2	1	$\bar{1}$	m
m	m	$\bar{1}$	1	2
$\bar{1}$	$\bar{1}$	m	2	1

Table 6(b)

C_{2h}	E	C_2	σ_h	i
E	E	C_2	σ_h	i
C_2	C_2	E	i	σ_h
σ_h	σ_h	i	E	C_2
i	i	σ_h	C_2	E

Table 7(a)

222	1	2_x	2_y	2_z
1	1	2_x	2_y	2_z
2_x	2_x	1	2_z	2_y
2_y	2_y	2_z	1	2_z
2_z	2_z	2_y	2_x	1

Table 7(b)

D_2	E	C_{2x}	C_{2y}	C_{2z}
E	E	C_{2x}	C_{2y}	C_{2z}
C_{2x}	C_{2x}	E	C_{2z}	C_{2y}
C_{2y}	C_{2y}	C_{2z}	E	C_{2x}
C_{2z}	C_{2z}	C_{2y}	C_{2x}	E

Table 8(a)

mmm	1	2_x	2_y	2_z	$\bar{1}$	m_x	m_y	m_z
1	1	2_x	2_y	2_z	$\bar{1}$	m_x	m_y	m_z
2_x	2_x	1	2_z	2_y	m_x	$\bar{1}$	m_z	m_y
2_y	2_y	2_z	1	2_x	m_y	m_z	$\bar{1}$	m_x
2_z	2_z	2_y	2_x	1	m_z	m_y	m_x	$\bar{1}$
$\bar{1}$	$\bar{1}$	m_x	m_y	m_z	1	2_x	2_y	2_z
m_x	m_x	$\bar{1}$	m_z	m_y	2_x	1	2_z	2_y
m_y	m_y	m_z	$\bar{1}$	m_x	2_y	2_z	1	2_x
m_z	m_z	m_y	m_x	$\bar{1}$	2_z	2_y	2_x	1

Table 8(b)

D_{2h}	E	C_{2x}	C_{2y}	C_{2z}	i	σ_x	σ_y	σ_z
E	E	C_{2x}	C_{2y}	C_{2z}	i	σ_x	σ_y	σ_z
C_{2x}	C_{2x}	E	C_{2z}	C_{2y}	σ_x	i	σ_z	σ_y
C_{2y}	C_{2y}	C_{2z}	E	C_{2x}	σ_y	σ_z	i	σ_x
C_{2z}	C_{2z}	C_{2y}	C_{2x}	E	σ_z	σ_y	σ_x	i
i	i	σ_x	σ_y	σ_z	E	C_{2x}	C_{2y}	C_{2z}
σ_x	σ_x	i	σ_z	σ_y	C_{2x}	E	C_{2z}	C_{2y}
σ_y	σ_y	σ_z	i	σ_x	C_{2y}	C_{2z}	E	C_{2x}
σ_z	σ_z	σ_y	σ_x	i	C_{2z}	C_{2y}	C_{2x}	E

Table 9(a)

4	$4^4(=1)$	4^3	$4^2(=2)$	4
1	1	4^3	2	4
4	4	1	4^3	2
2	2	4	1	4^3
4^3	4^3	2	4	1

Table 9(b)

C_4	E	C_4^3	C_2	C_4
E	E	C_4^3	C_2	C_4
C_4	C_4	E	C_4^3	C_2
C_2	C_2	C_4	E	C_4^3
C_4^3	C_4^3	C_2	C_4	E

Table 10(a)

$\bar{4}$	$\bar{4}^4(=1)$	$\bar{4}^3$	$\bar{4}^2(=2)$	$\bar{4}$
1	1	$\bar{4}^3$	2	$\bar{4}$
$\bar{4}$	$\bar{4}$	1	$\bar{4}^3$	2
2	2	$\bar{4}$	1	$\bar{4}^3$
$\bar{4}^3$	$\bar{4}^3$	2	$\bar{4}$	1

Table 10(b)

S_4	$S_4^4(=E)$	S_4^3	$S_4^2(=C_2)$	S_4
E	E	S_4^3	C_2	S_4
S_4	S_4	E	S_4^3	C_2
C_2	C_2	S_4	E	S_4^3
S_4^3	S_4^3	C_2	S_4	E

Table 11(a)

4/m	1	4^3	2	4	$\bar{1}$	$\bar{4}^3$	m	$\bar{4}$
1	1	4^3	2	4	$\bar{1}$	$\bar{4}^3$	m	$\bar{4}$
4	4	1	4^3	2	$\bar{4}$	$\bar{1}$	$\bar{4}^3$	m
2	2	4	1	4^3	m	$\bar{4}$	$\bar{1}$	$\bar{4}^3$
4^3	4^3	2	4	1	$\bar{4}^3$	m	$\bar{4}$	$\bar{1}$
$\bar{1}$	$\bar{1}$	$\bar{4}^3$	m	$\bar{4}$	1	4^3	2	4
$\bar{4}$	$\bar{4}$	$\bar{1}$	$\bar{4}^3$	m	4	1	4^3	2
m	m	$\bar{4}$	$\bar{1}$	$\bar{4}^3$	2	4	1	4^3
$\bar{4}^3$	$\bar{4}^3$	m	$\bar{4}$	$\bar{1}$	4^3	2	4	1

Table 11(b)

C_{4h}	E	C_4^3	C_2	C_4	i	S_4^3	σ_h	S_4
E	E	C_4^3	C_2	C_4	i	S_4^3	σ_h	S_4
C_4	C_4	E	C_4^3	C_2	S_4	i	S_4^3	σ_h
C_2	C_2	C_4	E	C_4^3	σ_h	S_4	i	S_4^3
C_4^3	C_4^3	C_2	C_4	E	S_4^3	σ_h	S_4	i
i	i	S_4^3	σ_h	S_4	E	C_4^3	C_2	C_4
S_4	S_4	i	S_4^3	σ_h	C_4	E	C_4^3	C_2
σ_h	σ_h	S_4	i	S_4^3	C_2	C_4	E	C_4^3
S_4^3	S_4^3	σ_h	S_4	i	C_4^3	C_2	C_4	E

Table 12(a)

422	1	4^3	2	4	2_x	2_y	2_1	2_2
1	1	4^3	2	4	2_x	2_y	2_1	2_2
4	4	1	4^3	2	2_1	2_2	2_y	2_x
2	2	4	1	4^3	2_y	2_x	2_2	2_1
4^3	4^3	2	4	1	2_2	2_1	2_x	2_y
2_x	2_x	2_1	2_y	2_2	1	2	4^3	4
2_y	2_y	2_2	2_x	2_1	2	1	4	4^3
2_1	2_1	2_y	2_2	2_x	4	4^3	1	2
2_2	2_2	2_x	2_1	2_y	4^3	4	2	1

Table 12(b)

D_4	E	C_4^3	C_2	C_4	C_{2x}	C_{2y}	$C_2(1)$	$C_2(2)$
E	E	C_4^3	C_2	C_4	C_{2x}	C_{2y}	$C_2(1)$	$C_2(2)$
C_4	C_4	E	C_4^3	C_2	$C_2(1)$	$C_2(2)$	C_{2y}	C_{2x}
C_2	C_2	C_4	E	C_4^3	C_{2y}	C_{2x}	$C_2(2)$	$C_2(1)$
C_4^3	C_4^3	C_2	C_4	E	$C_2(2)$	$C_2(1)$	C_{2x}	C_{2y}
C_{2x}	C_{2x}	$C_2(1)$	C_{2y}	$C_2(2)$	E	C_2	C_4^3	C_4
C_{2y}	C_{2y}	$C_2(2)$	C_{2x}	$C_2(1)$	C_2	E	C_4	C_4^3
$C_2(1)$	$C_2(1)$	C_{2y}	$C_2(2)$	C_{2x}	C_4	C_4^3	E	C_2
$C_2(2)$	$C_2(2)$	C_{2x}	$C_2(1)$	C_{2y}	C_4^3	C_4	C_2	E

Table 13(a)

4mm	1	4^3	2	4	m_x	m_y	m_1	m_2
1	1	4^3	2	4	m_x	m_y	m_1	m_2
4	4	1	4^3	2	m_1	m_2	m_y	m_x
2	2	4	1	4^3	m_y	m_x	m_2	m_1
4^3	4^3	2	4	1	m_2	m_1	m_x	m_y
m_x	m_x	m_1	m_y	m_2	1	2	4^3	4
m_y	m_y	m_2	m_x	m_1	2	I	4	4^3
m_1	m_1	m_y	m_2	m_x	4	4^3	1	2
m_2	m_2	m_x	m_1	m_y	4^3	4	2	1

Table 13(b)

C_{4v}	E	C_4^3	C_2	C_4	σ_x	σ_y	σ_1	σ_2
E	E	C_4^3	C_2	C_4	σ_x	σ_y	σ_1	σ_2
C_4	C_4	E	C_4^3	C_2	σ_1	σ_2	σ_y	σ_x
C_2	C_2	C_4	E	C_4^3	σ_y	σ_x	σ_2	σ_1
C_4^3	C_4^3	C_2	C_4	E	σ_2	σ_1	σ_x	σ_y
σ_x	σ_x	σ_1	σ_y	σ_2	E	C_2	C_4^3	C_4
σ_y	σ_y	σ_2	σ_x	σ_1	C_2	E	C_4	C_4^3
σ_1	σ_1	σ_y	σ_2	σ_x	C_4	C_4^3	E	C_2
σ_2	σ_2	σ_x	σ_1	σ_y	C_4^3	C_4	C_2	E

Table 14(a)

$\bar{4}2m$	1	$\bar{4}^3$	2	$\bar{4}$	2_x	2_y	m_1	m_2
1	1	$\bar{4}^3$	2	$\bar{4}$	2_x	2_y	m_1	m_2
$\bar{4}$	$\bar{4}$	1	$\bar{4}^3$	2	m_1	m_2	2_y	2_x
2	2	$\bar{4}$	1	$\bar{4}^3$	2_y	2_x	m_2	m_1
$\bar{4}^3$	$\bar{4}^3$	2	$\bar{4}$	1	m_2	m_1	2_x	2_y
2_x	2_x	m_1	2_y	m_2	1	2	$\bar{4}^3$	$\bar{4}$
2_y	2_y	m_2	2_x	m_1	2	1	$\bar{4}$	$\bar{4}^3$
m_1	m_1	2_y	m_2	2_x	$\bar{4}$	$\bar{4}^3$	1	2
m_2	m_2	2_x	m_1	2_y	$\bar{4}^3$	$\bar{4}$	2	1

Table 14(b)

D_{2d}	E	S_4^3	C_2	S_4	C_{2x}	C_{2y}	σ_1	σ_2
E	E	S_4^3	C_2	S_4	C_{2x}	C_{2y}	σ_1	σ_2
S_4	S_4	E	S_4^3	C_2	σ_1	σ_2	C_{2y}	C_{2x}
C_2	C_2	S_4	E	S_4^3	C_{2y}	C_{2x}	σ_2	σ_1
S_4^3	S_4^3	C_2	S_4	E	σ_2	σ_1	C_{2x}	C_{2y}
C_{2x}	C_{2x}	σ_1	C_{2y}	σ_2	E	C_2	S_4^3	S_4
C_{2y}	C_{2y}	σ_2	C_{2x}	σ_1	C_2	E	S_4	S_4^3
σ_1	σ_1	C_{2y}	σ_2	C_{2x}	S_4	S_4^3	E	C_2
σ_2	σ_2	C_{2x}	σ_1	C_{2y}	S_4^3	S_4	C_2	E

Table 15(a)

$\frac{4}{mmm}$	1	4^3	2	4	2_x	2_y	2_1	2_2	$\bar{1}$	$\bar{4}^3$	m	$\bar{4}$	m_x	m_y	m_1	m_2
1	1	4^3	2	4	2_x	2_y	2_1	2_2	$\bar{1}$	$\bar{4}^3$	m	$\bar{4}$	m_x	m_y	m_1	m_2
4	4	1	4^3	2	2_1	2_2	2_y	2_x	$\bar{4}$	$\bar{1}$	$\bar{4}^3$	m	m_1	m_2	m_y	m_x
2	2	4	1	4^3	2_y	2_x	2_2	2_1	m	$\bar{4}$	$\bar{1}$	$\bar{4}^3$	m_y	m_x	m_2	m_1
4^3	4^3	2	4	1	2_2	2_1	2_x	2_y	$\bar{4}^3$	m	$\bar{4}$	$\bar{1}$	m_2	m_1	m_x	m_y
2_x	2_x	2_1	2_y	2_2	1	2	4^3	4	m_x	m_1	m_y	m_2	$\bar{1}$	m	$\bar{4}^3$	$\bar{4}$
2_y	2_y	2_2	2_x	2_1	2	1	4	4^3	m_y	m_2	m_x	m_1	m	$\bar{1}$	$\bar{4}$	$\bar{4}^3$
2_1	2_1	2_y	2_2	2_x	4	4^3	1	2	m_1	m_y	m_2	m_x	$\bar{4}$	$\bar{4}^3$	$\bar{1}$	m
2_2	2_2	2_x	2_1	2_y	4^3	4	2	1	m_2	m_x	m_1	m_y	$\bar{4}^3$	$\bar{4}$	m	$\bar{1}$
$\bar{1}$	$\bar{1}$	$\bar{4}^3$	m	$\bar{4}$	m_x	m_y	m_1	m_2	1	4^3	2	4	2_x	2_y	2_1	2_2
$\bar{4}$	$\bar{4}$	$\bar{1}$	$\bar{4}^3$	m	m_1	m_2	m_y	m_x	4	1	4^3	2	2_1	2_2	2_y	2_x
m	m	$\bar{4}$	$\bar{1}$	$\bar{4}^3$	m_y	m_x	m_2	m_1	2	4	1	4^3	2_y	2_x	2_2	2_1
$\bar{4}^3$	$\bar{4}^3$	m	$\bar{4}$	$\bar{1}$	m_2	m_1	m_x	m_y	4^3	2	4	1	2_2	2_1	2_x	2_y
m_x	m_x	m_1	m_y	m_2	$\bar{1}$	m	$\bar{4}^3$	$\bar{4}$	2_x	2_1	2_y	2_2	1	2	4^3	4
m_y	m_y	m_2	m_x	m_1	m	$\bar{1}$	4	$\bar{4}^3$	2_y	2_2	2_x	2_1	2	1	4	4^3
m_1	m_1	m_y	m_2	m_x	$\bar{4}$	$\bar{4}^3$	$\bar{1}$	m	2_1	2_y	2_2	2_x	4	4^3	1	2
m_2	m_2	m_x	m_1	m_y	$\bar{4}^3$	$\bar{4}$	m	$\bar{1}$	2_2	2_x	2_1	2_y	4^3	4	2	1

Table 15(b)

D_{4h}	E	C_4^3	C_2	C_4	C_{2x}	C_{2y}	$C_2(1)$	$C_2(2)$	i	S_4^3	σ	S_4	σ_x	σ_y	σ_1	σ_2
E	E	C_4^3	C_2	C_4	C_{2x}	C_{2y}	$C_2(1)$	$C_2(2)$	i	S_4^3	σ	S_4	σ_x	σ_y	σ_1	σ_2
C_4	C_4	E	C_4^3	C_2	$C_2(1)$	$C_2(2)$	C_{2y}	C_{2x}	S_4	i	S_4^3	σ	σ_1	σ_2	σ_y	σ_x
C_2	C_2	C_4	E	C_4^3	C_{2y}	C_{2x}	$C_2(2)$	$C_2(1)$	σ	S_4	i	S_4^3	σ_y	σ_x	σ_2	σ_1
C_4^3	C_4^3	C_2	C_4	E	$C_2(2)$	$C_2(1)$	C_{2x}	C_{2y}	S_4^3	σ	S_4	i	σ_2	σ_1	σ_x	σ_y
C_{2x}	C_{2x}	$C_2(1)$	C_{2y}	$C_2(2)$	E	C_2	C_4^3	C_4	σ_x	σ_1	σ_y	σ_2	i	σ	S_4^3	S_4
C_{2y}	C_{2y}	$C_2(2)$	C_{2x}	$C_2(1)$	C_2	E	C_4	C_4^3	σ_y	σ_2	σ_x	σ_1	σ	i	S_4	S_4^3
$C_2(1)$	$C_2(1)$	C_{2y}	$C_2(2)$	C_{2x}	C_4	C_4^3	E	C_2	σ_1	σ_y	σ_2	σ_x	S_4	S_4^3	i	σ
$C_2(2)$	$C_2(2)$	C_{2x}	$C_2(1)$	C_{2y}	C_4^3	C_4	C_2	E	σ_2	σ_x	σ_1	σ_y	S_4^3	S_4	σ	i
i	i	S_4^3	σ	S_4	σ_x	σ_y	σ_1	σ_2	E	C_4^3	C_2	C_4	C_{2x}	C_{2y}	$C_2(1)$	$C_2(2)$
S_4	S_4	i	S_4^3	σ	σ_1	σ_2	σ_y	σ_x	C_4	E	C_4^3	C_2	$C_2(1)$	$C_2(2)$	C_{2y}	C_{2x}
σ	σ	S_4	i	S_4^3	σ_y	σ_x	σ_2	σ_1	C_2	C_4	E	C_4^3	C_{2y}	C_{2x}	$C_2(2)$	$C_2(1)$
S_4^3	S_4^3	σ	S_4	i	σ_2	σ_1	σ_x	σ_y	C_4^3	C_2	C_4	E	$C_2(2)$	$C_2(1)$	C_{2x}	C_{2y}
σ_x	σ_x	σ_1	σ_y	σ_2	i	σ	S_4^3	S_4	C_{2x}	$C_2(1)$	C_{2y}	$C_2(2)$	E	C_2	C_4^3	C_4
σ_y	σ_y	σ_2	σ_x	σ_1	σ	i	S_4	S_4^3	C_{2y}	$C_2(2)$	C_{2x}	$C_2(1)$	C_2	E	C_4	C_4^3
σ_1	σ_1	σ_y	σ_2	σ_x	S_4	S_4^3	i	σ	$C_2(1)$	C_{2y}	$C_2(2)$	C_{2x}	C_4	C_4^3	E	C_2
σ_2	σ_2	σ_x	σ_1	σ_y	S_4^3	S_4	σ	i	$C_2(2)$	C_{2x}	$C_2(1)$	C_{2y}	C_4^3	C_4	C_2	E

Table 16(a)

3	$3^3(=1)$	3^2	3
1	1	3^2	3
3	3	1	3^2
3^2	3^2	3	1

Table 16(b)

C_3	E	C_3^2	C_3
E	E	C_3^2	C_3
C_3	C_3	E	C_3^2
C_3^2	C_3^2	C_3	E

Table 17(a)

$\bar{3}$	$\bar{3}^6(=1)$	$\bar{3}^2(=3^2)$	$\bar{3}^4(=3)$	$\bar{3}^3(=\bar{1})$	$\bar{3}^5$	$\bar{3}$
1	1	3^2	3	$\bar{1}$	$\bar{3}^5$	$\bar{3}$
3	3	1	3^2	$\bar{3}$	$\bar{1}$	$\bar{3}^5$
3^2	3^2	3	1	$\bar{3}^5$	3	$\bar{1}$
$\bar{1}$	$\bar{1}$	$\bar{3}^5$	$\bar{3}$	1	3^2	3
$\bar{3}$	$\bar{3}$	$\bar{1}$	$\bar{3}^5$	3	1	3^2
$\bar{3}^5$	$\bar{3}^5$	3	$\bar{1}$	3^2	3	1

Table 17(b)

C_{3i}	E	C_3^2	C_3	i	S_6	S_6^5
E	E	C_3^2	C_3	i	S_6	S_6^5
C_3	C_3	E	C_3^2	S_6^5	i	S_6
C_3^2	C_3^2	C_3	E	S_6	S_6^5	i
i	i	S_6	S_6^5	E	C_3^2	C_3
S_6^5	S_6^5	i	S_6	C_3	E	C_3^2
S_6	S_6	S_6^5	i	C_3^2	C_3	E

Table 18(a)

32	1	3^2	3	2_x	2_y	2_{xy}
1	1	3^2	3	2_x	2_y	2_{xy}
3	3	1	3^2	2_{xy}	2_x	2_y
3^2	3^2	3	1	2_y	2_{xy}	2_x
2_x	2_x	2_{xy}	2_y	1	3	3^2
2_y	2_y	2_x	2_{xy}	3^2	1	3
2_z	2_z	2_y	2_x	3	3^2	1

Table 18(b)

D_3	E	C_3^2	C_3	C_{2x}	C_{2y}	C_{2xy}
E	E	C_3^2	C_3	C_{2x}	C_{2y}	C_{2xy}
C_3	C_3	E	C_3^2	C_{2xy}	C_{2x}	C_{2y}
C_3^2	C_3^2	C_3	E	C_{2y}	C_{2xy}	C_{2x}
C_{2x}	C_{2x}	C_{2xy}	C_{2y}	E	C_3	C_3^2
C_{2y}	C_{2y}	C_{2x}	C_{2xy}	C_3^2	E	C_3
C_{2xy}	C_{2xy}	C_{2y}	C_{2x}	C_3	C_3^2	E

Table 19(a)

3m	1	3^2	3	m_x	m_y	m_{xy}
1	1	3^2	3	m_x	m_y	m_{xy}
3	3	1	3^2	m_{xy}	m_x	m_y
3^2	3^2	3	1	m_y	m_{xy}	m_x
m_x	m_x	m_{xy}	m_y	1	3	3^2
m_y	m_y	m_x	m_{xy}	3^2	1	3
m_{xy}	m_{xy}	m_y	m_x	3	3^2	1

Table 19(b)

C_{3v}	E	C_3^2	C_3	σ_x	σ_y	σ_{xy}
E	E	C_3^2	C_3	σ_x	σ_y	σ_{xy}
C_3	C_3	E	C_3^2	σ_{xy}	σ_x	σ_y
C_3^2	C_3^2	C_3	E	σ_y	σ_{xy}	σ_x
σ_x	σ_x	σ_{xy}	σ_y	E	C_3	C_3^2
σ_y	σ_y	σ_x	σ_{xy}	C_3^2	E	C_3
σ_{xy}	σ_{xy}	σ_y	σ_x	C_3	C_3^2	E

Table 20(a)

$\bar{3}m$	1	3^2	3	$\bar{1}$	$\bar{3}^5$	$\bar{3}$	2_x	2_y	2_{xy}	m_x	m_y	m_{xy}
1	1	3^2	3	$\bar{1}$	$\bar{3}^5$	$\bar{3}$	2_x	2_y	2_{xy}	m_x	m_y	m_{xy}
3	3	1	3^2	$\bar{3}$	$\bar{1}$	$\bar{3}^5$	2_{xy}	2_x	2_y	m_{xy}	m_x	m_y
3^2	3^2	3	1	$\bar{3}^5$	$\bar{3}$	$\bar{1}$	2_y	2_{xy}	2_x	m_y	m_{xy}	m_x
$\bar{1}$	$\bar{1}$	$\bar{3}^5$	$\bar{3}$	1	3^2	3	m_x	m_y	m_{xy}	2_x	2_y	2_{xy}
$\bar{3}$	$\bar{3}$	$\bar{1}$	$\bar{3}^5$	3	1	3^2	m_{xy}	m_x	m_y	2_{xy}	2_x	2_y
$\bar{3}^5$	$\bar{3}^5$	$\bar{3}$	$\bar{1}$	3^2	3	1	m_y	m_{xy}	m_x	2_y	2_{xy}	2_x
2_x	2_x	2_{xy}	2_y	m_x	m_{xy}	m_y	1	3	3^2	$\bar{1}$	$\bar{3}$	$\bar{3}^5$
2_y	2_y	2_x	2_{xy}	m_y	m_x	m_{xy}	3^2	1	3	$\bar{3}^5$	$\bar{1}$	$\bar{3}$
2_{xy}	2_{xy}	2_y	2_x	m_{xy}	m_y	m_x	3	3^2	1	$\bar{3}$	$\bar{3}^5$	$\bar{1}$
m_x	m_x	m_{xy}	m_y	2_x	2_{xy}	2_y	$\bar{1}$	$\bar{3}$	$\bar{3}^5$	1	3	3^2
m_y	m_y	m_x	m_{xy}	2_y	2_x	2_{xy}	$\bar{3}^5$	$\bar{1}$	$\bar{3}$	3^2	1	3
m_{xy}	m_{xy}	m_y	m_x	2_{xy}	2_y	2_x	$\bar{3}$	$\bar{3}^5$	$\bar{1}$	3	3^2	1

Table 20(b)

D_{3d}	E	C_3^2	C_3	i	S_6	S_6^5	C_{2x}	C_{2y}	C_{2xy}	σ_x	σ_y	σ_{xy}
E	E	C_3^2	C_3	i	S_6	S_6^5	C_{2x}	C_{2y}	C_{2xy}	σ_x	σ_y	σ_{xy}
C_3	C_3	E	C_3^2	S_6^5	i	S_6	C_{2xy}	C_{2x}	C_{2y}	σ_{xy}	σ_x	σ_y
C_3^2	C_3^2	C_3	E	S_6	S_6^5	i	C_{2y}	C_{2xy}	C_{2x}	σ_y	σ_{xy}	σ_x
i	i	S_6	S_6^5	E	C_3^2	C_3	σ_x	σ_y	σ_{xy}	C_{2x}	C_{2y}	C_{2xy}
S_6^5	S_6^5	i	S_6	C_3	E	C_3^2	σ_{xy}	σ_x	σ_y	C_{2xy}	C_{2x}	C_{2y}
S_6	S_6	S_6^5	i	C_3^2	C_3	E	σ_y	σ_{xy}	σ_x	C_{2y}	C_{2xy}	C_{2x}
C_{2x}	C_{2x}	C_{2xy}	C_{2y}	σ_x	σ_{xy}	σ_y	E	C_3	C_3^2	i	S_6^5	S_6
C_{2y}	C_{2y}	C_{2x}	C_{2xy}	σ_y	σ_x	σ_{xy}	C_3^2	E	C_3	S_6	i	S_6^5
C_{2xy}	C_{2xy}	C_{2y}	C_{2x}	σ_{xy}	σ_y	σ_x	C_3	C_3^2	E	S_6^5	S_6	i
σ_x	σ_x	σ_{xy}	σ_y	C_{2x}	C_{2xy}	C_{2y}	i	S_6^5	S_6	E	C_3	C_3^2
σ_y	σ_y	σ_x	σ_{xy}	C_{2y}	C_{2x}	C_{2xy}	S_6	i	S_6^5	C_3^2	E	C_3
σ_{xy}	σ_{xy}	σ_y	σ_x	C_{2xy}	C_{2y}	C_{2x}	S_6^5	S_6	i	C_3	C_3^2	E

Table 21(a)

$\bar{6}$	$\bar{6}^6(=1)$	$\bar{6}^4(=3^2)$	$\bar{6}^2(=3)$	$\bar{6}^3(=m_h)$	$\bar{6}$	$\bar{6}^5$
1	1	3^2	3	m_h	$\bar{6}$	$\bar{6}^5$
3	3	1	3^2	$\bar{6}^5$	m_h	$\bar{6}$
3^2	3^2	3	1	$\bar{6}$	$\bar{6}^5$	m_h
m_h	m_h	$\bar{6}$	$\bar{6}^5$	1	3^2	3
$\bar{6}^5$	$\bar{6}^5$	m_h	$\bar{6}$	3	1	3^2
$\bar{6}$	$\bar{6}$	$\bar{6}^5$	m_h	3^2	3	1

Table 21(b)

C_{3h}	E	C_3^2	C_3	σ_h	S_3^5	S_3
E	E	C_3^2	C_3	σ_h	S_3^5	S_3
C_3	C_3	E	C_3^2	S_3	σ_h	S_3^5
C_3^2	C_3^2	C_3	E	S_3^5	S_3	σ_h
σ_h	σ_h	S_3^5	S_3	E	C_3^2	C_3
S_3	S_3	σ_h	S_3^5	C_3	E	C_3^2
S_3^5	S_3^5	S_3	σ_h	C_3^2	C_3	E

Table 22(a)

$\bar{6}2m$	1	3^2	3	2_x	2_y	2_{xy}	m_1	m_2	m_3	m_h	$\bar{6}$	$\bar{6}^5$
1	1	3^2	3	2_x	2_y	2_{xy}	m_1	m_2	m_3	m_h	$\bar{6}$	$\bar{6}^5$
3	3	1	3^2	2_{xy}	2_x	2_y	m_3	m_1	m_2	$\bar{6}^5$	m_h	$\bar{6}$
3^2	3^2	3	1	2_y	2_{xy}	2_x	m_2	m_3	m_1	$\bar{6}$	$\bar{6}^5$	m_h
2_x	2_x	2_{xy}	2_y	1	3	3^2	m_h	$\bar{6}^5$	$\bar{6}$	m_1	m_3	m_2
2_y	2_y	2_x	2_{xy}	3^2	1	3	$\bar{6}$	m_h	$\bar{6}^5$	m_2	m_1	m_3
2_{xy}	2_{xy}	2_y	2_x	3	3^2	1	$\bar{6}^5$	$\bar{6}$	m_h	m_3	m_2	m_1
m_1	m_1	m_3	m_2	m_h	$\bar{6}^5$	$\bar{6}$	1	3^2	3	2_x	2_{xy}	2_y
m_2	m_2	m_1	m_3	$\bar{6}$	m_h	$\bar{6}^5$	3	1	3^2	2_y	2_x	2_{xy}
m_3	m_3	m_2	m_1	$\bar{6}^5$	$\bar{6}$	m_h	3^2	3	1	2_{xy}	2_y	2_x
m_h	m_h	$\bar{6}$	$\bar{6}^5$	m_1	m_2	m_3	2_x	2_y	2_{xy}	1	3^2	3
$\bar{6}^5$	$\bar{6}^5$	m_h	$\bar{6}$	m_3	m_1	m_2	2_{xy}	2_x	2_y	3	1	3^2
$\bar{6}$	$\bar{6}$	$\bar{6}^5$	m_h	m_2	m_3	m_1	2_y	2_{xy}	2_x	3^2	3	1

Table 22(b)

D_{3h}	E	C_3^2	C_3	C_{2x}	C_{2y}	C_{2xy}	σ_1	σ_2	σ_3	σ_h	S_3^5	S_3
E	E	C_3^2	C_3	C_{2x}	C_{2y}	C_{2xy}	σ_1	σ_2	σ_3	σ_h	S_3^5	S_3
C_3	C_3	E	C_3^2	C_{2xy}	C_{2x}	C_{2y}	σ_3	σ_1	σ_2	S_3	σ_h	S_3^5
C_3^2	C_3^2	C_3	E	C_{2y}	C_{2xy}	C_{2x}	σ_2	σ_3	σ_1	S_3^5	S_3	σ_h
C_{2x}	C_{2x}	C_{2xy}	C_{2y}	E	C_3	C_3^2	σ_h	S_3	S_3^5	σ_1	σ_3	σ_2
C_{2y}	C_{2y}	C_{2x}	C_{2xy}	C_3^2	E	C_3	S_3^5	σ_h	S_3	σ_2	σ_1	σ_3
C_{2xy}	C_{2xy}	C_{2y}	C_{2x}	C_3	C_3^2	E	S_3	S_3^5	σ_h	σ_3	σ_2	σ_1
σ_1	σ_1	σ_3	σ_2	σ_h	S_3	S_3^5	E	C_3^2	C_3	C_{2x}	C_{2xy}	C_{2y}
σ_2	σ_2	σ_1	σ_3	S_3^5	σ_h	S_3	C_3	E	C_3^2	C_{2y}	C_{2x}	C_{2xy}
σ_3	σ_3	σ_2	σ_1	S_3	S_3^5	σ_h	C_3^2	C_3	E	C_{2xy}	C_{2y}	C_{2x}
σ_h	σ_h	S_3^5	S_3	σ_1	σ_2	σ_3	C_{2x}	C_{2y}	C_{2xy}	E	C_3^2	C_3
S_3	S_3	σ_h	S_3^5	σ_3	σ_1	σ_2	C_{2xy}	C_{2x}	C_{2y}	C_3	E	C_3^2
S_3^5	S_3^5	S_3	σ_h	σ_2	σ_3	σ_1	C_{2y}	C_{2xy}	C_{2x}	C_3^2	C_3	E

Table 23(a)

6	$6^6(=1)$	$6^4(=3^2)$	$6^2(=3)$	$6^3(=2)$	6	6^5
1	1	3^2	3	2	6	6^5
3	3	1	3^2	6^5	2	6
3^2	3^2	3	1	6	6^5	2
2	2	6	6^5	1	3^2	3
6^5	6^5	2	6	3	1	3^2
6	6	6^5	2	3^2	3	1

Table 23(b)

C_6	E	C_3^2	C_3	C_2	C_6	C_6^5
E	E	C_3^2	C_3	C_2	C_6	C_6^5
C_3	C_3	E	C_3^2	C_6^5	C_2	C_6
C_3^2	C_3^2	C_3	E	C_6	C_6^5	C_2
C_2	C_2	C_6	C_6^5	E	C_3^2	C_3
C_6^5	C_6^5	C_2	C_6	C_3	E	C_3^2
C_6	C_6	C_6^5	C_2	C_3^2	C_3	E

Table 24(a)

6/m	1	3^2	3	2	6	6^5	$\bar{1}$	$\bar{3}^5$	$\bar{3}$	m_h	$\bar{6}$	$\bar{6}^5$
1	1	3^2	3	2	6	6^5	$\bar{1}$	$\bar{3}^5$	$\bar{3}$	m_h	$\bar{6}$	$\bar{6}^5$
3	3	1	3^2	6^5	2	6	$\bar{3}$	$\bar{1}$	$\bar{3}^5$	$\bar{6}^5$	m_h	$\bar{6}$
3^2	3^2	3	1	6	6^5	2	$\bar{3}^5$	$\bar{3}$	$\bar{1}$	$\bar{6}$	$\bar{6}^5$	m_h
2	2	6	6^5	1	3^2	3	m_h	$\bar{6}$	$\bar{6}^5$	$\bar{1}$	$\bar{3}^5$	$\bar{3}$
6^5	6^5	2	6	3	1	3^2	$\bar{6}^5$	m_h	$\bar{6}$	$\bar{3}$	$\bar{1}$	$\bar{3}^5$
6	6	$\bar{6}^5$	2	3^2	3	1	$\bar{6}$	$\bar{6}^5$	m_h	$\bar{3}^5$	$\bar{3}$	$\bar{1}$
$\bar{1}$	$\bar{1}$	$\bar{3}^5$	$\bar{3}$	m_h	$\bar{6}$	$\bar{6}^5$	1	3^2	3	2	6	6^5
$\bar{3}$	$\bar{3}$	$\bar{1}$	$\bar{3}^5$	$\bar{6}^5$	m_h	$\bar{6}$	3	1	3^2	6^5	2	6
$\bar{3}^5$	$\bar{3}^5$	$\bar{3}$	$\bar{1}$	$\bar{6}$	$\bar{6}^5$	m_h	3^2	3	1	6	6^5	2
m_h	m_h	$\bar{6}$	$\bar{6}^5$	$\bar{1}$	$\bar{3}^5$	$\bar{3}$	2	6	6^5	1	3^2	3
$\bar{6}^5$	$\bar{6}^5$	m_h	$\bar{6}$	$\bar{3}$	$\bar{1}$	$\bar{3}^5$	6^5	2	6	3	1	3^2
$\bar{6}$	$\bar{6}$	$\bar{6}^5$	m_h	$\bar{3}^5$	$\bar{3}$	$\bar{1}$	6	6^5	2	3^2	3	1

Table 24(b)

C_{6h}	E	C_3^2	C_3	C_2	C_6	C_6^5	i	S_6	S_6^5	σ_h	S_3^5	S_3
E	E	C_3^2	C_3	C_2	C_6	C_6^5	i	S_6	S_6^5	σ_h	S_3^5	S_3
C_3	C_3	E	C_3^2	C_6^5	C_2	C_6	S_6^5	i	S_6	S_3	σ_h	S_3^5
C_3^2	C_3^2	C_3	E	C_6	C_6^5	C_2	S_6	S_6^5	i	S_3^5	S_3	σ_h
C_2	C_2	C_6	C_6^5	E	C_3^2	C_3	σ_h	S_3^5	S_3	i	S_6	S_6^5
C_6^5	C_6^5	C_2	C_6	C_3	E	C_3^2	S_3	σ_h	S_3^5	S_6^5	i	S_6
C_6	C_6	C_6^5	C_2	C_3^2	C_3	E	S_3^5	S_3	σ_h	S_6	S_6^5	i
i	i	S_6	S_6^5	σ_h	S_3^5	S_3	E	C_3^2	C_3	C_2	C_6	C_6^5
S_6^5	S_6^5	i	S_6	S_3	σ_h	S_3^5	C_3	E	C_3^2	C_6^5	C_2	C_6
S_6	S_6	S_6^5	i	S_3^5	S_3	σ_h	C_3^2	C_3	E	C_6	C_6^5	C_2
σ_h	σ_h	S_3^5	S_3	i	S_6	S_6^5	C_2	C_6	C_6^5	E	C_3^2	C_3
S_3	S_3	σ_h	S_3^5	S_6^5	i	S_6	C_6^5	C_2	C_6	C_3	E	C_3^2
S_3^5	S_3^5	S_3	σ_h	S_6	S_6^5	i	C_6	C_6^5	C_2	C_3^2	C_3	E

Table 25(a)

622	1	3^2	3	2	6	6^5	2_1	2_2	2_3	2_x	2_y	2_{xy}
1	1	3^2	3	2	6	6^5	2_1	2_2	2_3	2_x	2_Y	2_{xy}
3	3	1	3^2	6^5	2	6	2_3	2_1	2_2	2_{xy}	2_x	2_y
3^2	3^2	3	1	6	6^5	2	2_2	2_3	2_1	2_y	2_{xy}	2_x
2	2	6	6^5	1	3^2	3	2_x	2_y	2_{xy}	2_1	2_2	2_3
6^5	6^5	2	6	3	1	3^2	2_{xy}	2_x	2_y	2_3	2_1	2_2
6	6	6^5	2	3^2	3	1	2_y	2_{xy}	2_x	2_2	2_3	2_1
2_1	2_1	2_3	2_2	2_x	2_{xy}	2_y	1	3	3^2	2	6^5	6
2_2	2_2	2_1	2_3	2_y	2_x	2_{xy}	3^2	1	3	6	2	6^5
2_3	2_3	2_2	2_1	2_{xy}	2_y	2_x	3	3^2	1	6^5	6	2
2_x	2_x	2_{xy}	2_y	2_1	2_3	2_2	2	6^5	6	1	3	3^2
2_y	2_y	2_x	2_{xy}	2_2	2_1	2_3	6	2	6^5	3^2	1	3
2_{xy}	2_{xy}	2_y	2_x	2_3	2_2	2_1	6^5	6	2	3	3^2	1

Table 25(b)

D_6	E	C_3^2	C_3	C_2	C_6	C_6^5	$C_2(1)$	$C_2(2)$	$C_2(3)$	C_{2x}	C_{2y}	C_{2xy}
E	E	C_3^2	C_3	C_2	C_6	C_6^5	$C_2(1)$	$C_2(2)$	$C_2(3)$	C_{2x}	C_{2y}	C_{2xy}
C_3	C_3	E	C_3^2	C_6^5	C_2	C_6	$C_2(3)$	$C_2(1)$	$C_2(2)$	C_{2xy}	C_{2x}	C_{2y}
C_3^2	C_3^2	C_3	E	C_6	C_6^5	C_2	$C_2(2)$	$C_2(3)$	$C_2(1)$	C_{2y}	C_{2xy}	C_{2x}
C_2	C_2	C_6	C_6^5	E	C_3^2	C_3	C_{2x}	C_{2y}	C_{2xy}	$C_2(1)$	$C_2(2)$	$C_2(3)$
C_6^5	C_6^5	C_2	C_6	C_3	E	C_3^2	C_{2xy}	C_{2x}	C_{2y}	$C_2(3)$	$C_2(1)$	$C_2(2)$
C_6	C_6	C_6^5	C_2	C_3^2	C_3	E	C_{2y}	C_{2xy}	C_{2x}	$C_2(2)$	$C_2(3)$	$C_2(1)$
$C_2(1)$	$C_2(1)$	$C_2(3)$	$C_2(2)$	C_{2x}	C_{2xy}	C_{2y}	E	C_3	C_3^2	C_2	C_6^5	C_6
$C_2(2)$	$C_2(2)$	$C_2(1)$	$C_2(3)$	C_{2y}	C_{2x}	C_{2xy}	C_3^2	E	C_3	C_6	C_2	C_6^5
$C_2(3)$	$C_2(3)$	$C_2(2)$	$C_2(1)$	C_{2xy}	C_{2y}	C_{2x}	C_3	C_3^2	E	C_6^5	C_6	C_2
C_{2x}	C_{2x}	C_{2xy}	C_{2y}	$C_2(1)$	$C_2(3)$	$C_2(2)$	C_2	C_6^5	C_6	E	C_3	C_3^2
C_{2y}	C_{2y}	C_{2x}	C_{2xy}	$C_2(2)$	$C_2(1)$	$C_2(3)$	C_6	C_2	C_6^5	C_3^2	E	C_3
C_{2xy}	C_{2xy}	C_{2y}	C_{2x}	$C_2(3)$	$C_2(2)$	$C_2(1)$	C_6^5	C_6	C_2	C_3	C_3^2	E

Table 26(a)

6mm	1	3^2	3	2	6	6^5	m_1	m_2	m_3	m_x	m_y	m_{xy}
1	1	3^2	3	2	6	6^5	m_1	m_2	m_3	m_x	m_y	m_{xy}
3	3	1	3^2	6^5	2	6	m_3	m_1	m_2	m_{xy}	m_x	m_y
3^2	3^2	3	1	6	6^5	2	m_2	m_3	m_1	m_y	m_{xy}	m_x
2	2	6	6^5	1	3^2	3	m_x	m_y	m_{xy}	m_1	m_2	m_3
6^5	6^5	2	6	3	1	3^2	m_{xy}	m_x	m_y	m_3	m_1	m_2
6	6	6^5	2	3^2	3	1	m_y	m_{xy}	m_x	m_2	m_3	m_1
m_1	m_1	m_3	m_2	m_x	m_{xy}	m_y	1	3	3^2	2	6^5	6
m_2	m_2	m_1	m_3	m_y	m_x	m_{xy}	3^2	1	3	6	2	6^5
m_3	m_3	m_2	m_1	m_{xy}	m_y	m_x	3	3^2	1	6^5	6	2
m_x	m_x	m_{xy}	m_y	m_1	m_3	m_2	2	6^5	6	1	3	3^2
m_y	m_y	m_x	m_{xy}	m_2	m_1	m_3	6	2	6^5	3^2	1	3
m_{xy}	m_{xy}	m_y	m_x	m_3	m_2	m_1	6^5	6	2	3	3^2	1

Table 26(b)

C_{6v}	E	C_3^2	C_3	C_2	C_6	C_6^5	σ_1	σ_2	σ_3	σ_x	σ_y	σ_{xy}
E	E	C_3^2	C_3	C_2	C_6	C_6^5	σ_1	σ_2	σ_3	σ_x	σ_y	σ_{xy}
C_3	C_3	E	C_3^2	C_6^5	C_2	C_6	σ_3	σ_1	σ_2	σ_{xy}	σ_x	σ_y
C_3^2	C_3^2	C_3	E	C_6	C_6^5	C_2	σ_2	σ_3	σ_1	σ_y	σ_{xy}	σ_x
C_2	C_2	C_3	C_6^5	E	C_3^2	C_3	σ_x	σ_y	σ_{xy}	σ_1	σ_2	σ_3
C_6^5	C_6^5	C_2	C_6	C_3	E	C_3^2	σ_{xy}	σ_x	σ_y	σ_3	σ_1	σ_2
C_6	C_6	C_6^5	C_2	C_3^2	C_3	E	σ_y	σ_{xy}	σ_x	σ_2	σ_3	σ_1
σ_1	σ_1	σ_3	σ_2	σ_x	σ_{xy}	σ_y	E	C_3	C_3^2	C_2	C_6^5	C_6
σ_2	σ_2	σ_1	σ_3	σ_y	σ_x	σ_{xy}	C_3^2	E	C_3	C_6	C_2	C_6^5
σ_3	σ_3	σ_2	σ_1	σ_{xy}	σ_y	σ_x	C_3	C_3^2	E	C_6^5	C_6	C_2
σ_x	σ_x	σ_{xy}	σ_y	σ_1	σ_3	σ_2	C_2	C_6^5	C_6	E	C_3	C_3^2
σ_y	σ_y	σ_x	σ_{xy}	σ_2	σ_1	σ_3	C_6	C_2	C_6^5	C_3^2	E	C_3
σ_{xy}	σ_{xy}	σ_y	σ_x	σ_3	σ_2	σ_1	C_6^5	C_6	C_2	C_3	C_3^2	E

Table 27(a)i

$\dfrac{6}{mmm}$	1	3^2	3	2	6	6^5	$\bar{1}$	$\bar{3}^5$	$\bar{3}$	m_h	$\bar{6}$	$\bar{6}^5$
1	1	3^2	3	2	6	6^5	$\bar{1}$	$\bar{3}^5$	$\bar{3}$	m_h	$\bar{6}$	$\bar{6}^5$
3	3	1	3^2	6^5	2	6	$\bar{3}$	$\bar{1}$	$\bar{3}^5$	$\bar{6}^5$	m_h	$\bar{6}$
3^2	3^2	3	1	6	6^5	2	$\bar{3}^5$	$\bar{3}$	$\bar{1}$	$\bar{6}$	$\bar{6}^5$	m_h
2	2	6	6^5	1	3^2	3	m_h	$\bar{6}$	$\bar{6}^5$	$\bar{1}$	$\bar{3}^5$	$\bar{3}$
6^5	6^5	2	6	3	1	3^2	$\bar{6}^5$	m_h	$\bar{6}$	$\bar{3}$	$\bar{1}$	$\bar{3}^5$
6	6	$\bar{6}^5$	2	3^2	3	1	$\bar{6}$	$\bar{6}^5$	m_h	$\bar{3}^5$	$\bar{3}$	$\bar{1}$
$\bar{1}$	$\bar{1}$	$\bar{3}^5$	$\bar{3}$	m_h	$\bar{6}$	$\bar{6}^5$	1	3^2	3	2	6	6^5
$\bar{3}$	$\bar{3}$	$\bar{1}$	$\bar{3}^5$	$\bar{6}^5$	m_h	$\bar{6}$	3	1	3^2	6^5	2	6
$\bar{3}^5$	$\bar{3}^5$	$\bar{3}$	$\bar{1}$	$\bar{6}$	$\bar{6}^5$	m_h	3^2	3	1	6	6^5	2
m_h	m_h	$\bar{6}$	$\bar{6}^5$	$\bar{1}$	$\bar{3}^5$	$\bar{3}$	2	6	6^5	1	3^2	3
$\bar{6}^5$	$\bar{6}^5$	m_h	$\bar{6}$	$\bar{3}$	$\bar{1}$	$\bar{3}^5$	6^5	2	6	3	1	3^2
$\bar{6}$	$\bar{6}$	$\bar{6}^5$	m_h	$\bar{3}^5$	$\bar{3}$	$\bar{1}$	6	6^5	2	3^2	3	1
2_1	2_1	2_3	2_2	2_x	2_{xy}	2_y	m_1	m_3	m_2	m_x	m_{xy}	m_y
2_2	2_2	2_1	2_3	2_y	2_x	2_{xy}	m_2	m_1	m_3	m_y	m_x	m_{xy}
2_3	2_3	2_2	2_1	2_{xy}	2_y	2_x	m_3	m_2	m_1	m_{xy}	m_y	m_x
2_x	2_x	2_{xy}	2_y	2_1	2_3	2_2	m_x	m_{xy}	m_y	m_1	m_3	m_2
2_y	2_y	2_x	2_{xy}	2_2	2_1	2_3	m_y	m_x	m_{xy}	m_2	m_1	m_3
2_{xy}	2_{xy}	2_y	2_x	2_3	2_2	2_1	m_{xy}	m_y	m_x	m_3	m_2	m_1
m_1	m_1	m_3	m_2	m_x	m_{xy}	m_y	2_1	2_3	2_2	2_x	2_{xy}	2_y
m_2	m_2	m_1	m_3	m_y	m_x	m_{xy}	2_2	2_1	2_3	2_y	2_x	2_{xy}
m_3	m_3	m_2	m_1	m_{xy}	m_y	m_x	2_3	2_2	2_1	2_{xy}	2_y	2_x
m_x	m_x	m_{xy}	m_y	m_1	m_3	m_2	2_x	2_{xy}	2_y	2_1	2_3	2_2
m_y	m_y	m_x	m_{xy}	m_2	m_1	m_3	2_y	2_x	2_{xy}	2_2	2_1	2_3
m_{xy}	m_{xy}	m_y	m_x	m_3	m_2	m_1	2_{xy}	2_y	2_x	2_3	2_2	2_1

Table 27(a)ii

$\dfrac{6}{mmm}$	2_1	2_2	2_3	2_x	2_y	2_{xy}	m_1	m_2	m_3	m_x	m_y	m_{xy}
1	2_1	2_2	2_3	2_x	2_y	2_{xy}	m_1	m_2	m_3	m_x	m_y	m_{xy}
3	2_3	2_1	2_2	2_{xy}	2_x	2_y	m_3	m_1	m_2	m_{xy}	m_x	m_y
3^2	2_2	2_3	2_1	2_y	2_{xy}	2_x	m_2	m_3	m_1	m_y	m_{xy}	m_x
2	2_x	2_y	2_{xy}	2_1	2_2	2_3	m_x	m_y	m_{xy}	m_1	m_2	m_3
6^5	2_{xy}	2_x	2_y	2_3	2_1	2_2	m_{xy}	m_x	m_y	m_3	m_1	m_2
6	2_y	2_{xy}	2_x	2_2	2_3	2_1	m_y	m_{xy}	m_x	m_2	m_3	m_1
$\bar{1}$	m_1	m_2	m_3	m_x	m_y	m_{xy}	2_1	2_2	2_3	2_x	2_y	2_{xy}
$\bar{3}$	m_3	m_1	m_2	m_{xy}	m_x	m_y	2_3	2_1	2_2	2_{xy}	2_x	2_y
$\bar{3}^5$	m_2	m_3	m_1	m_y	m_{xy}	m_x	2_2	2_3	2_1	2_y	2_{xy}	2_x
m_h	m_x	m_y	m_{xy}	m_1	m_2	m_3	2_x	2_y	2_{xy}	2_1	2_2	2_3
$\bar{6}^5$	m_{xy}	m_x	m_y	m_3	m_1	m_2	2_{xy}	2_x	2_y	2_3	2_1	2_2
$\bar{6}$	m_y	m_{xy}	m_x	m_2	m_3	m_1	2_y	2_{xy}	2_x	2_2	2_3	2_1
2_1	1	3	3^2	2	6^5	6	$\bar{1}$	$\bar{3}$	$\bar{3}^5$	m_h	$\bar{6}^5$	$\bar{6}$
2_2	3^2	1	3	6	2	6^5	$\bar{3}^5$	$\bar{1}$	$\bar{3}$	$\bar{6}$	m_h	$\bar{6}^5$
2_3	3	3^2	1	6^5	6	2	$\bar{3}$	$\bar{3}^5$	$\bar{1}$	$\bar{6}^5$	$\bar{6}$	m_h
2_x	2	6^5	6	1	3	3^2	m_h	$\bar{6}^5$	$\bar{6}$	$\bar{1}$	$\bar{3}$	$\bar{3}^5$
2_y	6	2	6^5	3^2	1	3	$\bar{6}$	m_h	$\bar{6}^5$	$\bar{3}^5$	$\bar{1}$	$\bar{3}$
2_{xy}	6^5	6	2	3	3^2	1	$\bar{6}^5$	$\bar{6}$	m_h	$\bar{3}$	$\bar{3}^5$	$\bar{1}$
m_1	$\bar{1}$	$\bar{3}$	$\bar{3}^5$	m_h	$\bar{6}^5$	$\bar{6}$	1	3	3^2	2	6^5	6
m_2	$\bar{3}^5$	$\bar{1}$	$\bar{3}$	$\bar{6}$	m_h	$\bar{6}^5$	3^2	1	3	6	2	6^5
m_3	$\bar{3}$	$\bar{3}^5$	$\bar{1}$	$\bar{6}^5$	$\bar{6}$	m_h	3	3^2	1	6^5	6	2
m_x	m_h	$\bar{6}^5$	$\bar{6}$	$\bar{1}$	$\bar{3}$	$\bar{3}^5$	2	6^5	6	1	3	3^2
m_y	$\bar{6}$	m_h	$\bar{6}^5$	$\bar{3}^5$	$\bar{1}$	$\bar{3}$	6	2	6^5	3^2	1	3
m_{xy}	$\bar{6}^5$	$\bar{6}$	m_h	$\bar{3}$	$\bar{3}^5$	$\bar{1}$	6^5	6	2	3	3^2	1

Table 27(b)i

D_{6h}	E	C_3^2	C_3	C_2	C_6	C_6^5	i	S_6	S_6^5	σ_h	S_3^5	S_3
E	E	C_3^2	C_3	C_2	C_6	C_6^5	i	S_6	S_6^5	σ_h	S_3^5	S_3
C_3	C_3	E	C_3^2	C_6^5	C_2	C_6	S_6^5	i	S_6	S_3	σ_h	S_3^5
C_3^2	C_3^2	C_3	E	C_6	C_6^5	C_2	S_6	S_6^5	i	S_3^5	S_3	σ_h
C_2	C_2	C_6	C_6^5	E	C_3^2	C_3	σ_h	S_3^5	S_3	i	S_6	S_6^5
C_6^5	C_6^5	C_2	C_6	C_3	E	C_3^2	S_3	σ_h	S_3^5	S_6^5	i	S_6
C_6	C_6	C_6^5	C_2	C_3^2	C_3	E	S_3^5	S_3	σ_h	S_6	S_6^5	i
i	i	S_6	S_6^5	σ_h	S_3^5	S_3	E	C_3^2	C_3	C_2	C_6	C_6^5
S_6^5	S_6^5	i	S_6	S_3	σ_h	S_3^5	C_3	E	C_3^2	C_6^5	C_2	C_6
S_6	S_6	S_6^5	i	S_3^5	S_3	σ_h	C_3^2	C_3	E	C_6	C_6^5	C_2
σ_h	σ_h	S_3^5	S_3	i	S_6	S_6^5	C_2	C_6	C_6^5	E	C_3^2	C_3
S_3	S_3	σ_h	S_3^5	S_6^5	i	S_6	C_6^5	C_2	C_6	C_3	E	C_3^2
S_3^5	S_3^5	S_3	σ_h	S_6	S_6^5	i	C_6	C_6^5	C_2	C_3^2	C_3	E
$C_2(1)$	$C_2(1)$	$C_2(3)$	$C_2(2)$	C_{2x}	C_{2xy}	C_{2y}	σ_1	σ_3	σ_2	σ_x	σ_{xy}	σ_y
$C_2(2)$	$C_2(2)$	$C_2(1)$	$C_2(3)$	C_{2y}	C_{2x}	C_{2xy}	σ_2	σ_1	σ_3	σ_y	σ_x	σ_{xy}
$C_2(3)$	$C_2(3)$	$C_2(2)$	$C_2(1)$	C_{2xy}	C_{2y}	C_{2x}	σ_3	σ_2	σ_1	σ_{xy}	σ_y	σ_x
C_{2x}	C_{2x}	C_{2xy}	C_{2y}	$C_2(1)$	$C_2(3)$	$C_2(2)$	σ_x	σ_{xy}	σ_y	σ_1	σ_3	σ_2
C_{2y}	C_{2y}	C_{2x}	C_{2xy}	$C_2(2)$	$C_2(1)$	$C_2(3)$	σ_y	σ_x	σ_{xy}	σ_2	σ_1	σ_3
C_{2xy}	C_{2xy}	C_{2y}	C_{2x}	$C_2(3)$	$C_2(2)$	$C_2(1)$	σ_{xy}	σ_y	σ_x	σ_3	σ_2	σ_1
σ_1	σ_1	σ_3	σ_2	σ_x	σ_{xy}	σ_y	$C_2(1)$	$C_2(3)$	$C_2(2)$	C_{2x}	C_{2xy}	C_{2y}
σ_2	σ_2	σ_1	σ_3	σ_y	σ_x	σ_{xy}	$C_2(2)$	$C_2(1)$	$C_2(3)$	C_{2y}	C_{2x}	C_{2xy}
σ_3	σ_3	σ_2	σ_1	σ_{xy}	σ_y	σ_x	$C_2(3)$	$C_2(2)$	$C_2(1)$	C_{2xy}	C_{2y}	C_{2x}
σ_x	σ_x	σ_{xy}	σ_y	σ_1	σ_3	σ_2	C_{2x}	C_{2xy}	C_{2y}	$C_2(1)$	$C_2(3)$	$C_2(2)$
σ_y	σ_y	σ_x	σ_{xy}	σ_2	σ_1	σ_3	C_{2y}	C_{2x}	C_{2xy}	$C_2(2)$	$C_2(1)$	$C_2(3)$
σ_{xy}	σ_{xy}	σ_y	σ_x	σ_3	σ_2	σ_1	C_{2xy}	C_{2y}	C_{2x}	$C_2(3)$	$C_2(2)$	$C_2(1)$

Table 27(b)ii

D_{6h}	$C_2(1)$	$C_2(2)$	$C_2(3)$	C_{2x}	C_{2y}	C_{2xy}	σ_1	σ_2	σ_3	σ_x	σ_y	σ_{xy}
E	$C_2(1)$	$C_2(2)$	$C_2(3)$	C_{2x}	C_{2y}	C_{2xy}	σ_1	σ_2	σ_3	σ_x	σ_y	σ_{xy}
C_3	$C_2(3)$	$C_2(1)$	$C_2(2)$	C_{2xy}	C_{2x}	C_{2y}	σ_3	σ_1	σ_2	σ_{xy}	σ_x	σ_y
C_3^2	$C_2(2)$	$C_2(3)$	$C_2(1)$	C_{2y}	C_{2xy}	C_{2x}	σ_2	σ_3	σ_1	σ_y	σ_{xy}	σ_x
C_2	C_{2x}	C_{2y}	C_{2xy}	$C_2(1)$	$C_2(2)$	$C_2(3)$	σ_x	σ_y	σ_{xy}	σ_1	σ_2	σ_3
C_6^5	C_{2xy}	C_{2x}	C_{2y}	$C_2(3)$	$C_2(1)$	$C_2(2)$	σ_{xy}	σ_x	σ_y	σ_3	σ_1	σ_2
C_6	C_{2y}	C_{2xy}	C_{2x}	$C_2(2)$	$C_2(3)$	$C_2(1)$	σ_y	σ_{xy}	σ_x	σ_2	σ_3	σ_1
i	σ_1	σ_2	σ_3	σ_x	σ_y	σ_{xy}	$C_2(1)$	$C_2(2)$	$C_2(3)$	C_{2x}	C_{2y}	C_{2xy}
S_6^5	σ_3	σ_1	σ_2	σ_{xy}	σ_x	σ_y	$C_2(3)$	$C_2(1)$	$C_2(2)$	C_{2xy}	C_{2x}	C_{2y}
S_6	σ_2	σ_3	σ_1	σ_y	σ_{xy}	σ_x	$C_2(2)$	$C_2(3)$	$C_2(1)$	C_{2y}	C_{2xy}	C_{2x}
σ_h	σ_x	σ_y	σ_{xy}	σ_1	σ_2	σ_3	C_{2x}	C_{2y}	C_{2xy}	$C_2(1)$	$C_2(2)$	$C_2(3)$
S_3	σ_{xy}	σ_x	σ_y	σ_3	σ_1	σ_2	C_{2xy}	C_{2x}	C_{2y}	$C_2(3)$	$C_2(1)$	$C_2(2)$
S_3^5	σ_y	σ_{xy}	σ_x	σ_2	σ_3	σ_1	C_{2y}	C_{2xy}	C_{2x}	$C_2(2)$	$C_2(3)$	$C_2(1)$
$C_2(1)$	E	C_3	C_3^2	C_2	C_6^5	C_6	i	S_6^5	S_6	σ_h	S_3	S_3^5
$C_2(2)$	C_3^2	E	C_3	C_6	C_2	C_6^5	S_6	i	S_6^5	S_3^5	σ_h	S_3
$C_2(3)$	C_3	C_3^2	E	C_6^5	C_6	C_2	S_6^5	S_6	i	S_3	S_3^5	σ_h
C_{2x}	C_2	C_6^5	C_6	E	C_3	C_3^2	σ_h	S_3	S_3^5	i	S_6^5	S_6
C_{2y}	C_6	C_2	C_6^5	C_3^2	E	C_3	S_3^5	σ_h	S_3	S_6	i	S_6^5
C_{2xy}	C_6^5	C_6	C_2	C_3	C_3^2	E	S_3	S_3^5	σ_h	S_6^5	S_6	i
σ_1	i	S_6^5	S_6	σ_h	S_3	S_3^5	E	C_3	C_3^2	C_2	C_6^5	C_6
σ_2	S_6	i	S_6^5	S_3^5	σ_h	S_3	C_3^2	E	C_3	C_6	C_2	C_6^5
σ_3	S_6^5	S_6	i	S_3	S_3^5	σ_h	C_3	C_3^2	E	C_6^5	C_6	C_2
σ_x	σ_h	S_3	S_3^5	i	S_6^5	S_6	C_2	C_6^5	C_6	E	C_3	C_3^2
σ_y	S_3^5	σ_h	S_3	S_6	i	S_6^5	C_6	C_2	C_6^5	C_3^2	E	C_3
σ_{xy}	S_3	S_3^5	σ_h	S_6^5	S_6	i	C_6^5	C_6	C_2	C_3	C_3^2	E

Table 28(a)

23	1	2_x	2_y	2_z	3_1	3_2	3_3	3_4	3_1^2	3_2^2	3_3^2	3_4^2
1	1	2_x	2_y	2_z	3_1	3_2	3_3	3_4	3_1^2	3_2^2	3_3^2	3_4^2
2_x	2_x	1	2_z	2_y	3_3^2	3_3	3_2	3_1^2	3_4	3_4^2	3_1	3_2^2
2_y	2_y	2_z	1	2_x	3_4^2	3_1^2	3_4	3_3	3_2	3_3^2	3_2^2	3_1
2_z	2_z	2_y	2_x	1	3_2^2	3_4	3_1^2	3_2	3_3	3_1	3_4^2	3_3^2
3_1^2	3_1^2	3_3	3_4	3_2	1	3_3^2	3_4^2	3_2^2	3_1	2_y	2_z	2_x
3_2^2	3_2^2	3_3^2	3_1	3_4^2	3_3	1	2_y	2_x	2_z	3_2	3_4	3_1^2
3_3^2	3_3^2	3_2^2	3_4^2	3_1	3_4	2_y	1	2_z	2_x	3_1^2	3_3	3_2
3_4^2	3_4^2	3_1	3_3^2	3_2^2	3_2	2_x	2_z	1	2_y	3_3	3_1^2	3_4
3_1	3_1	3_4^2	3_2^2	3_3^2	3_1^2	2_z	2_x	2_y	1	3_4	3_2	3_3
3_2	3_2	3_4	3_3	3_1^2	2_y	3_2^2	3_1	3_3^2	3_4^2	1	2_x	2_z
3_3	3_3	3_1^2	3_2	3_4	2_z	3_4^2	3_3^2	3_1	3_2^2	2_x	1	2_y
3_4	3_4	3_2	3_1^2	3_3	2_x	3_1	3_2^2	3_4^2	3_3^2	2_z	2_y	1

Table 28(b)

T	E	C_{2x}	C_{2y}	C_{2z}	$C_3(1)$	$C_3(2)$	$C_3(3)$	$C_3(4)$	$C_3^2(1)$	$C_3^2(2)$	$C_3^2(3)$	$C_3^2(4)$
E	E	C_{2x}	C_{2y}	C_{2z}	$C_3(1)$	$C_3(2)$	$C_3(3)$	$C_3(4)$	$C_3^2(1)$	$C_3^2(2)$	$C_3^2(3)$	$C_3^2(4)$
C_{2x}	C_{2x}	E	C_{2z}	C_{2y}	$C_3^2(3)$	$C_3(3)$	$C_3(2)$	$C_3^2(1)$	$C_3(4)$	$C_3^2(4)$	$C_3(1)$	$C_3^2(2)$
C_{2y}	C_{2y}	C_{2z}	E	C_{2x}	$C_3^2(4)$	$C_3^2(1)$	$C_3(4)$	$C_3(3)$	$C_3(2)$	$C_3^2(3)$	$C_3^2(2)$	$C_3(1)$
C_{2z}	C_{2z}	C_{2y}	C_{2x}	E	$C_3^2(2)$	$C_3(4)$	$C_3^2(1)$	$C_3(2)$	$C_3(3)$	$C_3(1)$	$C_3^2(4)$	$C_3^2(3)$
$C_3^2(1)$	$C_3^2(1)$	$C_3(3)$	$C_3(4)$	$C_3(2)$	E	$C_3^2(3)$	$C_3^2(4)$	$C_3^2(2)$	$C_3(1)$	C_{2y}	C_{2z}	C_{2x}
$C_3^2(2)$	$C_3^2(2)$	$C_3^2(3)$	$C_3(1)$	$C_3^2(4)$	$C_3(3)$	E	C_{2y}	C_{2x}	C_{2z}	$C_3(2)$	$C_3(4)$	$C_3^2(1)$
$C_3^2(3)$	$C_3^2(3)$	$C_3^2(2)$	$C_3^2(4)$	$C_3(1)$	$C_3(4)$	C_{2y}	E	C_{2z}	C_{2x}	$C_3^2(1)$	$C_3(3)$	$C_3(2)$
$C_3^2(4)$	$C_3^2(4)$	$C_3(1)$	$C_3^2(3)$	$C_3^2(2)$	$C_3(2)$	C_{2x}	C_{2z}	E	C_{2y}	$C_3(3)$	$C_3^2(1)$	$C_3(4)$
$C_3(1)$	$C_3(1)$	$C_3^2(4)$	$C_3^2(2)$	$C_3^2(3)$	$C_3^2(1)$	C_{2z}	C_{2x}	C_{2y}	E	$C_3(4)$	$C_3(2)$	$C_3(3)$
$C_3(2)$	$C_3(2)$	$C_3(4)$	$C_3(3)$	$C_3^2(1)$	C_{2y}	$C_3^2(2)$	$C_3(1)$	$C_3^2(3)$	$C_3^2(4)$	E	C_{2x}	C_{2z}
$C_3(3)$	$C_3(3)$	$C_3^2(1)$	$C_3(2)$	$C_3(4)$	C_{2z}	$C_3^2(4)$	$C_3^2(3)$	$C_3(1)$	$C_3^2(2)$	C_{2x}	E	C_{2y}
$C_3(4)$	$C_3(4)$	$C_3(2)$	$C_3^2(1)$	$C_3(3)$	C_{2x}	$C_3(1)$	$C_3^2(2)$	$C_3^2(4)$	$C_3^2(3)$	C_{2z}	C_{2y}	E

Table 29(a)i

m3	1	2_x	2_y	2_z	3_1	3_2	3_3	3_4	3_1^2	3_2^2	3_3^2	3_4^2
1	1	2_x	2_y	2_z	3_1	3_2	3_3	3_4	3_1^2	3_2^2	3_3^2	3_4^2
2_x	2_x	1	2_z	2_y	3_3^2	3_3	3_2	3_1^2	3_4	3_4^2	3_1	3_2^2
2_y	2_y	2_z	1	2_x	3_4^2	3_1^2	3_4	3_3	3_2	3_3^2	3_2^2	3_1
2_z	2_z	2_y	2_x	1	3_2^2	3_4	3_1^2	3_2	3_3	3_1	3_4^2	3_3^2
3_1^2	3_1^2	3_3	3_4	3_2	1	3_3^2	3_4^2	3_2^2	3_1	2_y	2_z	2_x
3_2^2	3_2^2	3_3^2	3_1	3_4^2	3_3	1	2_y	2_x	2_z	3_2	3_4	3_1^2
3_3^2	3_3^2	3_2^2	3_4^2	3_1	3_4	2_y	1	2_z	2_x	3_1^2	3_3	3_2
3_4^2	3_4^2	3_1	3_3^2	3_2^2	3_2	2_x	2_z	1	2_y	3_3	3_1^2	3_4
3_1	3_1	3_4^2	3_2^2	3_3^2	3_1^2	2_z	2_x	2_y	1	3_4	3_2	3_3
3_2	3_2	3_4	3_3	3_1^2	2_y	3_2^2	3_1	3_3^2	3_4^2	1	2_x	2_z
3_3	3_3	3_1^2	3_2	3_4	2_z	3_4^2	3_3^2	3_1	3_2^2	2_x	1	2_y
3_4	3_4	3_2	3_1^2	3_3	2_x	3_1	3_2^2	3_4^2	3_3^2	2_z	2_y	1
$\bar{1}$	$\bar{1}$	m_x	m_y	m_z	$\bar{3}_1$	$\bar{3}_2$	$\bar{3}_3$	$\bar{3}_4$	$\bar{3}_1^2$	$\bar{3}_2^2$	$\bar{3}_3^2$	$\bar{3}_4^2$
m_x	m_x	$\bar{1}$	m_z	m_y	$\bar{3}_3^2$	$\bar{3}_3$	$\bar{3}_2$	$\bar{3}_1^2$	$\bar{3}_4$	$\bar{3}_4^2$	$\bar{3}_1$	$\bar{3}_2^2$
m_y	m_y	m_z	$\bar{1}$	m_x	$\bar{3}_4^2$	$\bar{3}_1^2$	$\bar{3}_4$	$\bar{3}_3$	$\bar{3}_2$	$\bar{3}_3^2$	$\bar{3}_2^2$	$\bar{3}_1$
m_z	m_z	m_y	m_x	$\bar{1}$	$\bar{3}_2^2$	$\bar{3}_4$	$\bar{3}_1^2$	$\bar{3}_2$	$\bar{3}_3$	$\bar{3}_1$	$\bar{3}_4^2$	$\bar{3}_3^2$
$\bar{3}_1^2$	$\bar{3}_1^2$	$\bar{3}_3$	$\bar{3}_4$	$\bar{3}_2$	$\bar{1}$	$\bar{3}_3^2$	$\bar{3}_4^2$	$\bar{3}_2^2$	$\bar{3}_1$	m_y	m_z	m_x
$\bar{3}_2^2$	$\bar{3}_2^2$	$\bar{3}_3^2$	$\bar{3}_1$	$\bar{3}_4^2$	$\bar{3}_3$	$\bar{1}$	m_y	m_x	m_z	$\bar{3}_2$	$\bar{3}_4$	$\bar{3}_1^2$
$\bar{3}_3^2$	$\bar{3}_3^2$	$\bar{3}_2^2$	$\bar{3}_4^2$	$\bar{3}_1$	$\bar{3}_4$	m_y	$\bar{1}$	m_z	m_x	$\bar{3}_1^2$	$\bar{3}_3$	$\bar{3}_2$
$\bar{3}_4^2$	$\bar{3}_4^2$	$\bar{3}_1$	$\bar{3}_3^2$	$\bar{3}_2^2$	$\bar{3}_2$	m_x	m_z	$\bar{1}$	m_y	$\bar{3}_3$	$\bar{3}_1^2$	$\bar{3}_4$
$\bar{3}_1$	$\bar{3}_1$	$\bar{3}_4^2$	$\bar{3}_2^2$	$\bar{3}_3^2$	$\bar{3}_1^2$	m_z	m_x	m_y	$\bar{1}$	$\bar{3}_4$	$\bar{3}_2$	$\bar{3}_3$
$\bar{3}_2$	$\bar{3}_2$	$\bar{3}_4$	$\bar{3}_3$	$\bar{3}_1^2$	m_y	$\bar{3}_2^2$	$\bar{3}_1$	$\bar{3}_3^2$	$\bar{3}_4^2$	$\bar{1}$	m_x	m_z
$\bar{3}_3$	$\bar{3}_3$	$\bar{3}_1^2$	$\bar{3}_2$	$\bar{3}_4$	m_z	$\bar{3}_4^2$	$\bar{3}_3^2$	$\bar{3}_1$	$\bar{3}_2^2$	m_x	$\bar{1}$	m_y
$\bar{3}_4$	$\bar{3}_4$	$\bar{3}_2$	$\bar{3}_1^2$	$\bar{3}_3$	m_x	$\bar{3}_1$	$\bar{3}_2^2$	$\bar{3}_4^2$	$\bar{3}_3^2$	m_z	m_y	$\bar{1}$

Table 29(a)ii

$m3$	$\bar{1}$	m_x	m_y	m_z	$\bar{3}_1$	$\bar{3}_2$	$\bar{3}_3$	$\bar{3}_4$	$\bar{3}_1^2$	$\bar{3}_2^2$	$\bar{3}_3^2$	$\bar{3}_4^2$
1	$\bar{1}$	m_x	m_y	m_z	$\bar{3}_1$	$\bar{3}_2$	$\bar{3}_3$	$\bar{3}_4$	$\bar{3}_1^2$	$\bar{3}_2^2$	$\bar{3}_3^2$	$\bar{3}_4^2$
2_x	m_x	$\bar{1}$	m_z	m_y	$\bar{3}_3^2$	$\bar{3}_3$	$\bar{3}_2$	$\bar{3}_1^2$	$\bar{3}_4$	$\bar{3}_4^2$	$\bar{3}_1$	$\bar{3}_2^2$
2_y	m_y	m_z	$\bar{1}$	m_x	$\bar{3}_4^2$	$\bar{3}_1^2$	$\bar{3}_4$	$\bar{3}_3$	$\bar{3}_2$	$\bar{3}_3^2$	$\bar{3}_2^2$	$\bar{3}_1$
2_z	m_z	m_y	m_x	$\bar{1}$	$\bar{3}_2^2$	$\bar{3}_4$	$\bar{3}_1^2$	$\bar{3}_2$	$\bar{3}_3$	$\bar{3}_1$	$\bar{3}_4^2$	$\bar{3}_3^2$
3_1^2	$\bar{3}_1^2$	$\bar{3}_3$	$\bar{3}_4$	$\bar{3}_2$	$\bar{1}$	$\bar{3}_3^2$	$\bar{3}_4^2$	$\bar{3}_2^2$	$\bar{3}_1$	m_y	m_z	m_x
3_2^2	$\bar{3}_2^2$	$\bar{3}_3^2$	$\bar{3}_1$	$\bar{3}_4^2$	$\bar{3}_3$	$\bar{1}$	m_y	m_x	m_z	$\bar{3}_2$	$\bar{3}_4$	$\bar{3}_1^2$
3_3^2	$\bar{3}_3^2$	$\bar{3}_2^2$	$\bar{3}_4^2$	$\bar{3}_1$	$\bar{3}_4$	m_y	$\bar{1}$	m_z	m_x	$\bar{3}_1^2$	$\bar{3}_3$	$\bar{3}_2$
3_4^2	$\bar{3}_4^2$	$\bar{3}_1$	$\bar{3}_3^2$	$\bar{3}_2^2$	$\bar{3}_2$	m_x	m_z	$\bar{1}$	m_y	$\bar{3}_3$	$\bar{3}_1^2$	$\bar{3}_4$
3_1	$\bar{3}_1$	$\bar{3}_4^2$	$\bar{3}_2^2$	$\bar{3}_3^2$	$\bar{3}_1^2$	m_z	m_x	m_y	$\bar{1}$	$\bar{3}_4$	$\bar{3}_2$	$\bar{3}_3$
3_2	$\bar{3}_2$	$\bar{3}_4$	$\bar{3}_3$	$\bar{3}_1^2$	m_y	$\bar{3}_2^2$	$\bar{3}_1$	$\bar{3}_3^2$	$\bar{3}_4^2$	$\bar{1}$	m_x	m_z
3_3	$\bar{3}_3$	$\bar{3}_1^2$	$\bar{3}_2$	$\bar{3}_4$	m_z	$\bar{3}_4^2$	$\bar{3}_3^2$	$\bar{3}_1$	$\bar{3}_2^2$	m_x	$\bar{1}$	m_y
3_4	$\bar{3}_4$	$\bar{3}_2$	$\bar{3}_1^2$	$\bar{3}_3$	m_x	$\bar{3}_1$	$\bar{3}_2^2$	$\bar{3}_4^2$	$\bar{3}_3^2$	m_z	m_y	$\bar{1}$
$\bar{1}$	1	2_x	2_y	2_z	3_1	3_2	3_3	3_4	3_1^2	3_2^2	3_3^2	3_4^2
m_x	2_x	1	2_z	2_y	3_3^2	3_3	3_2	3_1^2	3_4	3_4^2	3_1	3_2^2
m_y	2_y	2_z	1	2_x	3_4^2	3_1^2	3_4	3_3	3_2	3_3^2	3_2^2	3_1
m_z	2_z	2_y	2_x	1	3_2^2	3_4	3_1^2	3_2	3_3	3_1	3_4^2	3_3^2
$\bar{3}_1^2$	3_1^2	3_3	3_4	3_2	1	3_3^2	3_4^2	3_2^2	3_1	2_y	2_z	2_x
$\bar{3}_2^2$	3_2^2	3_3^2	3_1	3_4^2	3_3	1	2_y	2_x	2_z	3_2	3_4	3_1^2
$\bar{3}_3^2$	3_3^2	3_2^2	3_4^2	3_1	3_4	2_y	1	2_z	2_x	3_1^2	3_3	3_2
$\bar{3}_4^2$	3_4^2	3_1	3_3^2	3_2^2	3_2	2_x	2_z	1	2_y	3_3	3_1^2	3_4
$\bar{3}_1$	3_1	3_4^2	3_2^2	3_3^2	3_1^2	2_z	2_x	2_y	1	3_4	3_2	3_3
$\bar{3}_2$	3_2	3_4	3_3	3_1^2	2_y	3_2^2	3_1	3_3^2	3_4^2	1	2_x	2_z
$\bar{3}_3$	3_3	3_1^2	3_2	3_4	2_z	3_4^2	3_3^2	3_1	3_2^2	2_x	1	2_y
$\bar{3}_4$	3_4	3_2	3_1^2	3_3	2_x	3_1	3_2^2	3_4^2	3_3^2	2_z	2_y	1

Table 29(b)i

T_h	E	C_{2x}	C_{2y}	C_{2z}	$C_3(1)$	$C_3(2)$	$C_3(3)$	$C_3(4)$	$C_3^2(1)$	$C_3^2(2)$	$C_3^2(3)$	$C_3^2(4)$
E	E	C_{2x}	C_{2y}	C_{2z}	$C_3(1)$	$C_3(2)$	$C_3(3)$	$C_3(4)$	$C_3^2(1)$	$C_3^2(2)$	$C_3^2(3)$	$C_3^2(4)$
C_{2x}	C_{2x}	E	C_{2z}	C_{2y}	$C_3^2(3)$	$C_3(3)$	$C_3(2)$	$C_3^2(1)$	$C_3(4)$	$C_3^2(4)$	$C_3(1)$	$C_3^2(2)$
C_{2y}	C_{2y}	C_{2z}	E	C_{2x}	$C_3^2(4)$	$C_3^2(1)$	$C_3(4)$	$C_3(3)$	$C_3(2)$	$C_3^2(3)$	$C_3^2(2)$	$C_3(1)$
C_{2z}	C_{2z}	C_{2y}	C_{2x}	E	$C_3^2(2)$	$C_3(4)$	$C_3^2(1)$	$C_3(2)$	$C_3(3)$	$C_3(1)$	$C_3^2(4)$	$C_3^2(3)$
$C_3^2(1)$	$C_3^2(1)$	$C_3(3)$	$C_3(4)$	$C_3(2)$	E	$C_3^2(3)$	$C_3^2(4)$	$C_3^2(2)$	$C_3(1)$	C_{2y}	C_{2z}	C_{2x}
$C_3^2(2)$	$C_3^2(2)$	$C_3^2(3)$	$C_3(1)$	$C_3^2(4)$	$C_3(3)$	E	C_{2y}	C_{2x}	C_{2z}	$C_3(2)$	$C_3(4)$	$C_3^2(1)$
$C_3^2(3)$	$C_3^2(3)$	$C_3^2(2)$	$C_3^2(4)$	$C_3(1)$	$C_3(4)$	C_{2y}	E	C_{2z}	C_{2x}	$C_3^2(1)$	$C_3(3)$	$C_3(2)$
$C_3^2(4)$	$C_3^2(4)$	$C_3(1)$	$C_3^2(3)$	$C_3^2(2)$	$C_3(2)$	C_{2x}	C_{2z}	E	C_{2y}	$C_3(3)$	$C_3^2(1)$	$C_3(4)$
$C_3(1)$	$C_3(1)$	$C_3^2(4)$	$C_3^2(2)$	$C_3^2(3)$	$C_3^2(1)$	C_{2z}	C_{2x}	C_{2y}	E	$C_3(4)$	$C_3(2)$	$C_3(3)$
$C_3(2)$	$C_3(2)$	$C_3(4)$	$C_3(3)$	$C_3^2(1)$	C_{2y}	$C_3^2(2)$	$C_3(1)$	$C_3^2(3)$	$C_3^2(4)$	E	C_{2x}	C_{2z}
$C_3(3)$	$C_3(3)$	$C_3^2(1)$	$C_3(2)$	$C_3(4)$	C_{2z}	$C_3^2(4)$	$C_3^2(3)$	$C_3(1)$	$C_3^2(2)$	C_{2x}	E	C_{2y}
$C_3(4)$	$C_3(4)$	$C_3(2)$	$C_3^2(1)$	$C_3(3)$	C_{2x}	$C_3(1)$	$C_3^2(2)$	$C_3^2(4)$	$C_3^2(3)$	C_{2z}	C_{2y}	E
i	i	σ_x	σ_y	σ_z	$S_6^5(1)$	$S_6^5(2)$	$S_6^5(3)$	$S_6^5(4)$	$S_6(1)$	$S_6(2)$	$S_6(3)$	$S_6(4)$
σ_x	σ_x	i	σ_z	σ_y	$S_6(3)$	$S_6^5(3)$	$S_6^5(2)$	$S_6(1)$	$S_6^5(4)$	$S_6(4)$	$S_6^5(1)$	$S_6(2)$
σ_y	σ_y	σ_z	i	σ_x	$S_6(4)$	$S_6(1)$	$S_6^5(4)$	$S_6^5(3)$	$S_6^5(2)$	$S_6(3)$	$S_6(2)$	$S_6^5(1)$
σ_z	σ_z	σ_y	σ_x	i	$S_6(2)$	$S_6^5(4)$	$S_6(1)$	$S_6^5(2)$	$S_6^5(3)$	$S_6^5(1)$	$S_6(4)$	$S_6(3)$
$S_6(1)$	$S_6(1)$	$S_6^5(3)$	$S_6^5(4)$	$S_6^5(2)$	i	$S_6(3)$	$S_6(4)$	$S_6(2)$	$S_6^5(1)$	σ_y	σ_z	σ_x
$S_6(2)$	$S_6(2)$	$S_6(3)$	$S_6^5(1)$	$S_6(4)$	$S_6^5(3)$	i	σ_y	σ_x	σ_z	$S_6^5(2)$	$S_6^5(4)$	$S_6(1)$
$S_6(3)$	$S_6(3)$	$S_6(2)$	$S_6(4)$	$S_6^5(1)$	$S_6^5(4)$	σ_y	i	σ_z	σ_x	$S_6(1)$	$S_6^5(3)$	$S_6^5(2)$
$S_6(4)$	$S_6(4)$	$S_6^5(1)$	$S_6(3)$	$S_6(2)$	$S_6^5(2)$	σ_x	σ_z	i	σ_y	$S_6^5(3)$	$S_6(1)$	$S_6^5(4)$
$S_6^5(1)$	$S_6^5(1)$	$S_6(4)$	$S_6(2)$	$S_6(3)$	$S_6(1)$	σ_z	σ_x	σ_y	i	$S_6^5(4)$	$S_6^5(2)$	$S_6^5(3)$
$S_6^5(2)$	$S_6^5(2)$	$S_6^5(4)$	$S_6^5(3)$	$S_6(1)$	σ_y	$S_6(2)$	$S_6^5(1)$	$S_6(3)$	$S_6(4)$	i	σ_x	σ_z
$S_6^5(3)$	$S_6^5(3)$	$S_6(1)$	$S_6^5(2)$	$S_6^5(4)$	σ_z	$S_6(4)$	$S_6(3)$	$S_6^5(1)$	$S_6(2)$	σ_x	i	σ_y
$S_6^5(4)$	$S_6^5(4)$	$S_6^5(2)$	$S_6(1)$	$S_6^5(3)$	σ_x	$S_6^5(1)$	$S_6(2)$	$S_6(4)$	$S_6(3)$	σ_z	σ_y	i

Table 29(b)ii

T_h	i	σ_x	σ_y	σ_z	$S_6^5(1)$	$S_6^5(2)$	$S_6^5(3)$	$S_6^5(4)$	$S_6(1)$	$S_6(2)$	$S_6(3)$	$S_6(4)$
E	i	σ_x	σ_y	σ_z	$S_6^5(1)$	$S_6^5(2)$	$S_6^5(3)$	$S_6^5(4)$	$S_6(1)$	$S_6(2)$	$S_6(3)$	$S_6(4)$
C_{2x}	σ_x	i	σ_z	σ_y	$S_6(3)$	$S_6^5(3)$	$S_6^5(2)$	$S_6(1)$	$S_6^5(4)$	$S_6(4)$	$S_6^5(1)$	$S_6(2)$
C_{2y}	σ_y	σ_z	i	σ_x	$S_6(4)$	$S_6(1)$	$S_6^5(4)$	$S_6^5(3)$	$S_6^5(2)$	$S_6(3)$	$S_6(2)$	$S_6^5(1)$
C_{2z}	σ_z	σ_y	σ_x	i	$S_6(2)$	$S_6^5(4)$	$S_6(1)$	$S_6^5(2)$	$S_6^5(3)$	$S_6^5(1)$	$S_6(4)$	$S_6(3)$
$C_3^2(1)$	$S_6(1)$	$S_6^5(3)$	$S_6^5(4)$	$S_6^5(2)$	i	$S_6(3)$	$S_6(4)$	$S_6(2)$	$S_6^5(1)$	σ_y	σ_z	σ_x
$C_3^2(2)$	$S_6(2)$	$S_6(3)$	$S_6^5(1)$	$S_6(4)$	$S_6^5(3)$	i	σ_y	σ_x	σ_z	$S_6^5(2)$	$S_6^5(4)$	$S_6(1)$
$C_3^2(3)$	$S_6(3)$	$S_6(2)$	$S_6(4)$	$S_6^5(1)$	$S_6^5(4)$	σ_y	i	σ_z	σ_x	$S_6(1)$	$S_6^5(3)$	$S_6^5(2)$
$C_3^2(4)$	$S_6(4)$	$S_6^5(1)$	$S_6(3)$	$S_6(2)$	$S_6^5(2)$	σ_x	σ_z	i	σ_y	$S_6^5(3)$	$S_6(1)$	$S_6^5(4)$
$C_3(1)$	$S_6^5(1)$	$S_6(4)$	$S_6(2)$	$S_6(3)$	$S_6(1)$	σ_z	σ_x	σ_y	i	$S_6^5(4)$	$S_6^5(2)$	$S_6^5(3)$
$C_3(2)$	$S_6^5(2)$	$S_6^5(4)$	$S_6^5(3)$	$S_6(1)$	σ_y	$S_6(2)$	$S_6^5(1)$	$S_6(3)$	$S_6(4)$	i	σ_x	σ_z
$C_3(3)$	$S_6^5(3)$	$S_6(1)$	$S_6^5(2)$	$S_6^5(4)$	σ_z	$S_6(4)$	$S_6(3)$	$S_6^5(1)$	$S_6(2)$	σ_x	i	σ_y
$C_3(4)$	$S_6^5(4)$	$S_6^5(2)$	$S_6(1)$	$S_6^5(3)$	σ_x	$S_6^5(1)$	$S_6(2)$	$S_6(4)$	$S_6(3)$	σ_z	σ_y	i
i	E	C_{2x}	C_{2y}	C_{2z}	$C_3(1)$	$C_3(2)$	$C_3(3)$	$C_3(4)$	$C_3^2(1)$	$C_3^2(2)$	$C_3^2(3)$	$C_3^2(4)$
σ_x	C_{2x}	E	C_{2z}	C_{2y}	$C_3^2(3)$	$C_3(3)$	$C_3(2)$	$C_3^2(1)$	$C_3(4)$	$C_3^2(4)$	$C_3(1)$	$C_3^2(2)$
σ_y	C_{2y}	C_{2z}	E	C_{2x}	$C_3^2(4)$	$C_3^2(1)$	$C_3(4)$	$C_3(3)$	$C_3(2)$	$C_3^2(3)$	$C_3^2(2)$	$C_3(1)$
σ_z	C_{2z}	C_{2y}	C_{2x}	E	$C_3^2(2)$	$C_3(4)$	$C_3^2(1)$	$C_3(2)$	$C_3(3)$	$C_3(1)$	$C_3^2(4)$	$C_3^2(3)$
$S_6(1)$	$C_3^2(1)$	$C_3(3)$	$C_3(4)$	$C_3(2)$	E	$C_3^2(3)$	$C_3^2(4)$	$C_3^2(2)$	$C_3(1)$	C_{2y}	C_{2z}	C_{2x}
$S_6(2)$	$C_3^2(2)$	$C_3^2(3)$	$C_3(1)$	$C_3^2(4)$	$C_3(3)$	E	C_{2y}	C_{2x}	C_{2z}	$C_3(2)$	$C_3(4)$	$C_3^2(1)$
$S_6(3)$	$C_3^2(3)$	$C_3^2(2)$	$C_3^2(4)$	$C_3(1)$	$C_3(4)$	C_{2y}	E	C_{2z}	C_{2x}	$C_3^2(1)$	$C_3(3)$	$C_3(2)$
$S_6(4)$	$C_3^2(4)$	$C_3(1)$	$C_3^2(3)$	$C_3^2(2)$	$C_3(2)$	C_{2x}	C_{2z}	E	C_{2y}	$C_3(3)$	$C_3^2(1)$	$C_3(4)$
$S_6^5(1)$	$C_3(1)$	$C_3^2(4)$	$C_3^2(2)$	$C_3^2(3)$	$C_3^2(1)$	C_{2z}	C_{2x}	C_{2y}	E	$C_3(4)$	$C_3(2)$	$C_3(3)$
$S_6^5(2)$	$C_3(2)$	$C_3(4)$	$C_3(3)$	$C_3^2(1)$	C_{2y}	$C_3^2(2)$	$C_3(1)$	$C_3^2(3)$	$C_3^2(4)$	E	C_{2x}	C_{2z}
$S_6^5(3)$	$C_3(3)$	$C_3^2(1)$	$C_3(2)$	$C_3(4)$	C_{2z}	$C_3^2(4)$	$C_3^2(3)$	$C_3(1)$	$C_3^2(2)$	C_{2x}	E	C_{2y}
$S_6^5(4)$	$C_3(4)$	$C_3(2)$	$C_3^2(1)$	$C_3(3)$	C_{2x}	$C_3(1)$	$C_3^2(2)$	$C_3^2(4)$	$C_3(3)$	C_{2z}	C_{2y}	E

Table 30(a)i

432	1	2_x	2_y	2_z	3_1	3_2	3_3	3_4	3_1^2	3_2^2	3_3^2	3_4^2
1	1	2_x	2_y	2_z	3_1	3_2	3_3	3_4	3_1^2	3_2^2	3_3^2	3_4^2
2_x	2_x	1	2_z	2_y	3_3^2	3_3	3_2	3_1^2	3_4	3_4^2	3_1	3_2^2
2_y	2_y	2_z	1	2_x	3_4^2	3_1^2	3_4	3_3	3_2	3_3^2	3_2^2	3_1
2_z	2_z	2_y	2_x	1	3_2^2	3_4	3_1^2	3_2	3_3	3_1	3_4^2	3_3^2
3_1^2	3_1^2	3_3	3_4	3_2	1	3_3^2	3_4^2	3_2^2	3_1	2_y	2_z	2_x
3_2^2	3_2^2	3_3^2	3_1	3_4^2	3_3	1	2_y	2_x	2_z	3_2	3_4	3_1^2
3_3^2	3_3^2	3_2^2	3_4^2	3_1	3_4	2_y	1	2_z	2_x	3_1^2	3_3	3_2
3_4^2	3_4^2	3_1	3_3^2	3_2^2	3_2	2_x	2_z	1	2_y	3_3	3_1^2	3_4
3_1	3_1	3_4^2	3_2^2	3_3^2	3_1^2	2_z	2_x	2_y	1	3_4	3_2	3_3
3_2	3_2	3_4	3_3	3_1^2	2_y	3_2^2	3_1	3_3^2	3_4^2	1	2_x	2_z
3_3	3_3	3_1^2	3_2	3_4	2_z	3_4^2	3_3^2	3_1	3_2^2	2_x	1	2_y
3_4	3_4	3_2	3_1^2	3_3	2_x	3_1	3_2^2	3_4^2	3_3^2	2_z	2_y	1
4_x^3	4_x^3	4_x	2_{yz}	$2_{y\bar{z}}$	2_{zx}	4_z	$2_{x\bar{y}}$	2_{xy}	4_z^3	$2_{z\bar{x}}$	4_y^3	4_y
4_y^3	4_y^3	$2_{z\bar{x}}$	4_y	2_{zx}	2_{xy}	2_{yz}	4_x^3	$2_{y\bar{z}}$	4_x	4_z^3	$2_{x\bar{y}}$	4_z
4_z^3	4_z^3	$2_{x\bar{y}}$	2_{xy}	4_z	4_x^3	4_y^3	4_y	$2_{z\bar{x}}$	2_{zx}	2_{yz}	$2_{y\bar{z}}$	4_x
4_x	4_x	4_x^3	$2_{y\bar{z}}$	2_{yz}	4_y^3	$2_{x\bar{y}}$	4_z	4_z^3	2_{xy}	4_y	2_{zx}	$2_{z\bar{x}}$
4_y	4_y	2_{zx}	4_y^3	$2_{z\bar{x}}$	4_z	4_x	$2_{y\bar{z}}$	4_x^3	2_{yz}	$2_{x\bar{y}}$	4_z^3	2_{xy}
4_z	4_z	2_{xy}	$2_{x\bar{y}}$	4_z^3	2_{yz}	$2_{z\bar{x}}$	2_{zx}	4_y^3	4_y	4_x^3	4_x	$2_{y\bar{z}}$
2_{xy}	2_{xy}	4_z	4_z^3	$2_{x\bar{y}}$	4_x	2_{zx}	$2_{z\bar{x}}$	4_y	4_y^3	$2_{y\bar{z}}$	2_{yz}	4_x^3
2_{yz}	2_{yz}	$2_{y\bar{z}}$	4_x^3	4_x	4_y	4_z^3	2_{xy}	$2_{x\bar{y}}$	4_z	4_y^3	$2_{z\bar{x}}$	2_{zx}
2_{zx}	2_{zx}	4_y	$2_{z\bar{x}}$	4_y^3	4_z^3	$2_{y\bar{z}}$	4_x	2_{yz}	4_x^3	2_{xy}	4_z	$2_{x\bar{y}}$
$2_{x\bar{y}}$	$2_{x\bar{y}}$	4_z^3	4_z	2_{xy}	$2_{y\bar{z}}$	4_y	4_y^3	2_{zx}	$2_{z\bar{x}}$	4_x	4_x^3	2_{yz}
$2_{y\bar{z}}$	$2_{y\bar{z}}$	2_{yz}	4_x	4_x^3	$2_{z\bar{x}}$	2_{xy}	4_z^3	4_z	$2_{x\bar{y}}$	2_{zx}	4_y	4_y^3
$2_{z\bar{x}}$	$2_{z\bar{x}}$	4_y^3	2_{zx}	4_y	$2_{x\bar{y}}$	4_x^3	2_{yz}	4_x	$2_{y\bar{z}}$	4_z	2_{xy}	4_z^3

Table 30(a)ii

432	4_x	4_y	4_z	4_x^3	4_y^3	4_z^3	2_{xy}	2_{yz}	2_{zx}	$2_{x\bar{y}}$	$2_{y\bar{z}}$	$2_{z\bar{x}}$
1	4_x	4_y	4_z	4_x^3	4_y^3	4_z^3	2_{xy}	2_{yz}	2_{zx}	$2_{x\bar{y}}$	$2_{y\bar{z}}$	$2_{z\bar{x}}$
2_x	4_x^3	$2_{z\bar{x}}$	$2_{x\bar{y}}$	4_x	2_{zx}	2_{xy}	4_z^3	$2_{y\bar{z}}$	4_y^3	4_z	2_{yz}	4_y
2_y	2_{yz}	4_y^3	2_{xy}	$2_{y\bar{z}}$	4_y	$2_{x\bar{y}}$	4_z	4_x	$2_{z\bar{x}}$	4_z^3	4_x^3	2_{zx}
2_z	$2_{y\bar{z}}$	2_{zx}	4_z^3	2_{yz}	$2_{z\bar{x}}$	4_z	$2_{x\bar{y}}$	4_x^3	4_y	2_{xy}	4_x	4_y^3
3_1^2	2_{zx}	2_{xy}	4_x	4_y	4_z^3	2_{yz}	4_x^3	4_y^3	4_z	$2_{y\bar{z}}$	$2_{z\bar{x}}$	$2_{x\bar{y}}$
3_2^3	4_z^3	2_{yz}	4_y	$2_{x\bar{y}}$	4_x^3	$2_{z\bar{x}}$	2_{zx}	4_z	$2_{y\bar{z}}$	4_y^3	2_{xy}	4_x
3_3^2	$2_{x\bar{y}}$	4_x	4_y^3	4_z^3	$2_{y\bar{z}}$	2_{zx}	$2_{z\bar{x}}$	2_{xy}	4_x^3	4_y	4_z	2_{yz}
3_4^2	2_{xy}	$2_{y\bar{z}}$	$2_{z\bar{x}}$	4_z	4_x	4_y	4_y^3	$2_{x\bar{y}}$	2_{yz}	2_{zx}	4_z^3	4_x^3
3_1	4_z	4_x^3	2_{zx}	2_{xy}	2_{yz}	4_y^3	4_y	4_z^3	4_x	$2_{z\bar{x}}$	$2_{x\bar{y}}$	$2_{y\bar{z}}$
3_2	$2_{z\bar{x}}$	4_z	2_{yz}	4_y^3	$2_{x\bar{y}}$	4_x	$2_{y\bar{z}}$	4_y	2_{xy}	4_x^3	2_{zx}	4_z^3
3_3	4_y	$2_{x\bar{y}}$	$2_{y\bar{z}}$	2_{zx}	4_z	4_x^3	2_{yz}	$2_{z\bar{x}}$	4_z^3	4_x	4_y^3	2_{xy}
3_4	4_y^3	4_z^3	4_x^3	$2_{z\bar{x}}$	2_{xy}	$2_{y\bar{z}}$	4_x	2_{zx}	$2_{x\bar{y}}$	2_{yz}	4_y	4_z
4_x^3	1	3_2^2	3_3	2_x	3_1	3_4	3_1^2	2_z	3_3^2	3_2	2_y	3_4^2
4_y^3	3_2	1	3_1	3_4	2_y	3_3^2	3_4^2	3_1^2	2_x	3_2^2	3_3	2_z
4_z^3	3_3^2	3_1^2	1	3_2^2	3_4	2_z	2_x	3_1	3_3	2_y	3_4^2	3_2
4_x	2_x	3_4^2	3_2	1	3_3^2	3_1^2	3_4	2_y	3_1	3_3	2_z	3_2^2
4_y	3_1^2	2_y	3_4^2	3_3	1	3_2^2	3_1	3_2	2_z	3_3^2	3_4	2_x
4_z	3_4^2	3_3	2_z	3_1	3_2	1	2_y	3_2^2	3_1^2	2_x	3_3^2	3_4
2_{xy}	3_1	3_4	2_x	3_4^2	3_1^2	2_y	1	3_3^2	3_2	2_z	3_2^2	3_3
2_{yz}	2_z	3_1	3_1^2	2_y	3_2^2	3_2	3_3	1	3_4^2	3_4	2_x	3_3^2
2_{zx}	3_3	2_x	3_3^2	3_1^2	2_z	3_1	3_2^2	3_4	1	3_4^2	3_2	2_y
$2_{x\bar{y}}$	3_2^2	3_2	2_y	3_3^2	3_3	2_x	2_z	3_4^2	3_4	1	3_1	3_1^2
$2_{y\bar{z}}$	2_y	3_3^2	3_4	2_z	3_4^2	3_3	3_2	2_x	3_2^2	3_1^2	1	3_1
$2_{z\bar{x}}$	3_4	2_z	3_2^2	3_2	2_x	3_4^2	3_3^2	3_3	2_y	3_1	3_1^2	1

Table 30(b)i

O	E	C_{2x}	C_{2y}	C_{2z}	$C_3(1)$	$C_3(2)$	$C_3(3)$	$C_3(4)$	$C_3^2(1)$	$C_3^2(2)$	$C_3^2(3)$	$C_3^2(4)$
E	E	C_{2x}	C_{2y}	C_{2z}	$C_3(1)$	$C_3(2)$	$C_3(3)$	$C_3(4)$	$C_3^2(1)$	$C_3^2(2)$	$C_3^2(3)$	$C_3^2(4)$
C_{2x}	C_{2x}	E	C_{2z}	C_{2y}	$C_3^2(3)$	$C_3(3)$	$C_3(2)$	$C_3^2(1)$	$C_3(4)$	$C_3^2(4)$	$C_3(1)$	$C_3^2(2)$
C_{2y}	C_{2y}	C_{2z}	E	C_{2x}	$C_3^2(4)$	$C_3^2(1)$	$C_3(4)$	$C_3(3)$	$C_3(2)$	$C_3^2(3)$	$C_3^2(2)$	$C_3(1)$
C_{2z}	C_{2z}	C_{2y}	C_{2x}	E	$C_3^2(2)$	$C_3(4)$	$C_3^2(1)$	$C_3(2)$	$C_3(3)$	$C_3(1)$	$C_3^2(4)$	$C_3^2(3)$
$C_3^2(1)$	$C_3^2(1)$	$C_3(3)$	$C_3(4)$	$C_3(2)$	E	$C_3^2(3)$	$C_3^2(4)$	$C_3^2(2)$	$C_3(1)$	C_{2y}	C_{2z}	C_{2x}
$C_3^2(2)$	$C_3^2(2)$	$C_3^2(3)$	$C_3(1)$	$C_3^2(4)$	$C_3(3)$	E	C_{2y}	C_{2x}	C_{2z}	$C_3(2)$	$C_3(4)$	$C_3^2(1)$
$C_3^2(3)$	$C_3^2(3)$	$C_3^2(2)$	$C_3^2(4)$	$C_3(1)$	$C_3(4)$	C_{2y}	E	C_{2z}	C_{2x}	$C_3^2(1)$	$C_3(3)$	$C_3(2)$
$C_3^2(4)$	$C_3^2(4)$	$C_3(1)$	$C_3^2(3)$	$C_3^2(2)$	$C_3(2)$	C_{2x}	C_{2z}	E	C_{2y}	$C_3(3)$	$C_3^2(1)$	$C_3(4)$
$C_3(1)$	$C_3(1)$	$C_3^2(4)$	$C_3^2(2)$	$C_3^2(3)$	$C_3^2(1)$	C_{2z}	C_{2x}	C_{2y}	E	$C_3(4)$	$C_3(2)$	$C_3(3)$
$C_3(2)$	$C_3(2)$	$C_3(4)$	$C_3(3)$	$C_3^2(1)$	C_{2y}	$C_3^2(2)$	$C_3(1)$	$C_3^2(3)$	$C_3^2(4)$	E	C_{2x}	C_{2z}
$C_3(3)$	$C_3(3)$	$C_3^2(1)$	$C_3(2)$	$C_3(4)$	C_{2z}	$C_3^2(4)$	$C_3^2(3)$	$C_3(1)$	$C_3^2(2)$	C_{2x}	E	C_{2y}
$C_3(4)$	$C_3(4)$	$C_3(2)$	$C_3^2(1)$	$C_3(3)$	C_{2x}	$C_3(1)$	$C_3^2(2)$	$C_3^2(4)$	$C_3^2(3)$	C_{2z}	C_{2y}	E
C_{4x}^3	C_{4x}^3	C_{4x}	C_{2yz}	$C_{2y\bar{z}}$	C_{2zx}	C_{4z}	$C_{2x\bar{y}}$	C_{2xy}	C_{4z}^3	$C_{2z\bar{x}}$	C_{4y}^3	C_{4y}
C_{4y}^3	C_{4y}^3	$C_{2z\bar{x}}$	C_{4y}	C_{2zx}	C_{2xy}	C_{2yz}	C_{4x}^3	$C_{2y\bar{z}}$	C_{4x}	C_{4z}^3	$C_{2x\bar{y}}$	C_{4z}
C_{4z}^3	C_{4z}^3	$C_{2x\bar{y}}$	C_{2xy}	C_{4z}	C_{4x}^3	C_{4y}^3	C_{4y}	$C_{2z\bar{x}}$	C_{2zx}	C_{2yz}	$C_{2y\bar{z}}$	C_{4x}
C_{4x}	C_{4x}	C_{4x}^3	$C_{2y\bar{z}}$	C_{2yz}	C_{4y}^3	$C_{2x\bar{y}}$	C_{4z}	C_{4z}^3	C_{2xy}	C_{4y}	C_{2zx}	$C_{2z\bar{x}}$
C_{4y}	C_{4y}	C_{2zx}	C_{4y}^3	$C_{2z\bar{x}}$	C_{4z}	C_{4x}	$C_{2y\bar{z}}$	C_{4x}^3	C_{2yz}	$C_{2x\bar{y}}$	C_{4z}^3	C_{2xy}
C_{4z}	C_{4z}	C_{2xy}	$C_{2x\bar{y}}$	C_{4z}^3	C_{2yz}	$C_{2y\bar{z}}$	C_{2zx}	C_{4y}^3	C_{4y}	C_{4x}^3	C_{4x}	$C_{2y\bar{z}}$
C_{2xy}	C_{2xy}	C_{4z}	C_{4z}^3	$C_{2x\bar{y}}$	C_{4x}	C_{2zx}	$C_{2z\bar{x}}$	C_{4y}	C_{4y}^3	$C_{2y\bar{z}}$	C_{2yz}	C_{4x}^3
C_{2yz}	C_{2yz}	$C_{2y\bar{z}}$	C_{4x}^3	C_{4x}	C_{4y}	C_{4z}^3	C_{2xy}	$C_{2x\bar{y}}$	C_{4z}	C_{4y}^3	$C_{2z\bar{x}}$	C_{2zx}
C_{2zx}	C_{2zx}	C_{4y}	$C_{2z\bar{x}}$	C_{4y}^3	C_{4z}^3	$C_{2y\bar{z}}$	C_{4x}	C_{2yz}	C_{4x}^3	C_{2xy}	C_{4z}	$C_{2x\bar{y}}$
$C_{2x\bar{y}}$	$C_{2x\bar{y}}$	C_{4z}^3	C_{4z}	C_{2xy}	$C_{2y\bar{z}}$	C_{4y}	C_{4y}^3	C_{2zx}	$C_{2z\bar{x}}$	C_{4x}	C_{4x}^3	C_{2yz}
$C_{2y\bar{z}}$	$C_{2y\bar{z}}$	C_{2yz}	C_{4x}	C_{4x}^3	$C_{2z\bar{x}}$	C_{2xy}	C_{4z}^3	C_{4z}	$C_{2x\bar{y}}$	C_{2zx}	C_{4y}	C_{4y}^3
$C_{2z\bar{x}}$	$C_{2z\bar{x}}$	C_{4y}^3	C_{2zx}	C_{4y}	$C_{2x\bar{y}}$	C_{4x}^3	C_{2yz}	C_{4x}	$C_{2y\bar{z}}$	C_{4z}	C_{2xy}	C_{4z}^3

Table 30(b)ii

O	C_{4x}	C_{4y}	C_{4z}	C_{4x}^3	C_{4y}^3	C_{4z}^3	C_{2xy}	C_{2yz}	C_{2zx}	$C_{2x\bar{y}}$	$C_{2y\bar{z}}$	$C_{2\bar{z}x}$
E	C_{4x}	C_{4y}	C_{4z}	C_{4x}^3	C_{4y}^3	C_{4z}^3	C_{2xy}	C_{2yz}	C_{2zx}	$C_{2x\bar{y}}$	$C_{2y\bar{z}}$	$C_{2\bar{z}x}$
C_{2x}	C_{4x}^3	$C_{2\bar{z}x}$	$C_{2x\bar{y}}$	C_{4x}	C_{2zx}	C_{2xy}	C_{4z}^3	$C_{2y\bar{z}}$	C_{4y}^3	C_{4z}	C_{2yz}	C_{4y}
C_{2y}	C_{2yz}	C_{4y}^3	C_{2xy}	$C_{2y\bar{z}}$	C_{4y}	$C_{2x\bar{y}}$	C_{4z}	C_{4x}	$C_{2\bar{z}x}$	C_{4z}^3	C_{4x}^3	C_{2zx}
C_{2z}	$C_{2y\bar{z}}$	C_{2zx}	C_{4z}^3	C_{2yz}	$C_{2\bar{z}x}$	C_{4z}	$C_{2x\bar{y}}$	C_{4x}^3	C_{4y}	C_{2xy}	C_{4x}	C_{4y}^3
$C_3^2(1)$	C_{2zx}	C_{2xy}	C_{4x}	C_{4y}	C_{4z}^3	C_{2yz}	C_{4x}^3	C_{4y}^3	C_{4z}	$C_{2y\bar{z}}$	$C_{2\bar{z}x}$	$C_{2x\bar{y}}$
$C_3^2(2)$	C_{4z}^3	C_{2yz}	C_{4y}	$C_{2x\bar{y}}$	C_{4x}^3	$C_{2\bar{z}x}$	C_{2zx}	C_{4z}	$C_{2y\bar{z}}$	C_{4y}^3	C_{2xy}	C_{4x}
$C_3^2(3)$	$C_{2x\bar{y}}$	C_{4x}	C_{4y}^3	C_{4z}^3	$C_{2y\bar{z}}$	C_{2zx}	$C_{2\bar{z}x}$	C_{2xy}	C_{4x}^3	C_{4y}	C_{4z}	C_{2yz}
$C_3^2(4)$	C_{2xy}	$C_{2y\bar{z}}$	$C_{2\bar{z}x}$	C_{4z}	C_{4x}	C_{4y}	C_{4y}^3	$C_{2x\bar{y}}$	C_{2yz}	C_{2zx}	C_{4z}^3	C_{4x}^3
$C_3(1)$	C_{4z}	C_{4x}^3	C_{2zx}	C_{2xy}	C_{2yz}	C_{4y}^3	C_{4y}	C_{4z}^3	C_{4x}	$C_{2\bar{z}x}$	$C_{2x\bar{y}}$	$C_{2y\bar{z}}$
$C_3(2)$	$C_{2\bar{z}x}$	C_{4z}	C_{2yz}	C_{4y}^3	$C_{2x\bar{y}}$	C_{4x}	$C_{2y\bar{z}}$	C_{4y}	C_{2xy}	C_{4x}^3	C_{2zx}	C_{4z}^3
$C_3(3)$	C_{4y}	$C_{2x\bar{y}}$	$C_{2y\bar{z}}$	C_{2zx}	C_{4z}	C_{4x}^3	C_{2yz}	$C_{2\bar{z}x}$	C_{4z}^3	C_{4x}	C_{4y}^3	C_{2xy}
$C_3(4)$	C_{4y}^3	C_{4z}^3	C_{4x}^3	$C_{2\bar{z}x}$	C_{2xy}	$C_{2y\bar{z}}$	C_{4x}	C_{2zx}	$C_{2x\bar{y}}$	C_{2yz}	C_{4y}	C_{4z}
C_{4x}^3	E	$C_3^2(2)$	$C_3(3)$	C_{2x}	$C_3(1)$	$C_3(4)$	$C_3^2(1)$	C_{2z}	$C_3^2(3)$	$C_3(2)$	C_{2y}	$C_3^2(4)$
C_{4y}^3	$C_3(2)$	E	$C_3(1)$	$C_3(4)$	C_{2y}	$C_3^2(3)$	$C_3^2(4)$	$C_3^2(1)$	C_{2x}	$C_3^2(2)$	$C_3(3)$	C_{2z}
C_{4z}^3	$C_3^2(3)$	$C_3^2(1)$	E	$C_3^2(2)$	$C_3(4)$	C_{2z}	C_{2x}	$C_3(1)$	$C_3(3)$	C_{2y}	$C_3^2(4)$	$C_3(2)$
C_{4x}	C_{2x}	$C_3^2(4)$	$C_3(2)$	E	$C_3^2(3)$	$C_3^2(1)$	$C_3(4)$	C_{2y}	$C_3(1)$	$C_3(3)$	C_{2z}	$C_3^2(2)$
C_{4y}	$C_3^2(1)$	C_{2y}	$C_3^2(4)$	$C_3(3)$	E	$C_3^2(2)$	$C_3(1)$	$C_3(2)$	C_{2z}	$C_3^2(3)$	$C_3(4)$	C_{2x}
C_{4z}	$C_3^2(4)$	$C_3(3)$	C_{2z}	$C_3(1)$	$C_3(2)$	E	C_{2y}	$C_3^2(2)$	$C_3^2(1)$	C_{2x}	$C_3^2(3)$	$C_3(4)$
C_{2xy}	$C_3(1)$	$C_3(4)$	C_{2x}	$C_3^2(4)$	$C_3^2(1)$	C_{2y}	E	$C_3^2(3)$	$C_3(2)$	C_{2z}	$C_3^2(2)$	$C_3(3)$
C_{2yz}	C_{2z}	$C_3(1)$	$C_3^2(1)$	C_{2y}	$C_3^2(2)$	$C_3(2)$	$C_3(3)$	E	$C_3^2(4)$	$C_3(4)$	C_{2x}	$C_3^2(3)$
C_{2zx}	$C_3(3)$	C_{2x}	$C_3^2(3)$	$C_3^2(1)$	C_{2z}	$C_3(1)$	$C_3^2(2)$	$C_3(4)$	E	$C_3^2(4)$	$C_3(2)$	C_{2y}
$C_{2x\bar{y}}$	$C_3^2(2)$	$C_3(2)$	C_{2y}	$C_3^2(3)$	$C_3(3)$	C_{2x}	C_{2z}	$C_3^2(4)$	$C_3(4)$	E	$C_3(1)$	$C_3^2(1)$
$C_{2y\bar{z}}$	C_{2y}	$C_3^2(3)$	$C_3(4)$	C_{2z}	$C_3^2(4)$	$C_3(3)$	$C_3(2)$	C_{2x}	$C_3^2(2)$	$C_3^2(1)$	E	$C_3(1)$
$C_{2\bar{z}x}$	$C_3(4)$	C_{2z}	$C_3^2(2)$	$C_3(2)$	C_{2x}	$C_3^2(4)$	$C_3^2(3)$	$C_3(3)$	C_{2y}	$C_3(1)$	$C_3^2(1)$	E

Table 31(a)i

$\bar{4}3m$	1	2_x	2_y	2_z	3_1	3_2	3_3	3_4	3_1^2	3_2^2	3_3^2	3_4^2
1	1	2_x	2_y	2_z	3_1	3_2	3_3	3_4	3_1^2	3_2^2	3_3^2	3_4^2
2_x	2_x	1	2_z	2_y	$3_{\bar{3}}^2$	3_3	3_2	3_1^2	3_4	3_4^2	3_1	3_2^2
2_y	2_y	2_z	1	2_x	3_4^2	3_1^2	3_4	3_3	3_2	3_3^2	3_2^2	3_1
2_z	2_z	2_y	2_x	1	3_2^2	3_4	3_1^2	3_2	3_3	3_1	3_4^2	3_3^2
3_1^2	3_1^2	3_3	3_4	3_2	1	3_3^2	3_4^2	3_2^2	3_1	2_y	2_z	2_x
3_2^2	3_2^2	3_3^2	3_1	3_4^2	3_3	1	2_y	2_x	2_z	3_2	3_4	3_1^2
3_3^2	3_3^2	3_2^2	3_4^2	3_1	3_4	2_y	1	2_z	2_x	3_1^2	3_3	3_2
3_4^2	3_4^2	3_1	3_3^2	3_2^2	3_2	2_x	2_z	1	2_y	3_3	3_1^2	3_4
3_1	3_1	3_4^2	3_2^2	3_3^2	3_1^2	2_z	2_x	2_y	1	3_4	3_2	3_3
3_2	3_2	3_4	3_3	3_1^2	2_y	3_2^2	3_1	3_3^2	3_4^2	1	2_x	2_z
3_3	3_3	3_1^2	3_2	3_4	2_z	3_4^2	3_3^2	3_1	3_2^2	2_x	1	2_y
3_4	3_4	3_2	3_1^2	3_3	2_x	3_1	3_2^2	3_4^2	3_3^2	2_z	2_y	1
$\bar{4}_x^3$	$\bar{4}_x^3$	$\bar{4}_x$	m_{yz}	$m_{y\bar{z}}$	m_{zx}	$\bar{4}_z$	$m_{x\bar{y}}$	m_{xy}	$\bar{4}_z^3$	$m_{z\bar{x}}$	$\bar{4}_y^3$	$\bar{4}_y$
$\bar{4}_y^3$	$\bar{4}_y^3$	$m_{z\bar{x}}$	$\bar{4}_y$	m_{zx}	m_{xy}	m_{yz}	$\bar{4}_x^3$	$m_{y\bar{z}}$	$\bar{4}_x$	$\bar{4}_z^3$	$m_{x\bar{y}}$	$\bar{4}_z$
$\bar{4}_z^3$	$\bar{4}_z^3$	$m_{x\bar{y}}$	m_{xy}	$\bar{4}_z$	$\bar{4}_x^3$	$\bar{4}_y^3$	$\bar{4}_y$	$m_{z\bar{x}}$	m_{zx}	m_{yz}	$m_{y\bar{z}}$	$\bar{4}_x$
$\bar{4}_x$	$\bar{4}_x$	$\bar{4}_x^3$	$m_{y\bar{z}}$	m_{yz}	$\bar{4}_y^3$	$m_{x\bar{y}}$	$\bar{4}_z$	$\bar{4}_z^3$	m_{xy}	$\bar{4}_y$	m_{zx}	$m_{z\bar{x}}$
$\bar{4}_y$	$\bar{4}_y$	m_{zx}	$\bar{4}_y^3$	$m_{z\bar{x}}$	$\bar{4}_z$	$\bar{4}_x$	$m_{y\bar{z}}$	$\bar{4}_x^3$	m_{yz}	$m_{x\bar{y}}$	$\bar{4}_z^3$	m_{xy}
$\bar{4}_z$	$\bar{4}_z$	m_{xy}	$m_{x\bar{y}}$	$\bar{4}_z^3$	m_{yz}	$m_{z\bar{x}}$	m_{zx}	$\bar{4}_y^3$	$\bar{4}_y$	$\bar{4}_x^3$	$\bar{4}_x$	$m_{y\bar{z}}$
m_{xy}	m_{xy}	$\bar{4}_z$	$\bar{4}_z^3$	$m_{x\bar{y}}$	$\bar{4}_x$	m_{zx}	$m_{z\bar{x}}$	$\bar{4}_y$	$\bar{4}_y^3$	$m_{y\bar{z}}$	m_{yz}	$\bar{4}_x^3$
m_{yz}	m_{yz}	$m_{y\bar{z}}$	$\bar{4}_x^3$	$\bar{4}_x$	$\bar{4}_y$	$\bar{4}_z^3$	m_{xy}	$m_{x\bar{y}}$	$\bar{4}_z$	$\bar{4}_y^3$	$m_{z\bar{x}}$	m_{zx}
m_{zx}	m_{zx}	$\bar{4}_y$	$m_{z\bar{x}}$	$\bar{4}_y^3$	$\bar{4}_z^3$	$m_{y\bar{z}}$	$\bar{4}_x$	m_{yz}	$\bar{4}_x^3$	m_{xy}	$\bar{4}_z$	$m_{x\bar{y}}$
$m_{x\bar{y}}$	$m_{x\bar{y}}$	$\bar{4}_z^3$	$\bar{4}_z$	m_{xy}	$m_{y\bar{z}}$	$\bar{4}_y$	$\bar{4}_y^3$	m_{zx}	$m_{z\bar{x}}$	$\bar{4}_x$	$\bar{4}_x^3$	m_{yz}
$m_{y\bar{z}}$	$m_{y\bar{z}}$	m_{yz}	$\bar{4}_x$	$\bar{4}_x^3$	$m_{z\bar{x}}$	m_{xy}	$\bar{4}_z^3$	$\bar{4}_z$	$m_{x\bar{y}}$	m_{zx}	$\bar{4}_y$	$\bar{4}_y^3$
$m_{z\bar{x}}$	$m_{z\bar{x}}$	$\bar{4}_y^3$	m_{zx}	$\bar{4}_y$	$m_{x\bar{y}}$	$\bar{4}_x^3$	m_{yz}	$\bar{4}_x$	$m_{y\bar{z}}$	$\bar{4}_z$	m_{xy}	$\bar{4}_z^3$

Table 31(a)ii

$\bar{4}3m$	$\bar{4}_x$	$\bar{4}_y$	$\bar{4}_z$	$\bar{4}_x^3$	$\bar{4}_y^3$	$\bar{4}_z^3$	m_{xy}	m_{yz}	m_{zx}	$m_{x\bar{y}}$	$m_{y\bar{z}}$	$m_{z\bar{x}}$
1	$\bar{4}_x$	$\bar{4}_y$	$\bar{4}_z$	$\bar{4}_x^3$	$\bar{4}_y^3$	$\bar{4}_z^3$	m_{xy}	m_{yz}	m_{zx}	$m_{x\bar{y}}$	$m_{y\bar{z}}$	$m_{z\bar{x}}$
2_x	$\bar{4}_x^3$	$m_{z\bar{x}}$	$m_{x\bar{y}}$	$\bar{4}_x$	m_{zx}	m_{xy}	$\bar{4}_z^3$	$m_{y\bar{z}}$	$\bar{4}_y^3$	$\bar{4}_z$	m_{yz}	$\bar{4}_y$
2_y	m_{yz}	$\bar{4}_y^3$	m_{xy}	$m_{y\bar{z}}$	$\bar{4}_y$	$m_{x\bar{y}}$	$\bar{4}_z$	$\bar{4}_x$	$m_{z\bar{x}}$	$\bar{4}_z^3$	$\bar{4}_x^3$	m_{zx}
2_z	$m_{y\bar{z}}$	m_{zx}	$\bar{4}_z^3$	m_{yz}	$m_{z\bar{x}}$	$\bar{4}_z$	$m_{x\bar{y}}$	$\bar{4}_x^3$	$\bar{4}_y$	m_{xy}	$\bar{4}_x$	$\bar{4}_y^3$
3_1^2	m_{zx}	m_{xy}	$\bar{4}_x$	$\bar{4}_y$	$\bar{4}_z^3$	m_{yz}	$\bar{4}_x^3$	$\bar{4}_y^3$	$\bar{4}_z$	$m_{y\bar{z}}$	$m_{z\bar{x}}$	$m_{x\bar{y}}$
3_2^2	$\bar{4}_z^3$	m_{yz}	$\bar{4}_y$	$m_{x\bar{y}}$	$\bar{4}_x^3$	$m_{z\bar{x}}$	m_{zx}	$\bar{4}_z$	$m_{y\bar{z}}$	$\bar{4}_y^3$	m_{xy}	$\bar{4}_x$
3_3^2	$m_{x\bar{y}}$	$\bar{4}_x$	$\bar{4}_y^3$	$\bar{4}_z^3$	$m_{y\bar{z}}$	m_{zx}	$m_{z\bar{x}}$	m_{xy}	$\bar{4}_x^3$	$\bar{4}_y$	$\bar{4}_z$	m_{yz}
3_4^2	m_{xy}	$m_{y\bar{z}}$	$m_{z\bar{x}}$	$\bar{4}_z$	$\bar{4}_x$	$\bar{4}_y$	$\bar{4}_y^3$	$m_{x\bar{y}}$	m_{yz}	m_{zx}	$\bar{4}_z^3$	$\bar{4}_x^3$
3_1	$\bar{4}_z$	$\bar{4}_x^3$	m_{zx}	m_{xy}	m_{yz}	$\bar{4}_y^3$	$\bar{4}_y$	$\bar{4}_z^3$	$\bar{4}_x$	$m_{z\bar{x}}$	$m_{x\bar{y}}$	$m_{y\bar{z}}$
3_2	$m_{z\bar{x}}$	$\bar{4}_z$	m_{yz}	$\bar{4}_y^3$	$m_{x\bar{y}}$	$\bar{4}_x$	$m_{y\bar{z}}$	$\bar{4}_y$	m_{xy}	$\bar{4}_x^3$	m_{zx}	$\bar{4}_z^3$
3_3	$\bar{4}_y$	$m_{x\bar{y}}$	$m_{y\bar{z}}$	m_{zx}	$\bar{4}_z$	$\bar{4}_x^3$	m_{yz}	$m_{z\bar{x}}$	$\bar{4}_z^3$	$\bar{4}_x$	$\bar{4}_y^3$	m_{xy}
3_4	$\bar{4}_y^3$	$\bar{4}_z^3$	$\bar{4}_x^3$	$m_{z\bar{x}}$	m_{xy}	$m_{y\bar{z}}$	$\bar{4}_x$	m_{zx}	$m_{x\bar{y}}$	m_{yz}	$\bar{4}_y$	$\bar{4}_z$
$\bar{4}_x^3$	1	3_2^2	3_3	2_x	3_1	3_4	3_1^2	2_z	3_3^2	3_2	2_y	3_4^2
$\bar{4}_y^3$	3_2	1	3_1	3_4	2_y	3_3^2	3_4^2	3_1^2	2_x	3_2^2	3_3	2_z
$\bar{4}_z^3$	3_3^2	3_1^2	1	3_2^2	3_4	2_z	2_x	3_1	3_3	2_y	3_4^2	3_2
$\bar{4}_x$	2_x	3_4^2	3_2	1	3_3^2	3_1^2	3_4	2_y	3_1	3_3	2_z	3_2^2
$\bar{4}_y$	3_1^2	2_y	3_4^2	3_3	1	3_2^2	3_1	3_2	2_z	3_3^2	3_4	2_x
$\bar{4}_z$	3_4^2	3_3	2_z	3_1	3_2	1	2_y	3_2^2	3_1^2	2_x	3_3^2	3_4
m_{xy}	3_1	3_4	2_x	3_4^2	3_1^2	2_y	1	3_3^2	3_2	2_z	3_2^2	3_3
m_{yz}	2_z	3_1	3_1^2	2_y	3_2^2	3_2	3_3	1	3_4^2	3_4	2_x	3_3^2
m_{zx}	3_3	2_x	3_3^2	3_1^2	2_z	3_1	3_2^2	3_4	1	3_4^2	3_2	2_y
$m_{x\bar{y}}$	3_2^2	3_2	2_y	3_3^2	3_3	2_x	2_z	3_4^2	3_4	1	3_1	3_1^2
$m_{y\bar{z}}$	2_y	3_3^2	3_4	2_z	3_4^2	3_3	3_2	2_x	3_2^2	3_1^2	1	3_1
$m_{z\bar{x}}$	3_4	2_z	3_2^2	3_2	2_x	3_4^2	3_3^2	3_3	2_y	3_1	3_1^2	1

Table 31(b)i

T_d	E	C_{2x}	C_{2y}	C_{2z}	$C_3(1)$	$C_3(2)$	$C_3(3)$	$C_3(4)$	$C_3^2(1)$	$C_3^2(2)$	$C_3^2(3)$	$C_3^2(4)$
E	E	C_{2x}	C_{2y}	C_{2z}	$C_3(1)$	$C_3(2)$	$C_3(3)$	$C_3(4)$	$C_3^2(1)$	$C_3^2(2)$	$C_3^2(3)$	$C_3^2(4)$
C_{2x}	C_{2x}	E	C_{2z}	C_{2y}	$C_3^2(3)$	$C_3(3)$	$C_3(2)$	$C_3^2(1)$	$C_3(4)$	$C_3^2(4)$	$C_3(1)$	$C_3^2(2)$
C_{2y}	C_{2y}	C_{2z}	E	C_{2x}	$C_3^2(4)$	$C_3^2(1)$	$C_3(4)$	$C_3(3)$	$C_3(2)$	$C_3^2(3)$	$C_3^2(2)$	$C_3(1)$
C_{2z}	C_{2z}	C_{2y}	C_{2x}	E	$C_3^2(2)$	$C_3(4)$	$C_3^2(1)$	$C_3(2)$	$C_3(3)$	$C_3(1)$	$C_3^2(4)$	$C_3^2(3)$
$C_3^2(1)$	$C_3^2(1)$	$C_3(3)$	$C_3(4)$	$C_3(2)$	E	$C_3^2(3)$	$C_3^2(4)$	$C_3^2(2)$	$C_3(1)$	C_{2y}	C_{2z}	C_{2x}
$C_3^2(2)$	$C_3^2(2)$	$C_3^2(3)$	$C_3(1)$	$C_3^2(4)$	$C_3(3)$	E	C_{2y}	C_{2x}	C_{2z}	$C_3(2)$	$C_3(4)$	$C_3^2(1)$
$C_3^2(3)$	$C_3^2(3)$	$C_3^2(2)$	$C_3^2(4)$	$C_3(1)$	$C_3(4)$	C_{2y}	E	C_{2z}	C_{2x}	$C_3^2(1)$	$C_3(3)$	$C_3(2)$
$C_3^2(4)$	$C_3^2(4)$	$C_3(1)$	$C_3^2(3)$	$C_3^2(2)$	$C_3(2)$	C_{2x}	C_{2z}	E	C_{2y}	$C_3(3)$	$C_3^2(1)$	$C_3(4)$
$C_3(1)$	$C_3(1)$	$C_3^2(4)$	$C_3^2(2)$	$C_3^2(3)$	$C_3^2(1)$	C_{2z}	C_{2x}	C_{2y}	E	$C_3(4)$	$C_3(2)$	$C_3(3)$
$C_3(2)$	$C_3(2)$	$C_3(4)$	$C_3(3)$	$C_3^2(1)$	C_{2y}	$C_3^2(2)$	$C_3(1)$	$C_3^2(3)$	$C_3^2(4)$	E	C_{2x}	C_{2z}
$C_3(3)$	$C_3(3)$	$C_3^2(1)$	$C_3(2)$	$C_3(4)$	C_{2z}	$C_3^2(4)$	$C_3^2(3)$	$C_3(1)$	$C_3^2(2)$	C_{2x}	E	C_{2y}
$C_3(4)$	$C_3(4)$	$C_3(2)$	$C_3^2(1)$	$C_3(3)$	C_{2x}	$C_3(1)$	$C_3^2(2)$	$C_3^2(4)$	$C_3^2(3)$	C_{2z}	C_{2y}	E
S_{4x}^3	S_{4x}^3	S_{4x}	σ_{yz}	$\sigma_{y\bar{z}}$	σ_{zx}	S_{4z}	$\sigma_{x\bar{y}}$	σ_{xy}	S_{4z}^3	$\sigma_{z\bar{x}}$	S_{4y}^3	S_{4y}
S_{4y}^3	S_{4y}^3	$\sigma_{z\bar{x}}$	S_{4y}	σ_{zx}	σ_{xy}	σ_{yz}	S_{4x}^3	$\sigma_{y\bar{z}}$	S_{4x}	S_{4z}^3	$\sigma_{x\bar{y}}$	S_{4z}
S_{4z}^3	S_{4z}^3	$\sigma_{x\bar{y}}$	σ_{xy}	S_{4z}	S_{4x}^3	S_{4y}^3	S_{4y}	$\sigma_{z\bar{x}}$	σ_{zx}	σ_{yz}	$\sigma_{y\bar{z}}$	S_{4x}
S_{4x}	S_{4x}	S_{4x}^3	$\sigma_{y\bar{z}}$	σ_{yz}	S_{4y}^3	$\sigma_{x\bar{y}}$	S_{4z}	S_{4z}^3	σ_{xy}	S_{4y}	σ_{zx}	$\sigma_{z\bar{x}}$
S_{4y}	S_{4y}	σ_{zx}	S_{4y}^3	$\sigma_{z\bar{x}}$	S_{4z}	S_{4x}	$\sigma_{y\bar{z}}$	S_{4x}^3	σ_{yz}	$\sigma_{x\bar{y}}$	S_{4z}^3	σ_{xy}
S_{4z}	S_{4z}	σ_{xy}	$\sigma_{x\bar{y}}$	S_{4z}^3	σ_{yz}	$\sigma_{z\bar{x}}$	σ_{zx}	S_{4y}^3	S_{4y}	S_{4x}^3	S_{4x}	$\sigma_{y\bar{z}}$
σ_{xy}	σ_{xy}	S_{4z}	S_{4z}^3	$\sigma_{x\bar{y}}$	S_{4x}	σ_{zx}	$\sigma_{z\bar{x}}$	S_{4y}	S_{4y}^3	$\sigma_{y\bar{z}}$	σ_{yz}	S_{4x}^3
σ_{yz}	σ_{yz}	$\sigma_{y\bar{z}}$	S_{4x}^3	S_{4x}	S_{4y}	S_{4z}^3	σ_{xy}	$\sigma_{x\bar{y}}$	S_{4z}	S_{4y}^3	$\sigma_{z\bar{x}}$	σ_{zx}
σ_{zx}	σ_{zx}	S_{4y}	$\sigma_{z\bar{x}}$	S_{4y}^3	S_{4z}^3	$\sigma_{y\bar{z}}$	S_{4x}	σ_{yz}	S_{4x}^3	σ_{xy}	S_{4z}	$\sigma_{x\bar{y}}$
$\sigma_{x\bar{y}}$	$\sigma_{x\bar{y}}$	S_{4z}^3	S_{4z}	σ_{xy}	$\sigma_{y\bar{z}}$	S_{4y}	S_{4y}^3	σ_{zx}	$\sigma_{z\bar{x}}$	S_{4x}	S_{4x}^3	σ_{yz}
$\sigma_{y\bar{z}}$	$\sigma_{y\bar{z}}$	σ_{yz}	S_{4x}	S_{4x}^3	$\sigma_{z\bar{x}}$	σ_{xy}	S_{4z}^3	S_{4z}	$\sigma_{x\bar{y}}$	σ_{zx}	S_{4y}	S_{4y}^3
$\sigma_{z\bar{x}}$	$\sigma_{z\bar{x}}$	S_{4y}^3	σ_{zx}	S_{4y}	$\sigma_{x\bar{y}}$	S_{4x}^3	σ_{yz}	S_{4x}	$\sigma_{y\bar{z}}$	S_{4z}	σ_{xy}	S_{4z}^3

Table 31(b)ii

T_d	S_{4x}	S_{4y}	S_{4z}	S_{4x}^3	S_{4y}^3	S_{4z}^3	σ_{xy}	σ_{yz}	σ_{zx}	$\sigma_{x\bar{y}}$	$\sigma_{y\bar{z}}$	$\sigma_{z\bar{x}}$
E	S_{4x}	S_{4y}	S_{4z}	S_{4x}^3	S_{4y}^3	S_{4z}^3	σ_{xy}	σ_{yz}	σ_{zx}	$\sigma_{x\bar{y}}$	$\sigma_{y\bar{z}}$	$\sigma_{z\bar{x}}$
C_{2x}	S_{4x}^3	$\sigma_{z\bar{x}}$	$\sigma_{x\bar{y}}$	S_{4x}	σ_{zx}	σ_{xy}	S_{4z}^3	$\sigma_{y\bar{z}}$	S_{4y}^3	S_{4z}	σ_{yz}	S_{4y}
C_{2y}	σ_{yz}	S_{4y}^3	σ_{xy}	$\sigma_{y\bar{z}}$	S_{4y}	$\sigma_{x\bar{y}}$	S_{4z}	S_{4x}	$\sigma_{z\bar{x}}$	S_{4z}^3	S_{4x}^3	σ_{zx}
C_{2z}	$\sigma_{y\bar{z}}$	σ_{zx}	S_{4z}^3	σ_{yz}	$\sigma_{z\bar{x}}$	S_{4z}	$\sigma_{x\bar{y}}$	S_{4x}^3	S_{4y}	σ_{xy}	S_{4x}	S_{4y}^3
$C_3^2(1)$	σ_{zx}	σ_{xy}	S_{4x}	S_{4y}	S_{4z}^3	σ_{yz}	S_{4x}^3	S_{4y}^3	S_{4z}	$\sigma_{y\bar{z}}$	$\sigma_{z\bar{x}}$	$\sigma_{x\bar{y}}$
$C_3^2(2)$	S_{4z}^3	σ_{yz}	S_{4y}	$\sigma_{x\bar{y}}$	S_{4x}^3	$\sigma_{z\bar{x}}$	σ_{zx}	S_{4z}	$\sigma_{y\bar{z}}$	S_{4y}^3	σ_{xy}	S_{4x}
$C_3^2(3)$	$\sigma_{x\bar{y}}$	S_{4x}	S_{4y}^3	S_{4z}^3	$\sigma_{y\bar{z}}$	σ_{zx}	$\sigma_{z\bar{x}}$	σ_{xy}	S_{4x}^3	S_{4y}	S_{4z}	σ_{yz}
$C_3^2(4)$	σ_{xy}	$\sigma_{y\bar{z}}$	$\sigma_{z\bar{x}}$	S_{4z}	S_{4x}	S_{4y}	S_{4y}^3	$\sigma_{x\bar{y}}$	σ_{yz}	σ_{zx}	S_{4z}^3	S_{4x}^3
$C_3(1)$	S_{4z}	S_{4x}^3	σ_{zx}	σ_{xy}	σ_{yz}	S_{4y}^3	S_{4y}	S_{4z}^3	S_{4x}	$\sigma_{z\bar{x}}$	$\sigma_{x\bar{y}}$	$\sigma_{y\bar{z}}$
$C_3(2)$	$\sigma_{z\bar{x}}$	S_{4z}	σ_{yz}	S_{4y}^3	$\sigma_{x\bar{y}}$	S_{4x}	$\sigma_{y\bar{z}}$	S_{4y}	σ_{xy}	S_{4x}^3	σ_{zx}	S_{4z}^3
$C_3(3)$	S_{4y}	$\sigma_{x\bar{y}}$	$\sigma_{y\bar{z}}$	σ_{zx}	S_{4z}	S_{4x}^3	σ_{yz}	$\sigma_{z\bar{x}}$	S_{4z}^3	S_{4x}	S_{4y}^3	σ_{xy}
$C_3(4)$	S_{4y}^3	S_{4z}^3	S_{4x}^3	$\sigma_{z\bar{x}}$	σ_{xy}	$\sigma_{y\bar{z}}$	S_{4x}	σ_{zx}	$\sigma_{x\bar{y}}$	σ_{yz}	S_{4y}	S_{4z}
S_{4x}^3	E	$C_3^2(2)$	$C_3(3)$	C_{2x}	$C_3(1)$	$C_3(4)$	$C_3^2(1)$	C_{2z}	$C_3^2(3)$	$C_3(2)$	C_{2y}	$C_3^2(4)$
S_{4y}^3	$C_3(2)$	E	$C_3(1)$	$C_3(4)$	C_{2y}	$C_3^2(3)$	$C_3^2(4)$	$C_3^2(1)$	C_{2x}	$C_3^2(2)$	$C_3(3)$	C_{2z}
S_{4z}^3	$C_3^2(3)$	$C_3^2(1)$	E	$C_3^2(2)$	$C_3(4)$	C_{2z}	C_{2x}	$C_3(1)$	$C_3(3)$	C_{2y}	$C_3^2(4)$	$C_3(2)$
S_{4x}	C_{2x}	$C_3^2(4)$	$C_3(2)$	E	$C_3^2(3)$	$C_3^2(1)$	$C_3(4)$	C_{2y}	$C_3(1)$	$C_3(3)$	C_{2z}	$C_3^2(2)$
S_{4y}	$C_3^2(1)$	C_{2y}	$C_3^2(4)$	$C_3(3)$	E	$C_3^2(2)$	$C_3(1)$	$C_3(2)$	C_{2z}	$C_3^2(3)$	$C_3(4)$	C_{2x}
S_{4z}	$C_3^2(4)$	$C_3(3)$	C_{2z}	$C_3(1)$	$C_3(2)$	E	C_{2y}	$C_3^2(2)$	$C_3^2(1)$	C_{2x}	$C_3^2(3)$	$C_3(4)$
σ_{xy}	$C_3(1)$	$C_3(4)$	C_{2x}	$C_3^2(4)$	$C_3^2(1)$	C_{2y}	E	$C_3^2(3)$	$C_3(2)$	C_{2z}	$C_3^2(2)$	$C_3(3)$
σ_{yz}	C_{2z}	$C_3(1)$	$C_3^2(1)$	C_{2y}	$C_3^2(2)$	$C_3(2)$	$C_3(3)$	E	$C_3^2(4)$	$C_3(4)$	C_{2x}	$C_3^2(3)$
σ_{zx}	$C_3(3)$	C_{2x}	$C_3^2(3)$	$C_3^2(1)$	C_{2z}	$C_3(1)$	$C_3^2(2)$	$C_3(4)$	E	$C_3^2(4)$	$C_3(2)$	C_{2y}
$\sigma_{x\bar{y}}$	$C_3^2(2)$	$C_3(2)$	C_{2y}	$C_3^2(3)$	$C_3(3)$	C_{2x}	C_{2z}	$C_3^2(4)$	$C_3(4)$	E	$C_3(1)$	$C_3^2(1)$
$\sigma_{y\bar{z}}$	C_{2y}	$C_3^2(3)$	$C_3(4)$	C_{2z}	$C_3^2(4)$	$C_3(3)$	$C_3(2)$	C_{2x}	$C_3^2(2)$	$C_3^2(1)$	E	$C_3(1)$
$\sigma_{z\bar{x}}$	$C_3(4)$	C_{2z}	$C_3^2(2)$	$C_3(2)$	C_{2x}	$C_3^2(4)$	$C_3^2(3)$	$C_3(3)$	C_{2y}	$C_3(1)$	$C_3^2(1)$	E

Table 32(a)1i

m3m	1	2_x	2_y	2_z	3_1	3_2	3_3	3_4	3_1^2	3_2^2	3_3^2	3_4^2
1	1	2_x	2_y	2_z	3_1	3_2	3_3	3_4	3_1^2	3_2^2	3_3^2	3_4^2
2_x	2_x	1	2_z	2_y	3_3^2	3_3	3_2	3_1^2	3_4	3_4^2	3_1	3_2^2
2_y	2_y	2_z	1	2_x	3_4^2	3_1^2	3_4	3_3	3_2	3_3^2	3_2^2	3_1
2_z	2_z	2_y	2_x	1	3_2^2	3_4	3_1^2	3_2	3_3	3_1	3_4^2	3_3^2
3_1^2	3_1^2	3_3	3_4	3_2	1	3_3^2	3_4^2	3_2^2	3_1	2_y	2_z	2_x
3_2^2	3_2^2	3_3^2	3_1	3_4^2	3_3	1	2_y	2_x	2_z	3_2	3_4	3_1^2
3_3^2	3_3^2	3_2^2	3_4^2	3_1	3_4	2_y	1	2_z	2_x	3_1^2	3_3	3_2
3_4^2	3_4^2	3_1	3_3^2	3_2^2	3_2	2_x	2_z	1	2_y	3_3	3_1^2	3_4
3_1	3_1	3_4^2	3_2^2	3_3^2	3_1^2	2_z	2_x	2_y	1	3_4	3_2	3_3
3_2	3_2	3_4	3_3	3_1^2	2_y	3_2^2	3_1	3_3^2	3_4^2	1	2_x	2_z
3_3	3_3	3_1^2	3_2	3_4	2_z	3_4^2	3_3^2	3_1	3_2^2	2_x	1	2_y
3_4	3_4	3_2	3_1^2	3_3	2_x	3_1	3_2^2	3_4^2	3_3^2	2_z	2_y	1
$\bar{1}$	$\bar{1}$	m_x	m_y	m_z	$\bar{3}_1$	$\bar{3}_2$	$\bar{3}_3$	$\bar{3}_4$	$\bar{3}_1^2$	$\bar{3}_2^2$	$\bar{3}_3^2$	$\bar{3}_4^2$
m_x	m_x	$\bar{1}$	m_z	m_y	$\bar{3}_3^2$	$\bar{3}_3$	$\bar{3}_2$	$\bar{3}_1^2$	$\bar{3}_4$	$\bar{3}_4^2$	$\bar{3}_1$	$\bar{3}_2^2$
m_y	m_y	m_z	$\bar{1}$	m_x	$\bar{3}_4^2$	$\bar{3}_1^2$	$\bar{3}_4$	$\bar{3}_3$	$\bar{3}_2$	$\bar{3}_3^2$	$\bar{3}_2^2$	$\bar{3}_1$
m_z	m_z	m_y	m_x	$\bar{1}$	$\bar{3}_2^2$	$\bar{3}_4$	$\bar{3}_1^2$	$\bar{3}_2$	$\bar{3}_3$	$\bar{3}_1$	$\bar{3}_4^2$	$\bar{3}_3^2$
$\bar{3}_1^2$	$\bar{3}_1^2$	$\bar{3}_3$	$\bar{3}_4$	$\bar{3}_2$	$\bar{1}$	$\bar{3}_3^2$	$\bar{3}_4^2$	$\bar{3}_2^2$	$\bar{3}_1$	m_y	m_z	m_x
$\bar{3}_2^2$	$\bar{3}_2^2$	$\bar{3}_3^2$	$\bar{3}_1$	$\bar{3}_4^2$	$\bar{3}_3$	$\bar{1}$	m_y	m_x	m_z	$\bar{3}_2$	$\bar{3}_4$	$\bar{3}_1^2$
$\bar{3}_3^2$	$\bar{3}_3^2$	$\bar{3}_2^2$	$\bar{3}_4^2$	$\bar{3}_1$	$\bar{3}_4$	m_y	$\bar{1}$	m_z	m_x	$\bar{3}_1^2$	$\bar{3}_3$	$\bar{3}_2$
$\bar{3}_4^2$	$\bar{3}_4^2$	$\bar{3}_1$	$\bar{3}_3^2$	$\bar{3}_2^2$	$\bar{3}_2$	m_x	m_z	$\bar{1}$	m_y	$\bar{3}_3$	$\bar{3}_1^2$	$\bar{3}_4$
$\bar{3}_1$	$\bar{3}_1$	$\bar{3}_4^2$	$\bar{3}_2^2$	$\bar{3}_3^2$	$\bar{3}_1^2$	m_z	m_x	m_y	$\bar{1}$	$\bar{3}_4$	$\bar{3}_2$	$\bar{3}_3$
$\bar{3}_2$	$\bar{3}_2$	$\bar{3}_4$	$\bar{3}_3$	$\bar{3}_1^2$	m_y	$\bar{3}_2^2$	$\bar{3}_1$	$\bar{3}_3^2$	$\bar{3}_4^2$	$\bar{1}$	m_x	m_z
$\bar{3}_3$	$\bar{3}_3$	$\bar{3}_1^2$	$\bar{3}_2$	$\bar{3}_4$	m_z	$\bar{3}_4^2$	$\bar{3}_3^2$	$\bar{3}_1$	$\bar{3}_2^2$	m_x	$\bar{1}$	m_y
$\bar{3}_4$	$\bar{3}_4$	$\bar{3}_2$	$\bar{3}_1^2$	$\bar{3}_3$	m_x	$\bar{3}_1$	$\bar{3}_2^2$	$\bar{3}_4^2$	$\bar{3}_3^2$	m_z	m_y	$\bar{1}$

Table 32(a)1ii

m3m	$\bar{1}$	m_x	m_y	m_z	$\bar{3}_1$	$\bar{3}_2$	$\bar{3}_3$	$\bar{3}_4$	$\bar{3}_1^2$	$\bar{3}_2^2$	$\bar{3}_3^2$	$\bar{3}_4^2$
1	$\bar{1}$	m_x	m_y	m_z	$\bar{3}_1$	$\bar{3}_2$	$\bar{3}_3$	$\bar{3}_4$	$\bar{3}_1^2$	$\bar{3}_2^2$	$\bar{3}_3^2$	$\bar{3}_4^2$
2_x	m_x	$\bar{1}$	m_z	m_y	$\bar{3}_3^2$	$\bar{3}_3$	$\bar{3}_2$	$\bar{3}_1^2$	$\bar{3}_4$	$\bar{3}_4^2$	$\bar{3}_1$	$\bar{3}_2^2$
2_y	m_y	m_z	$\bar{1}$	m_x	$\bar{3}_4^2$	$\bar{3}_1^2$	$\bar{3}_4$	$\bar{3}_3$	$\bar{3}_2$	$\bar{3}_3^2$	$\bar{3}_2^2$	$\bar{3}_1$
2_z	m_z	m_y	m_x	$\bar{1}$	$\bar{3}_2^2$	$\bar{3}_4$	$\bar{3}_1^2$	$\bar{3}_2$	$\bar{3}_3$	$\bar{3}_1$	$\bar{3}_4^2$	$\bar{3}_3^2$
3_1^2	$\bar{3}_1^2$	$\bar{3}_3$	$\bar{3}_4$	$\bar{3}_2$	$\bar{1}$	$\bar{3}_3^2$	$\bar{3}_4^2$	$\bar{3}_2^2$	$\bar{3}_1$	m_y	m_z	m_x
3_2^2	$\bar{3}_2^2$	$\bar{3}_3^2$	$\bar{3}_1$	$\bar{3}_4^2$	$\bar{3}_3$	$\bar{1}$	m_y	m_x	m_z	$\bar{3}_2$	$\bar{3}_4$	$\bar{3}_1^2$
3_3^2	$\bar{3}_3^2$	$\bar{3}_2^2$	$\bar{3}_4^2$	$\bar{3}_1$	$\bar{3}_4$	m_y	$\bar{1}$	m_z	m_x	$\bar{3}_1^2$	$\bar{3}_3$	$\bar{3}_2$
3_4^2	$\bar{3}_4^2$	$\bar{3}_1$	$\bar{3}_3^2$	$\bar{3}_2^2$	$\bar{3}_2$	m_x	m_z	$\bar{1}$	m_y	$\bar{3}_3$	$\bar{3}_1^2$	$\bar{3}_4$
3_1	$\bar{3}_1$	$\bar{3}_4^2$	$\bar{3}_2^2$	$\bar{3}_3^2$	$\bar{3}_1^2$	m_z	m_x	m_y	$\bar{1}$	$\bar{3}_4$	$\bar{3}_2$	$\bar{3}_3$
3_2	$\bar{3}_2$	$\bar{3}_4$	$\bar{3}_3$	$\bar{3}_1^2$	m_y	$\bar{3}_2^2$	$\bar{3}_1$	$\bar{3}_3^2$	$\bar{3}_4^2$	$\bar{1}$	m_x	m_z
3_3	$\bar{3}_3$	$\bar{3}_1^2$	$\bar{3}_2$	$\bar{3}_4$	m_z	$\bar{3}_4^2$	$\bar{3}_3^2$	$\bar{3}_1$	$\bar{3}_2^2$	m_x	$\bar{1}$	m_y
3_4	$\bar{3}_4$	$\bar{3}_2$	$\bar{3}_1^2$	$\bar{3}_3$	m_x	$\bar{3}_1$	$\bar{3}_2^2$	$\bar{3}_4^2$	$\bar{3}_3^2$	m_z	m_y	$\bar{1}$
$\bar{1}$	1	2_x	2_y	2_z	3_1	3_2	3_3	3_4	3_1^2	3_2^2	3_3^2	3_4^2
m_x	2_x	1	2_z	2_y	3_3^2	3_3	3_2	3_1^2	3_4	3_4^2	3_1	3_2^2
m_y	2_y	2_z	1	2_x	3_4^2	3_1^2	3_4	3_3	3_2	3_3^2	3_2^2	3_1
m_z	2_z	2_y	2_x	1	3_2^2	3_4	3_1^2	3_2	3_3	3_1	3_4^2	3_3^2
$\bar{3}_1^2$	3_1^2	3_3	3_4	3_2	1	3_3^2	3_4^2	3_2^2	3_1	2_y	2_z	2_x
$\bar{3}_2^2$	3_2^2	3_3^2	3_1	3_4^2	3_3	1	2_y	2_x	2_z	3_2	3_4	3_1^2
$\bar{3}_3^2$	3_3^2	3_2^2	3_4^2	3_1	3_4	2_y	1	2_z	2_x	3_1^2	3_3	3_2
$\bar{3}_4^2$	3_4^2	3_1	3_3^2	3_2^2	3_2	2_x	2_z	1	2_y	3_3	3_1^2	3_4
$\bar{3}_1$	3_1	3_4^2	3_2^2	3_3^2	3_1^2	2_z	2_x	2_y	1	3_4	3_2	3_3
$\bar{3}_2$	3_2	3_4	3_3	3_1^2	2_y	3_2^2	3_1	3_3^2	3_4^2	1	2_x	2_z
$\bar{3}_3$	3_3	3_1^2	3_2	3_4	2_z	3_4^2	3_3^2	3_1	3_2^2	2_x	1	2_y
$\bar{3}_4$	3_4	3_2	3_1^2	3_3	2_x	3_1	3_2^2	3_4^2	3_3^2	2_z	2_y	1

Table 32(a)1iii

$m3m$	4_x	4_y	4_z	4_x^3	4_y^3	4_z^3	2_{xy}	2_{yz}	2_{zx}	$2_{x\bar{y}}$	$2_{y\bar{z}}$	$2_{z\bar{x}}$
1	4_x	4_y	4_z	4_x^3	4_y^3	4_z^3	2_{xy}	2_{yz}	2_{zx}	$2_{x\bar{y}}$	$2_{y\bar{z}}$	$2_{z\bar{x}}$
2_x	4_x^3	$2_{z\bar{x}}$	$2_{x\bar{y}}$	4_x	2_{zx}	2_{xy}	4_z^3	$2_{y\bar{z}}$	4_y^3	4_z	2_{yz}	4_y
2_y	2_{yz}	4_y^3	2_{xy}	$2_{y\bar{z}}$	4_y	$2_{x\bar{y}}$	4_z	4_x	$2_{z\bar{x}}$	4_z^3	4_x^3	2_{zx}
2_z	$2_{y\bar{z}}$	2_{zx}	4_z^3	2_{yz}	$2_{z\bar{x}}$	4_z	$2_{x\bar{y}}$	4_x^3	4_y	2_{xy}	4_x	4_y^3
3_1^2	2_{zx}	2_{xy}	4_x	4_y	4_z^3	2_{yz}	4_x^3	4_y^3	4_z	$2_{y\bar{z}}$	$2_{z\bar{x}}$	$2_{x\bar{y}}$
3_2^2	4_z^3	2_{yz}	4_y	$2_{x\bar{y}}$	4_x^3	$2_{z\bar{x}}$	2_{zx}	4_z	$2_{y\bar{z}}$	4_y^3	2_{xy}	4_x
3_3^2	$2_{x\bar{y}}$	4_x	4_y^3	4_z^3	$2_{y\bar{z}}$	2_{zx}	$2_{z\bar{x}}$	2_{xy}	4_x^3	4_y	4_z	2_{yz}
3_4^2	2_{xy}	$2_{y\bar{z}}$	$2_{z\bar{x}}$	4_z	4_x	4_y	4_y^3	$2_{x\bar{y}}$	2_{yz}	2_{zx}	4_z^3	4_x^3
3_1	4_z	4_x^3	2_{zx}	2_{xy}	2_{yz}	4_y^3	4_y	4_z^3	4_x	$2_{z\bar{x}}$	$2_{x\bar{y}}$	$2_{y\bar{z}}$
3_2	$2_{z\bar{x}}$	4_z	2_{yz}	4_y^3	$2_{x\bar{y}}$	4_x	$2_{y\bar{z}}$	4_y	2_{xy}	4_x^3	2_{zx}	4_z^3
3_3	4_y	$2_{x\bar{y}}$	$2_{y\bar{z}}$	2_{zx}	4_z	4_x^3	2_{yz}	$2_{z\bar{x}}$	4_z^3	4_x	4_y^3	2_{xy}
3_4	4_y^3	4_z^3	4_x^3	$2_{z\bar{x}}$	2_{xy}	$2_{y\bar{z}}$	4_x	2_{zx}	$2_{x\bar{y}}$	2_{yz}	4_y	4_z
$\bar{1}$	$\bar{4}_x$	$\bar{4}_y$	$\bar{4}_z$	$\bar{4}_x^3$	$\bar{4}_y^3$	$\bar{4}_z^3$	m_{xy}	m_{yz}	m_{zx}	$m_{x\bar{y}}$	$m_{y\bar{z}}$	$m_{z\bar{x}}$
m_x	$\bar{4}_x^3$	$m_{z\bar{x}}$	$m_{x\bar{y}}$	$\bar{4}_x$	m_{zx}	m_{xy}	$\bar{4}_z^3$	$m_{y\bar{z}}$	$\bar{4}_y^3$	$\bar{4}_z$	m_{yz}	$\bar{4}_y$
m_y	m_{yz}	$\bar{4}_y^3$	m_{xy}	$m_{y\bar{z}}$	$\bar{4}_y$	$m_{x\bar{y}}$	$\bar{4}_z$	$\bar{4}_x$	$m_{z\bar{x}}$	$\bar{4}_z^3$	$\bar{4}_x^3$	m_{zx}
m_z	$m_{y\bar{z}}$	m_{zx}	$\bar{4}_z^3$	m_{yz}	$m_{z\bar{x}}$	$\bar{4}_z$	$m_{x\bar{y}}$	$\bar{4}_x^3$	$\bar{4}_y$	m_{xy}	$\bar{4}_x$	$\bar{4}_y^3$
$\bar{3}_1^2$	m_{zx}	m_{xy}	$\bar{4}_x$	$\bar{4}_y$	$\bar{4}_z^3$	m_{yz}	$\bar{4}_x^3$	$\bar{4}_y^3$	$\bar{4}_z$	$m_{y\bar{z}}$	$m_{z\bar{x}}$	$m_{x\bar{y}}$
$\bar{3}_2^2$	$\bar{4}_z^3$	m_{yz}	$\bar{4}_y$	$m_{x\bar{y}}$	$\bar{4}_x^3$	$m_{z\bar{x}}$	m_{zx}	$\bar{4}_z$	$m_{y\bar{z}}$	$\bar{4}_y^3$	m_{xy}	$\bar{4}_x$
$\bar{3}_3^2$	$m_{x\bar{y}}$	$\bar{4}_x$	$\bar{4}_y^3$	$\bar{4}_z^3$	$m_{y\bar{z}}$	m_{zx}	$m_{z\bar{x}}$	m_{xy}	$\bar{4}_x^3$	$\bar{4}_y$	$\bar{4}_z$	m_{yz}
$\bar{3}_4^2$	m_{xy}	$m_{y\bar{z}}$	$m_{z\bar{x}}$	$\bar{4}_z$	$\bar{4}_x$	$\bar{4}_y$	$\bar{4}_y^3$	$m_{x\bar{y}}$	m_{yz}	m_{zx}	$\bar{4}_z^3$	$\bar{4}_x^3$
$\bar{3}_1$	$\bar{4}_z$	$\bar{4}_x^3$	m_{zx}	m_{xy}	m_{yz}	$\bar{4}_y^3$	$\bar{4}_y$	$\bar{4}_z^3$	$\bar{4}_x$	$m_{z\bar{x}}$	$m_{x\bar{y}}$	$m_{y\bar{z}}$
$\bar{3}_2$	$m_{z\bar{x}}$	$\bar{4}_z$	m_{yz}	$\bar{4}_y^3$	$m_{x\bar{y}}$	$\bar{4}_x$	$m_{y\bar{z}}$	$\bar{4}_y$	m_{xy}	$\bar{4}_x^3$	m_{zx}	$\bar{4}_z^3$
$\bar{3}_3$	$\bar{4}_y$	$m_{x\bar{y}}$	$m_{y\bar{z}}$	m_{zx}	$\bar{4}_z$	$\bar{4}_x^3$	m_{yz}	$m_{z\bar{x}}$	$\bar{4}_z^3$	$\bar{4}_x$	$\bar{4}_y^3$	m_{xy}
$\bar{3}_4$	$\bar{4}_y^3$	$\bar{4}_z^3$	$\bar{4}_x^3$	$m_{z\bar{x}}$	m_{xy}	$m_{y\bar{z}}$	$\bar{4}_x$	m_{zx}	$m_{x\bar{y}}$	m_{yz}	$\bar{4}_y$	$\bar{4}_z$

Table 32(a)1iv

m3m	$\bar{4}_x$	$\bar{4}_y$	$\bar{4}_z$	$\bar{4}_x^3$	$\bar{4}_y^3$	$\bar{4}_z^3$	m_{xy}	m_{yz}	m_{zx}	$m_{x\bar{y}}$	$m_{y\bar{z}}$	$m_{z\bar{x}}$
1	$\bar{4}_x$	$\bar{4}_y$	$\bar{4}_z$	$\bar{4}_x^3$	$\bar{4}_y^3$	$\bar{4}_z^3$	m_{xy}	m_{yz}	m_{zx}	$m_{x\bar{y}}$	$m_{y\bar{z}}$	$m_{z\bar{x}}$
2_x	$\bar{4}_x^3$	$m_{z\bar{x}}$	$m_{x\bar{y}}$	$\bar{4}_x$	m_{zx}	m_{xy}	$\bar{4}_z^3$	$m_{y\bar{z}}$	$\bar{4}_y^3$	$\bar{4}_z$	m_{yz}	$\bar{4}_y$
2_y	m_{yz}	$\bar{4}_y^3$	m_{xy}	$m_{y\bar{z}}$	$\bar{4}_y$	$m_{x\bar{y}}$	$\bar{4}_z$	$\bar{4}_x$	$m_{z\bar{x}}$	$\bar{4}_z^3$	$\bar{4}_x^3$	m_{zx}
2_z	$m_{y\bar{z}}$	m_{zx}	$\bar{4}_z^3$	m_{yz}	$m_{z\bar{x}}$	$\bar{4}_z$	$m_{x\bar{y}}$	$\bar{4}_x^3$	$\bar{4}_y$	m_{xy}	$\bar{4}_x$	$\bar{4}_y^3$
3_1^2	m_{zx}	m_{xy}	$\bar{4}_x$	$\bar{4}_y$	$\bar{4}_z^3$	m_{yz}	$\bar{4}_x^3$	$\bar{4}_y^3$	$\bar{4}_z$	$m_{y\bar{z}}$	$m_{z\bar{x}}$	$m_{x\bar{y}}$
3_2^2	$\bar{4}_z^3$	m_{yz}	$\bar{4}_y$	$m_{x\bar{y}}$	$\bar{4}_x^3$	$m_{z\bar{x}}$	m_{zx}	$\bar{4}_z$	$m_{y\bar{z}}$	$\bar{4}_y^3$	m_{xy}	$\bar{4}_x$
3_3^2	$m_{x\bar{y}}$	$\bar{4}_x$	$\bar{4}_y^3$	$\bar{4}_z^3$	$m_{y\bar{z}}$	m_{zx}	$m_{z\bar{x}}$	m_{xy}	$\bar{4}_x^3$	$\bar{4}_y$	$\bar{4}_z$	m_{yz}
3_4^2	m_{xy}	$m_{y\bar{z}}$	$m_{z\bar{x}}$	$\bar{4}_z$	$\bar{4}_x$	$\bar{4}_y$	$\bar{4}_y^3$	$m_{x\bar{y}}$	m_{yz}	m_{zx}	$\bar{4}_z^3$	$\bar{4}_x^3$
3_1	$\bar{4}_z$	$\bar{4}_x^3$	m_{zx}	m_{xy}	m_{yz}	$\bar{4}_y^3$	$\bar{4}_y$	$\bar{4}_z^3$	$\bar{4}_x$	$m_{z\bar{x}}$	$m_{x\bar{y}}$	$m_{y\bar{z}}$
3_2	$m_{z\bar{x}}$	$\bar{4}_z$	m_{yz}	$\bar{4}_y^3$	$m_{x\bar{y}}$	$\bar{4}_x$	$m_{y\bar{z}}$	$\bar{4}_y$	m_{xy}	$\bar{4}_x^3$	m_{zx}	$\bar{4}_z^3$
3_3	$\bar{4}_y$	$m_{x\bar{y}}$	$m_{y\bar{z}}$	m_{zx}	$\bar{4}_z$	$\bar{4}_x^3$	m_{yz}	$m_{z\bar{x}}$	$\bar{4}_z^3$	$\bar{4}_x$	$\bar{4}_y^3$	m_{xy}
3_4	$\bar{4}_y^3$	$\bar{4}_z^3$	$\bar{4}_x^3$	$m_{z\bar{x}}$	m_{xy}	$m_{y\bar{z}}$	$\bar{4}_x$	m_{zx}	$m_{x\bar{y}}$	m_{yz}	$\bar{4}_y$	$\bar{4}_z$
$\bar{1}$	4_x	4_y	4_z	4_x^3	4_y^3	4_z^3	2_{xy}	2_{yz}	2_{zx}	$2_{x\bar{y}}$	$2_{y\bar{z}}$	$2_{z\bar{x}}$
m_x	4_x^3	$2_{z\bar{x}}$	$2_{x\bar{y}}$	4_x	2_{zx}	2_{xy}	4_z^3	$2_{y\bar{z}}$	4_y^3	4_z	2_{yz}	4_y
m_y	2_{yz}	4_y^3	2_{xy}	$2_{y\bar{z}}$	4_y	$2_{x\bar{y}}$	4_z	4_x	$2_{z\bar{x}}$	4_z^3	4_x^3	2_{zx}
m_z	$2_{y\bar{z}}$	2_{zx}	4_z^3	2_{yz}	$2_{z\bar{x}}$	4_z	$2_{x\bar{y}}$	4_x^3	4_y	2_{xy}	4_x	4_y^3
$\bar{3}_1^2$	2_{zx}	2_{xy}	4_x	4_y	4_z^3	2_{yz}	4_x^3	4_y^3	4_z	$2_{y\bar{z}}$	$2_{z\bar{x}}$	$2_{x\bar{y}}$
$\bar{3}_2^2$	4_z^3	2_{yz}	4_y	$2_{x\bar{y}}$	4_x^3	$2_{z\bar{x}}$	2_{zx}	4_z	$2_{y\bar{z}}$	4_y^3	2_{xy}	4_x
$\bar{3}_3^2$	$2_{x\bar{y}}$	4_x	4_y^3	4_z^3	$2_{y\bar{z}}$	2_{zx}	$2_{z\bar{x}}$	2_{xy}	4_x^3	4_y	4_z	2_{yz}
$\bar{3}_4^2$	2_{xy}	$2_{y\bar{z}}$	$2_{z\bar{x}}$	4_z	4_x	4_y	4_y^3	$2_{x\bar{y}}$	2_{yz}	2_{zx}	4_z^3	4_x^3
$\bar{3}_1$	4_z	4_x^3	2_{zx}	2_{xy}	2_{yz}	4_y^3	4_y	4_z^3	4_x	$2_{z\bar{x}}$	$2_{x\bar{y}}$	$2_{y\bar{z}}$
$\bar{3}_2$	$2_{z\bar{x}}$	4_z	2_{yz}	4_y^3	$2_{x\bar{y}}$	4_x	$2_{y\bar{z}}$	4_y	2_{xy}	4_x^3	2_{zx}	4_z^3
$\bar{3}_3$	4_y	$2_{x\bar{y}}$	$2_{y\bar{z}}$	2_{zx}	4_z	4_x^3	2_{yz}	$2_{z\bar{x}}$	4_z^3	4_x	4_y^3	2_{xy}
$\bar{3}_4$	4_y^3	4_z^3	4_x^3	$2_{z\bar{x}}$	2_{xy}	$2_{y\bar{z}}$	4_x	2_{zx}	$2_{x\bar{y}}$	2_{yz}	4_y	4_z

Table 32(a)2i

m3m	1	2_x	2_y	2_z	3_1	3_2	3_3	3_4	3_1^2	3_2^2	3_3^2	3_4^2
4_x^3	4_x^3	4_x	2_{yz}	$2_{y\bar{z}}$	2_{zx}	4_z	$2_{x\bar{y}}$	2_{xy}	4_z^3	$2_{z\bar{x}}$	4_y^3	4_y
4_y^3	4_y^3	$2_{z\bar{x}}$	4_y	2_{zx}	2_{xy}	2_{yz}	4_x^3	$2_{y\bar{z}}$	4_x	4_z^3	$2_{x\bar{y}}$	4_z
4_z^3	4_z^3	$2_{x\bar{y}}$	2_{xy}	4_z	4_x^3	4_y^3	4_y	$2_{z\bar{x}}$	2_{zx}	2_{yz}	$2_{y\bar{z}}$	4_x
4_x	4_x	4_x^3	$2_{y\bar{z}}$	2_{yz}	4_y^3	$2_{x\bar{y}}$	4_z	4_z^3	2_{xy}	4_y	2_{zx}	$2_{z\bar{x}}$
4_y	4_y	2_{zx}	4_y^3	$2_{z\bar{x}}$	4_z	4_x	$2_{y\bar{z}}$	4_x^3	2_{yz}	$2_{x\bar{y}}$	4_z^3	2_{xy}
4_z	4_z	2_{xy}	$2_{x\bar{y}}$	4_z^3	2_{yz}	$2_{z\bar{x}}$	2_{zx}	4_y^3	4_y	4_x^3	4_x	$2_{y\bar{z}}$
2_{xy}	2_{xy}	4_z	4_z^3	$2_{x\bar{y}}$	4_x	2_{zx}	$2_{z\bar{x}}$	4_y	4_y^3	$2_{y\bar{z}}$	2_{yz}	4_x^3
2_{yz}	2_{yz}	$2_{y\bar{z}}$	4_x^3	4_x	4_y	4_z^3	2_{xy}	$2_{x\bar{y}}$	4_z	4_y^3	$2_{z\bar{x}}$	2_{zx}
2_{zx}	2_{zx}	4_y	$2_{z\bar{x}}$	4_y^3	4_z^3	$2_{y\bar{z}}$	4_x	2_{yz}	4_x^3	2_{xy}	4_z	$2_{x\bar{y}}$
$2_{x\bar{y}}$	$2_{x\bar{y}}$	4_z^3	4_z	2_{xy}	$2_{y\bar{z}}$	4_y	4_y^3	2_{zx}	$2_{z\bar{x}}$	4_x	4_x^3	2_{yz}
$2_{y\bar{z}}$	$2_{y\bar{z}}$	2_{yz}	4_x	4_x^3	$2_{z\bar{x}}$	2_{xy}	4_z^3	4_z	$2_{x\bar{y}}$	2_{zx}	4_y	4_y^3
$2_{z\bar{x}}$	$2_{z\bar{x}}$	4_y^3	2_{zx}	4_y	$2_{x\bar{y}}$	4_x^3	2_{yz}	4_x	$2_{y\bar{z}}$	4_z	2_{xy}	4_z^3
$\bar{4}_x^3$	$\bar{4}_x^3$	$\bar{4}_x$	m_{yz}	$m_{y\bar{z}}$	m_{zx}	$\bar{4}_z$	$m_{x\bar{y}}$	m_{xy}	$\bar{4}_z^3$	$m_{z\bar{x}}$	$\bar{4}_y^3$	$\bar{4}_y$
$\bar{4}_y^3$	$\bar{4}_y^3$	$m_{z\bar{x}}$	$\bar{4}_y$	m_{zx}	m_{xy}	m_{yz}	$\bar{4}_x^3$	$m_{y\bar{z}}$	$\bar{4}_x$	$\bar{4}_z^3$	$m_{x\bar{y}}$	$\bar{4}_z$
$\bar{4}_z^3$	$\bar{4}_z^3$	$m_{x\bar{y}}$	m_{xy}	$\bar{4}_z$	$\bar{4}_x^3$	$\bar{4}_y^3$	$\bar{4}_y$	$m_{z\bar{x}}$	m_{zx}	m_{yz}	$m_{y\bar{z}}$	$\bar{4}_x$
$\bar{4}_x$	$\bar{4}_x$	$\bar{4}_x^3$	$m_{y\bar{z}}$	m_{yz}	$\bar{4}_y^3$	$m_{x\bar{y}}$	$\bar{4}_z$	$\bar{4}_z^3$	m_{xy}	$\bar{4}_y$	m_{zx}	$m_{z\bar{x}}$
$\bar{4}_y$	$\bar{4}_y$	m_{zx}	$\bar{4}_y^3$	$m_{z\bar{x}}$	$\bar{4}_z$	$\bar{4}_x$	$m_{y\bar{z}}$	$\bar{4}_x^3$	m_{yz}	$m_{x\bar{y}}$	$\bar{4}_z^3$	m_{xy}
$\bar{4}_z$	$\bar{4}_z$	m_{xy}	$m_{x\bar{y}}$	$\bar{4}_z^3$	m_{yz}	$m_{z\bar{x}}$	m_{zx}	$\bar{4}_y^3$	$\bar{4}_y$	$\bar{4}_x^3$	$\bar{4}_x$	$m_{y\bar{z}}$
m_{xy}	m_{xy}	$\bar{4}_z$	$\bar{4}_z^3$	$m_{x\bar{y}}$	$\bar{4}_x$	m_{zx}	$m_{z\bar{x}}$	$\bar{4}_y$	$\bar{4}_y^3$	$m_{y\bar{z}}$	m_{yz}	$\bar{4}_x^3$
m_{yz}	m_{yz}	$m_{y\bar{z}}$	$\bar{4}_x^3$	$\bar{4}_x$	$\bar{4}_y$	$\bar{4}_z^3$	m_{xy}	$m_{x\bar{y}}$	$\bar{4}_z$	$\bar{4}_y^3$	$m_{z\bar{x}}$	m_{zx}
m_{zx}	m_{zx}	$\bar{4}_y$	$m_{z\bar{x}}$	$\bar{4}_y^3$	$\bar{4}_z^3$	$m_{y\bar{z}}$	$\bar{4}_x$	m_{yz}	$\bar{4}_x^3$	m_{xy}	$\bar{4}_z$	$m_{x\bar{y}}$
$m_{x\bar{y}}$	$m_{x\bar{y}}$	$\bar{4}_z^3$	$\bar{4}_z$	m_{xy}	$m_{y\bar{z}}$	$\bar{4}_y$	$\bar{4}_y^3$	m_{zx}	$m_{z\bar{x}}$	$\bar{4}_x$	$\bar{4}_x^3$	m_{yz}
$m_{y\bar{z}}$	$m_{y\bar{z}}$	m_{yz}	$\bar{4}_x$	$\bar{4}_x^3$	$m_{z\bar{x}}$	m_{xy}	$\bar{4}_z^3$	$\bar{4}_z$	$m_{x\bar{y}}$	m_{zx}	$\bar{4}_y$	$\bar{4}_y^3$
$m_{z\bar{x}}$	$m_{z\bar{x}}$	$\bar{4}_y^3$	m_{zx}	$\bar{4}_y$	$m_{x\bar{y}}$	$\bar{4}_x^3$	m_{yz}	$\bar{4}_x$	$m_{y\bar{z}}$	$\bar{4}_z$	m_{xy}	$\bar{4}_z^3$

Table 32(a)2ii

m3m	$\bar{1}$	m_x	m_y	m_z	$\bar{3}_1$	$\bar{3}_2$	$\bar{3}_3$	$\bar{3}_4$	$\bar{3}_1^2$	$\bar{3}_2^2$	$\bar{3}_3^2$	$\bar{3}_4^2$
4_x^3	$\bar{4}_x^3$	$\bar{4}_x$	m_{yz}	$m_{y\bar{z}}$	m_{zx}	$\bar{4}_z$	$m_{x\bar{y}}$	m_{xy}	$\bar{4}_z^3$	$m_{z\bar{x}}$	$\bar{4}_y^3$	$\bar{4}_y$
4_y^3	$\bar{4}_y^3$	$m_{z\bar{x}}$	$\bar{4}_y$	m_{zx}	m_{xy}	m_{yz}	$\bar{4}_x^3$	$m_{y\bar{z}}$	$\bar{4}_x$	$\bar{4}_z^3$	$m_{x\bar{y}}$	$\bar{4}_z$
4_z^3	$\bar{4}_z^3$	$m_{x\bar{y}}$	m_{xy}	$\bar{4}_z$	$\bar{4}_x^3$	$\bar{4}_y^3$	$\bar{4}_y$	$m_{z\bar{x}}$	m_{zx}	m_{yz}	$m_{y\bar{z}}$	$\bar{4}_x$
4_x	$\bar{4}_x$	$\bar{4}_x^3$	$m_{y\bar{z}}$	m_{yz}	$\bar{4}_y^3$	$m_{x\bar{y}}$	$\bar{4}_z$	$\bar{4}_z^3$	m_{xy}	$\bar{4}_y$	m_{zx}	$m_{z\bar{x}}$
4_y	$\bar{4}_y$	m_{zx}	$\bar{4}_y^3$	$m_{z\bar{x}}$	$\bar{4}_z$	$\bar{4}_x$	$m_{y\bar{z}}$	$\bar{4}_x^3$	m_{yz}	$m_{x\bar{y}}$	$\bar{4}_z^3$	m_{xy}
4_z	$\bar{4}_z$	m_{xy}	$m_{x\bar{y}}$	$\bar{4}_z^3$	m_{yz}	$m_{z\bar{x}}$	m_{zx}	$\bar{4}_y^3$	$\bar{4}_y$	$\bar{4}_x^3$	$\bar{4}_x$	$m_{y\bar{z}}$
2_{xy}	m_{xy}	$\bar{4}_z$	$\bar{4}_z^3$	$m_{x\bar{y}}$	$\bar{4}_x$	m_{zx}	$m_{z\bar{x}}$	$\bar{4}_y$	$\bar{4}_y^3$	$m_{y\bar{z}}$	m_{yz}	$\bar{4}_x^3$
2_{yz}	m_{yz}	$m_{y\bar{z}}$	$\bar{4}_x^3$	$\bar{4}_x$	$\bar{4}_y$	$\bar{4}_z^3$	m_{xy}	$m_{x\bar{y}}$	$\bar{4}_z$	$\bar{4}_y^3$	$m_{z\bar{x}}$	m_{zx}
2_{zx}	m_{zx}	$\bar{4}_y$	$m_{z\bar{x}}$	$\bar{4}_y^3$	$\bar{4}_z^3$	$m_{y\bar{z}}$	$\bar{4}_x$	m_{yz}	$\bar{4}_x^3$	m_{xy}	$\bar{4}_z$	$m_{x\bar{y}}$
$2_{x\bar{y}}$	$m_{x\bar{y}}$	$\bar{4}_z^3$	$\bar{4}_z$	m_{xy}	$m_{y\bar{z}}$	$\bar{4}_y$	$\bar{4}_y^3$	m_{zx}	$m_{z\bar{x}}$	$\bar{4}_x$	$\bar{4}_x^3$	m_{yz}
$2_{y\bar{z}}$	$m_{y\bar{z}}$	m_{yz}	$\bar{4}_x$	$\bar{4}_x^3$	$m_{z\bar{x}}$	m_{xy}	$\bar{4}_z^3$	$\bar{4}_z$	$m_{x\bar{y}}$	m_{zx}	$\bar{4}_y$	$\bar{4}_y^3$
$2_{z\bar{x}}$	$m_{z\bar{x}}$	$\bar{4}_y^3$	m_{zx}	$\bar{4}_y$	$m_{x\bar{y}}$	$\bar{4}_x^3$	m_{yz}	$\bar{4}_x$	$m_{y\bar{z}}$	$\bar{4}_z$	m_{xy}	$\bar{4}_z^3$
$\bar{4}_x^3$	4_x^3	4_x	2_{yz}	$2_{y\bar{z}}$	2_{zx}	4_z	$2_{x\bar{y}}$	2_{xy}	4_z^3	$2_{z\bar{x}}$	4_y^3	4_y
$\bar{4}_y^3$	4_y^3	$2_{z\bar{x}}$	4_y	2_{zx}	2_{xy}	2_{yz}	4_x^3	$2_{y\bar{z}}$	4_x	4_z^3	$2_{x\bar{y}}$	4_z
$\bar{4}_z^3$	4_z^3	$2_{x\bar{y}}$	2_{xy}	4_z	4_x^3	4_y^3	4_y	$2_{z\bar{x}}$	2_{zx}	2_{yz}	$2_{y\bar{z}}$	4_x
$\bar{4}_x$	4_x	4_x^3	$2_{y\bar{z}}$	2_{yz}	4_y^3	$2_{x\bar{y}}$	4_z	4_z^3	2_{xy}	4_y	2_{zx}	$2_{z\bar{x}}$
$\bar{4}_y$	4_y	2_{zx}	4_y^3	$2_{z\bar{x}}$	4_z	4_x	$2_{y\bar{z}}$	4_x^3	2_{yz}	$2_{x\bar{y}}$	4_z^3	2_{xy}
$\bar{4}_z$	4_z	2_{xy}	$2_{x\bar{y}}$	4_z^3	2_{yz}	$2_{z\bar{x}}$	2_{zx}	4_y^3	4_y	4_x^3	4_x	$2_{y\bar{z}}$
m_{xy}	2_{xy}	4_z	4_z^3	$2_{x\bar{y}}$	4_x	2_{zx}	$2_{z\bar{x}}$	4_y	4_y^3	$2_{y\bar{z}}$	2_{yz}	4_x^3
m_{yz}	2_{yz}	$2_{y\bar{z}}$	4_x^3	4_x	4_y	4_z^3	2_{xy}	$2_{x\bar{y}}$	4_z	4_y^3	$2_{z\bar{x}}$	2_{zx}
m_{zx}	2_{zx}	4_y	$2_{z\bar{x}}$	4_y^3	4_z^3	$2_{y\bar{z}}$	4_x	2_{yz}	4_x^3	2_{xy}	4_z	$2_{x\bar{y}}$
$m_{x\bar{y}}$	$2_{x\bar{y}}$	4_z^3	4_z	2_{xy}	$2_{y\bar{z}}$	4_y	4_y^3	2_{zx}	$2_{z\bar{x}}$	4_x	4_x^3	2_{yz}
$m_{y\bar{z}}$	$2_{y\bar{z}}$	2_{yz}	4_x	4_x^3	$2_{z\bar{x}}$	2_{xy}	4_z^3	4_z	$2_{x\bar{y}}$	2_{zx}	4_y	4_y^3
$m_{z\bar{x}}$	$2_{z\bar{x}}$	4_y^3	2_{zx}	4_y	$2_{x\bar{y}}$	4_x^3	2_{yz}	4_x	$2_{y\bar{z}}$	4_z	2_{xy}	4_z^3

Table 32(a)2iii

$m3m$	4_x	4_y	4_z	4_x^3	4_y^3	4_z^3	2_{xy}	2_{yz}	2_{zx}	$2_{x\bar{y}}$	$2_{y\bar{z}}$	$2_{z\bar{x}}$
4_x^3	1	3_2^2	3_3	2_x	3_1	3_4	3_1^2	2_z	3_3^2	3_2	2_y	3_4^2
4_y^3	3_2	1	3_1	3_4	2_y	3_3^2	3_4^2	3_1^2	2_x	3_2^2	3_3	2_z
4_z^3	3_3^2	3_1^2	1	3_2^2	3_4	2_z	2_x	3_1	3_3	2_y	3_4^2	3_2
4_x	2_x	3_4^2	3_2	1	3_3^2	3_1^2	3_4	2_y	3_1	3_3	2_z	3_2^2
4_y	3_1^2	2_y	3_4^2	3_3	1	3_2^2	3_1	3_2	2_z	3_3^2	3_4	2_x
4_z	3_4^2	3_3	2_z	3_1	3_2	1	2_y	3_2^2	3_1^2	2_x	3_3^2	3_4
2_{xy}	3_1	3_4	2_x	3_4^2	3_1^2	2_y	1	3_3^2	3_2	2_z	3_2^2	3_3
2_{yz}	2_z	3_1	3_1^2	2_y	3_2^2	3_2	3_3	1	3_4^2	3_4	2_x	3_3^2
2_{zx}	3_3	2_x	3_3^2	3_1^2	2_z	3_1	3_2^2	3_4	1	3_4^2	3_2	2_y
$2_{x\bar{y}}$	3_2^2	3_2	2_y	3_3^2	3_3	2_x	2_z	3_4^2	3_4	1	3_1	3_1^2
$2_{y\bar{z}}$	2_y	3_3^2	3_4	2_z	3_4^2	3_3	3_2	2_x	3_2^2	3_1^2	1	3_1
$2_{z\bar{x}}$	3_4	2_z	3_2^2	3_2	2_x	3_4^2	3_3^2	3_3	2_y	3_1	3_1^2	1
$\bar{4}_x^3$	$\bar{1}$	$\bar{3}_2^2$	$\bar{3}_3$	m_x	$\bar{3}_1$	$\bar{3}_4$	$\bar{3}_1^2$	m_z	$\bar{3}_3^2$	$\bar{3}_2$	m_y	$\bar{3}_4^2$
$\bar{4}_y^3$	$\bar{3}_2$	$\bar{1}$	$\bar{3}_1$	$\bar{3}_4$	m_y	$\bar{3}_3^2$	$\bar{3}_4^2$	$\bar{3}_1^2$	m_x	$\bar{3}_2^2$	$\bar{3}_3$	m_z
$\bar{4}_z^3$	$\bar{3}_3^2$	$\bar{3}_1^2$	$\bar{1}$	$\bar{3}_2^2$	$\bar{3}_4$	m_z	m_x	$\bar{3}_1$	$\bar{3}_3$	m_y	$\bar{3}_4^2$	$\bar{3}_2$
$\bar{4}_x$	m_x	$\bar{3}_4^2$	$\bar{3}_2$	$\bar{1}$	$\bar{3}_3^2$	$\bar{3}_1^2$	$\bar{3}_4$	m_y	$\bar{3}_1$	$\bar{3}_3$	m_z	$\bar{3}_2^2$
$\bar{4}_y$	$\bar{3}_1^2$	m_y	$\bar{3}_4^2$	$\bar{3}_3$	$\bar{1}$	$\bar{3}_2^2$	$\bar{3}_1$	$\bar{3}_2$	m_z	$\bar{3}_3^2$	$\bar{3}_4$	m_x
$\bar{4}_z$	$\bar{3}_4^2$	$\bar{3}_3$	m_z	$\bar{3}_1$	$\bar{3}_2$	$\bar{1}$	m_y	$\bar{3}_2^2$	$\bar{3}_1^2$	m_x	$\bar{3}_3^2$	$\bar{3}_4$
m_{xy}	$\bar{3}_1$	$\bar{3}_4$	m_x	$\bar{3}_4^2$	$\bar{3}_1^2$	m_y	$\bar{1}$	$\bar{3}_3^2$	$\bar{3}_2$	m_z	$\bar{3}_2^2$	$\bar{3}_3$
m_{yz}	m_z	$\bar{3}_1$	$\bar{3}_1^2$	m_y	$\bar{3}_2^2$	$\bar{3}_2$	$\bar{3}_3$	$\bar{1}$	$\bar{3}_4^2$	$\bar{3}_4$	m_x	$\bar{3}_3^2$
m_{zx}	$\bar{3}_3$	m_x	$\bar{3}_3^2$	$\bar{3}_1^2$	m_z	$\bar{3}_1$	$\bar{3}_2^2$	$\bar{3}_4$	$\bar{1}$	$\bar{3}_4^2$	$\bar{3}_2$	m_y
$m_{x\bar{y}}$	$\bar{3}_2^2$	$\bar{3}_2$	m_y	$\bar{3}_3^2$	$\bar{3}_3$	m_x	m_z	$\bar{3}_4^2$	$\bar{3}_4$	$\bar{1}$	$\bar{3}_1$	$\bar{3}_1^2$
$m_{y\bar{z}}$	m_y	$\bar{3}_3^2$	$\bar{3}_4$	m_z	$\bar{3}_4^2$	$\bar{3}_3$	$\bar{3}_2$	m_x	$\bar{3}_2^2$	$\bar{3}_1^2$	$\bar{1}$	$\bar{3}_1$
$m_{z\bar{x}}$	$\bar{3}_4$	m_z	$\bar{3}_2^2$	$\bar{3}_2$	m_x	$\bar{3}_4^2$	$\bar{3}_3^2$	$\bar{3}_3$	m_y	$\bar{3}_1$	$\bar{3}_1^2$	$\bar{1}$

Table 32(a)2iv

m3m	$\bar{4}_x$	$\bar{4}_y$	$\bar{4}_z$	$\bar{4}_x^3$	$\bar{4}_y^3$	$\bar{4}_z^3$	m_{xy}	m_{yz}	m_{zx}	$m_{x\bar{y}}$	$m_{y\bar{z}}$	$m_{z\bar{x}}$
4_x^3	$\bar{1}$	$\bar{3}_2^2$	$\bar{3}_3$	m_x	$\bar{3}_1$	$\bar{3}_4$	$\bar{3}_1^2$	m_z	$\bar{3}_3^2$	$\bar{3}_2$	m_y	$\bar{3}_4^2$
4_y^3	$\bar{3}_2$	$\bar{1}$	$\bar{3}_1$	$\bar{3}_4$	m_y	$\bar{3}_3^2$	$\bar{3}_4^2$	$\bar{3}_1^2$	m_x	$\bar{3}_2^2$	$\bar{3}_3$	m_z
4_z^3	$\bar{3}_3^2$	$\bar{3}_1^2$	$\bar{1}$	$\bar{3}_2^2$	$\bar{3}_4$	m_z	m_x	$\bar{3}_1$	$\bar{3}_3$	m_y	$\bar{3}_4^2$	$\bar{3}_2$
4_x	m_x	$\bar{3}_4^2$	$\bar{3}_2$	$\bar{1}$	$\bar{3}_3^2$	$\bar{3}_1^2$	$\bar{3}_4$	m_y	$\bar{3}_1$	$\bar{3}_3$	m_z	$\bar{3}_2^2$
4_y	$\bar{3}_1^2$	m_y	$\bar{3}_4^2$	$\bar{3}_3$	$\bar{1}$	$\bar{3}_2^2$	$\bar{3}_1$	$\bar{3}_2$	m_z	$\bar{3}_3^2$	$\bar{3}_4$	m_x
4_z	$\bar{3}_4^2$	$\bar{3}_3$	m_z	$\bar{3}_1$	$\bar{3}_2$	$\bar{1}$	m_y	$\bar{3}_2^2$	$\bar{3}_1^2$	m_x	$\bar{3}_3^2$	$\bar{3}_4$
2_{xy}	$\bar{3}_1$	$\bar{3}_4$	m_x	$\bar{3}_4^2$	$\bar{3}_1^2$	m_y	$\bar{1}$	$\bar{3}_3^2$	$\bar{3}_2$	m_z	$\bar{3}_2^2$	$\bar{3}_3$
2_{yz}	m_z	$\bar{3}_1$	$\bar{3}_1^2$	m_y	$\bar{3}_2^2$	$\bar{3}_2$	$\bar{3}_3$	$\bar{1}$	$\bar{3}_4^2$	$\bar{3}_4$	m_x	$\bar{3}_3^2$
2_{zx}	$\bar{3}_3$	m_x	$\bar{3}_3^2$	$\bar{3}_1^2$	m_z	$\bar{3}_1$	$\bar{3}_2^2$	$\bar{3}_4$	$\bar{1}$	$\bar{3}_4^2$	$\bar{3}_2$	m_y
$2_{x\bar{y}}$	$\bar{3}_2^2$	$\bar{3}_2$	m_y	$\bar{3}_3^2$	$\bar{3}_3$	m_x	m_z	$\bar{3}_4^2$	$\bar{3}_4$	$\bar{1}$	$\bar{3}_1$	$\bar{3}_1^2$
$2_{y\bar{z}}$	m_y	$\bar{3}_3^2$	$\bar{3}_4$	m_z	$\bar{3}_4^2$	$\bar{3}_3$	$\bar{3}_2$	m_x	$\bar{3}_2^2$	$\bar{3}_1^2$	$\bar{1}$	$\bar{3}_1$
$2_{z\bar{x}}$	$\bar{3}_4$	m_z	$\bar{3}_2^2$	$\bar{3}_2$	m_x	$\bar{3}_4^2$	$\bar{3}_3^2$	$\bar{3}_3$	m_y	$\bar{3}_1$	$\bar{3}_1^2$	$\bar{1}$
$\bar{4}_x^3$	1	3_2^2	3_3	2_x	3_1	3_4	3_1^2	2_z	3_3^2	3_2	2_y	3_4^2
$\bar{4}_y^3$	3_2	1	3_1	3_4	2_y	3_3^2	3_4^2	3_1^2	2_x	3_2^2	3_3	2_z
$\bar{4}_z^3$	3_3^2	3_1^2	1	3_2^2	3_4	2_z	2_x	3_1	3_3	2_y	3_4^2	3_2
$\bar{4}_x$	2_x	3_4^2	3_2	1	3_3^2	3_1^2	3_4	2_y	3_1	3_3	2_z	3_2^2
$\bar{4}_y$	3_1^2	2_y	3_4^2	3_3	1	3_2^2	3_1	3_2	2_z	3_3^2	3_4	2_x
$\bar{4}_z$	3_4^2	3_3	2_z	3_1	3_2	1	2_y	3_2^2	3_1^2	2_x	3_3^2	3_4
m_{xy}	3_1	3_4	2_x	3_4^2	3_1^2	2_y	1	3_3^2	3_2	2_z	3_2^2	3_3
m_{yz}	2_z	3_1	3_1^2	2_y	3_2^2	3_2	3_3	1	3_4^2	3_4	2_x	3_3^2
m_{zx}	3_3	2_x	3_3^2	3_1^2	2_z	3_1	3_2^2	3_4	1	3_4^2	3_2	2_y
$m_{x\bar{y}}$	3_2^2	3_2	2_y	3_3^2	3_3	2_x	2_z	3_4^2	3_4	1	3_1	3_1^2
$m_{y\bar{z}}$	2_y	3_3^2	3_4	2_z	3_4^2	3_3	3_2	2_x	3_2^2	3_1^2	1	3_1
$m_{z\bar{x}}$	3_4	2_z	3_2^2	3_2	2_x	3_4^2	3_3^2	3_3	2_y	3_1	3_1^2	1

Table 32(b)1i

O_h	E	C_{2x}	C_{2y}	C_{2z}	$C_3(1)$	$C_3(2)$	$C_3(3)$	$C_3(4)$	$C_3^2(1)$	$C_3^2(2)$	$C_3^2(3)$	$C_3^2(4)$
E	E	C_{2x}	C_{2y}	C_{2z}	$C_3(1)$	$C_3(2)$	$C_3(3)$	$C_3(4)$	$C_3^2(1)$	$C_3^2(2)$	$C_3^2(3)$	$C_3^2(4)$
C_{2x}	C_{2x}	E	C_{2z}	C_{2y}	$C_3^2(3)$	$C_3(3)$	$C_3(2)$	$C_3^2(1)$	$C_3(4)$	$C_3^2(4)$	$C_3(1)$	$C_3^2(2)$
C_{2y}	C_{2y}	C_{2z}	E	C_{2x}	$C_3^2(4)$	$C_3^2(1)$	$C_3(4)$	$C_3(3)$	$C_3(2)$	$C_3^2(3)$	$C_3^2(2)$	$C_3(1)$
C_{2z}	C_{2z}	C_{2y}	C_{2x}	E	$C_3^2(2)$	$C_3(4)$	$C_3^2(1)$	$C_3(2)$	$C_3(3)$	$C_3(1)$	$C_3^2(4)$	$C_3^2(3)$
$C_3^2(1)$	$C_3^2(1)$	$C_3(3)$	$C_3(4)$	$C_3(2)$	E	$C_3^2(3)$	$C_3^2(4)$	$C_3^2(2)$	$C_3(1)$	C_{2y}	C_{2z}	C_{2x}
$C_3^2(2)$	$C_3^2(2)$	$C_3^2(3)$	$C_3(1)$	$C_3^2(4)$	$C_3(3)$	E	C_{2y}	C_{2x}	C_{2z}	$C_3(2)$	$C_3(4)$	$C_3^2(1)$
$C_3^2(3)$	$C_3^2(3)$	$C_3^2(2)$	$C_3^2(4)$	$C_3(1)$	$C_3(4)$	C_{2y}	E	C_{2z}	C_{2x}	$C_3^2(1)$	$C_3(3)$	$C_3(2)$
$C_3^2(4)$	$C_3^2(4)$	$C_3(1)$	$C_3^2(3)$	$C_3^2(2)$	$C_3(2)$	C_{2x}	C_{2z}	E	C_{2y}	$C_3(3)$	$C_3^2(1)$	$C_3(4)$
$C_3(1)$	$C_3(1)$	$C_3^2(4)$	$C_3^2(2)$	$C_3^2(3)$	$C_3^2(1)$	C_{2z}	C_{2x}	C_{2y}	E	$C_3(4)$	$C_3(2)$	$C_3(3)$
$C_3(2)$	$C_3(2)$	$C_3(4)$	$C_3(3)$	$C_3^2(1)$	C_{2y}	$C_3^2(2)$	$C_3(1)$	$C_3^2(3)$	$C_3^2(4)$	E	C_{2x}	C_{2z}
$C_3(3)$	$C_3(3)$	$C_3^2(1)$	$C_3(2)$	$C_3(4)$	C_{2z}	$C_3^2(4)$	$C_3^2(3)$	$C_3(1)$	$C_3^2(2)$	C_{2x}	E	C_{2y}
$C_3(4)$	$C_3(4)$	$C_3(2)$	$C_3^2(1)$	$C_3(3)$	C_{2x}	$C_3(1)$	$C_3^2(2)$	$C_3^2(4)$	$C_3^2(3)$	C_{2z}	C_{2y}	E
i	i	σ_x	σ_y	σ_z	$S_6^5(1)$	$S_6^5(2)$	$S_6^5(3)$	$S_6^5(4)$	$S_6(1)$	$S_6(2)$	$S_6(3)$	$S_6(4)$
σ_x	σ_x	i	σ_z	σ_y	$S_6(3)$	$S_6^5(3)$	$S_6^5(2)$	$S_6(1)$	$S_6^5(4)$	$S_6(4)$	$S_6^5(1)$	$S_6(2)$
σ_y	σ_y	σ_z	i	σ_x	$S_6(4)$	$S_6(1)$	$S_6^5(4)$	$S_6^5(3)$	$S_6^5(2)$	$S_6(3)$	$S_6(2)$	$S_6^5(1)$
σ_z	σ_z	σ_y	σ_x	i	$S_6(2)$	$S_6^5(4)$	$S_6(1)$	$S_6^5(2)$	$S_6^5(3)$	$S_6^5(1)$	$S_6(4)$	$S_6(3)$
$S_6(1)$	$S_6(1)$	$S_6^5(3)$	$S_6^5(4)$	$S_6^5(2)$	i	$S_6(3)$	$S_6(4)$	$S_6(2)$	$S_6^5(1)$	σ_y	σ_z	σ_x
$S_6(2)$	$S_6(2)$	$S_6(3)$	$S_6^5(1)$	$S_6(4)$	$S_6^5(3)$	i	σ_y	σ_x	σ_z	$S_6^5(2)$	$S_6^5(4)$	$S_6(1)$
$S_6(3)$	$S_6(3)$	$S_6(2)$	$S_6(4)$	$S_6^5(1)$	$S_6^5(4)$	σ_y	i	σ_z	σ_x	$S_6(1)$	$S_6^5(3)$	$S_6^5(2)$
$S_6(4)$	$S_6(4)$	$S_6^5(1)$	$S_6(3)$	$S_6(2)$	$S_6^5(2)$	σ_x	σ_z	i	σ_y	$S_6^5(3)$	$S_6(1)$	$S_6^5(4)$
$S_6^5(1)$	$S_6^5(1)$	$S_6(4)$	$S_6(2)$	$S_6(3)$	$S_6(1)$	σ_z	σ_x	σ_y	i	$S_6^5(4)$	$S_6^5(2)$	$S_6^5(3)$
$S_6^5(2)$	$S_6^5(2)$	$S_6^5(4)$	$S_6^5(3)$	$S_6(1)$	σ_y	$S_6(2)$	$S_6^5(1)$	$S_6(3)$	$S_6(4)$	i	σ_x	σ_z
$S_6^5(3)$	$S_6^5(3)$	$S_6(1)$	$S_6^5(2)$	$S_6^5(4)$	σ_z	$S_6(4)$	$S_6(3)$	$S_6^5(1)$	$S_6(2)$	σ_x	i	σ_y
$S_6^5(4)$	$S_6^5(4)$	$S_6^5(2)$	$S_6(1)$	$S_6^5(3)$	σ_x	$S_6^5(1)$	$S_6(2)$	$S_6(4)$	$S_6(3)$	σ_z	σ_y	i

Table 32(b)1ii

O_h	i	σ_x	σ_y	σ_z	$S_6^5(1)$	$S_6^5(2)$	$S_6^5(3)$	$S_6^5(4)$	$S_6(1)$	$S_6(2)$	$S_6(3)$	$S_6(4)$
E	i	σ_x	σ_y	σ_z	$S_6^5(1)$	$S_6^5(2)$	$S_6^5(3)$	$S_6^5(4)$	$S_6(1)$	$S_6(2)$	$S_6(3)$	$S_6(4)$
C_{2x}	σ_x	i	σ_z	σ_y	$S_6(3)$	$S_6^5(3)$	$S_6^5(2)$	$S_6(1)$	$S_6^5(4)$	$S_6(4)$	$S_6^5(1)$	$S_6(2)$
C_{2y}	σ_y	σ_z	i	σ_x	$S_6(4)$	$S_6(1)$	$S_6^5(4)$	$S_6^5(3)$	$S_6^5(2)$	$S_6(3)$	$S_6(2)$	$S_6^5(1)$
C_{2z}	σ_z	σ_y	σ_x	i	$S_6(2)$	$S_6^5(4)$	$S_6(1)$	$S_6^5(2)$	$S_6^5(3)$	$S_6^5(1)$	$S_6(4)$	$S_6(3)$
$C_3^2(1)$	$S_6(1)$	$S_6^5(3)$	$S_6^5(4)$	$S_6^5(2)$	i	$S_6(3)$	$S_6(4)$	$S_6(2)$	$S_6^5(1)$	σ_y	σ_z	σ_x
$C_3^2(2)$	$S_6(2)$	$S_6(3)$	$S_6^5(1)$	$S_6(4)$	$S_6^5(3)$	i	σ_y	σ_x	σ_z	$S_6^5(2)$	$S_6^5(4)$	$S_6(1)$
$C_3^2(3)$	$S_6(3)$	$S_6(2)$	$S_6(4)$	$S_6^5(1)$	$S_6^5(4)$	σ_y	i	σ_z	σ_x	$S_6(1)$	$S_6^5(3)$	$S_6^5(2)$
$C_3^2(4)$	$S_6(4)$	$S_6^5(1)$	$S_6(3)$	$S_6(2)$	$S_6^5(2)$	σ_x	σ_z	i	σ_y	$S_6^5(3)$	$S_6(1)$	$S_6^5(4)$
$C_3(1)$	$S_6^5(1)$	$S_6(4)$	$S_6(2)$	$S_6(3)$	$S_6(1)$	σ_z	σ_x	σ_y	i	$S_6^5(4)$	$S_6^5(2)$	$S_6^5(3)$
$C_3(2)$	$S_6^5(2)$	$S_6^5(4)$	$S_6^5(3)$	$S_6(1)$	σ_y	$S_6(2)$	$S_6^5(1)$	$S_6(3)$	$S_6(4)$	i	σ_x	σ_z
$C_3(3)$	$S_6^5(3)$	$S_6(1)$	$S_6^5(2)$	$S_6^5(4)$	σ_z	$S_6(4)$	$S_6(3)$	$S_6^5(1)$	$S_6(2)$	σ_x	i	σ_y
$C_3(4)$	$S_6^5(4)$	$S_6^5(2)$	$S_6(1)$	$S_6^5(3)$	σ_x	$S_6^5(1)$	$S_6(2)$	$S_6(4)$	$S_6(3)$	σ_z	σ_y	i
i	E	C_{2x}	C_{2y}	C_{2z}	$C_3(1)$	$C_3(2)$	$C_3(3)$	$C_3(4)$	$C_3^2(1)$	$C_3^2(2)$	$C_3^2(3)$	$C_3^2(4)$
σ_x	C_{2x}	E	C_{2z}	C_{2y}	$C_3^2(3)$	$C_3(3)$	$C_3(2)$	$C_3^2(1)$	$C_3(4)$	$C_3^2(4)$	$C_3(1)$	$C_3^2(2)$
σ_y	C_{2y}	C_{2z}	E	C_{2x}	$C_3^2(4)$	$C_3^2(1)$	$C_3(4)$	$C_3(3)$	$C_3(2)$	$C_3^2(3)$	$C_3^2(2)$	$C_3(1)$
σ_z	C_{2z}	C_{2y}	C_{2x}	E	$C_3^2(2)$	$C_3(4)$	$C_3^2(1)$	$C_3(2)$	$C_3(3)$	$C_3(1)$	$C_3^2(4)$	$C_3^2(3)$
$S_6(1)$	$C_3^2(1)$	$C_3(3)$	$C_3(4)$	$C_3(2)$	E	$C_3^2(3)$	$C_3^2(4)$	$C_3^2(2)$	$C_3(1)$	C_{2y}	C_{2z}	C_{2x}
$S_6(2)$	$C_3^2(2)$	$C_3^2(3)$	$C_3(1)$	$C_3^2(4)$	$C_3(3)$	E	C_{2y}	C_{2x}	C_{2z}	$C_3(2)$	$C_3(4)$	$C_3^2(1)$
$S_6(3)$	$C_3^2(3)$	$C_3^2(2)$	$C_3^2(4)$	$C_3(1)$	$C_3(4)$	C_{2y}	E	C_{2z}	C_{2x}	$C_3^2(1)$	$C_3(3)$	$C_3(2)$
$S_6(4)$	$C_3^2(4)$	$C_3(1)$	$C_3^2(3)$	$C_3^2(2)$	$C_3(2)$	C_{2x}	C_{2z}	E	C_{2y}	$C_3(3)$	$C_3^2(1)$	$C_3(4)$
$S_6^5(1)$	$C_3(1)$	$C_3^2(4)$	$C_3^2(2)$	$C_3^2(3)$	$C_3^2(1)$	C_{2z}	C_{2x}	C_{2y}	E	$C_3(4)$	$C_3(2)$	$C_3(3)$
$S_6^5(2)$	$C_3(2)$	$C_3(4)$	$C_3(3)$	$C_3^2(1)$	C_{2y}	$C_3^2(2)$	$C_3(1)$	$C_3^2(3)$	$C_3^2(4)$	E	C_{2x}	C_{2z}
$S_6^5(3)$	$C_3(3)$	$C_3^2(1)$	$C_3(2)$	$C_3(4)$	C_{2z}	$C_3^2(4)$	$C_3^2(3)$	$C_3(1)$	$C_3^2(2)$	C_{2x}	E	C_{2y}
$S_6^5(4)$	$C_3(4)$	$C_3(2)$	$C_3^2(1)$	$C_3(3)$	C_{2x}	$C_3(1)$	$C_3^2(2)$	$C_3^2(4)$	$C_3^2(3)$	C_{2z}	C_{2y}	E

Table 32(b)1iii

O_h	C_{4x}	C_{4y}	C_{4z}	C_{4x}^3	C_{4y}^3	C_{4z}^3	C_{2xy}	C_{2yz}	C_{2zx}	$C_{2x\bar{y}}$	$C_{2y\bar{z}}$	$C_{2z\bar{x}}$
E	C_{4x}	C_{4y}	C_{4z}	C_{4x}^3	C_{4y}^3	C_{4z}^3	C_{2xy}	C_{2yz}	C_{2zx}	$C_{2x\bar{y}}$	$C_{2y\bar{z}}$	$C_{2z\bar{x}}$
C_{2x}	C_{4x}^3	$C_{2z\bar{x}}$	$C_{2x\bar{y}}$	C_{4x}	C_{2zx}	C_{2xy}	C_{4z}^3	C_{2yz}	C_{4y}^3	C_{4z}	C_{2yz}	C_{4y}
C_{2y}	C_{2yz}	C_{4y}^3	C_{2xy}	$C_{2y\bar{z}}$	C_{4y}	$C_{2x\bar{y}}$	C_{4z}	C_{4x}	$C_{2z\bar{x}}$	C_{4z}^3	C_{4x}^3	C_{2zx}
C_{2z}	$C_{2y\bar{z}}$	C_{2zx}	C_{4z}^3	C_{2yz}	$C_{2z\bar{x}}$	C_{4z}	$C_{2x\bar{y}}$	C_{4x}^3	C_{4y}	C_{2xy}	C_{4x}	C_{4y}^3
$C_3^2(1)$	C_{2zx}	C_{2xy}	C_{4x}	C_{4y}	C_{4z}^3	C_{2yz}	C_{4x}^3	C_{4y}^3	C_{4z}	$C_{2y\bar{z}}$	$C_{2z\bar{x}}$	$C_{2x\bar{y}}$
$C_3^2(2)$	C_{4z}^3	C_{2yz}	C_{4y}	$C_{2x\bar{y}}$	C_{4x}^3	$C_{2z\bar{x}}$	C_{2zx}	C_{4z}	$C_{2y\bar{z}}$	C_{4y}^3	C_{2xy}	C_{4x}
$C_3^2(3)$	$C_{2x\bar{y}}$	C_{4x}	C_{4y}^3	C_{4z}^3	$C_{2y\bar{z}}$	C_{2zx}	$C_{2z\bar{x}}$	C_{2xy}	C_{4x}^3	C_{4y}	C_{4z}	C_{2yz}
$C_3^2(4)$	C_{2xy}	$C_{2y\bar{z}}$	$C_{2z\bar{x}}$	C_{4z}	C_{4x}	C_{4y}	C_{4y}^3	$C_{2x\bar{y}}$	C_{2yz}	C_{2zx}	C_{4z}^3	C_{4x}^3
$C_3(1)$	C_{4z}	C_{4x}^3	C_{2zx}	C_{2xy}	C_{2yz}	C_{4y}^3	C_{4y}	C_{4z}^3	C_{4x}	$C_{2z\bar{x}}$	$C_{2x\bar{y}}$	$C_{2y\bar{z}}$
$C_3(2)$	$C_{2z\bar{x}}$	C_{4z}	C_{2yz}	C_{4y}^3	$C_{2x\bar{y}}$	C_{4x}	$C_{2y\bar{z}}$	C_{4y}	C_{2xy}	C_{4x}^3	C_{2zx}	C_{4z}^3
$C_3(3)$	C_{4y}	$C_{2x\bar{y}}$	$C_{2y\bar{z}}$	C_{2zx}	C_{4z}	C_{4x}^3	C_{2yz}	$C_{2z\bar{x}}$	C_{4z}^3	C_{4x}	C_{4y}^3	C_{2xy}
$C_3(4)$	C_{4y}^3	C_{4z}^3	C_{4x}^3	$C_{2z\bar{x}}$	C_{2xy}	$C_{2y\bar{z}}$	C_{4x}	C_{2zx}	$C_{2x\bar{y}}$	C_{2yz}	C_{4y}	C_{4z}
i	S_{4x}	S_{4y}	S_{4z}	S_{4x}^3	S_{4y}^3	S_{4z}^3	σ_{xy}	σ_{yz}	σ_{zx}	$\sigma_{x\bar{y}}$	$\sigma_{y\bar{z}}$	$\sigma_{z\bar{x}}$
σ_x	S_{4x}^3	$\sigma_{z\bar{x}}$	$\sigma_{x\bar{y}}$	S_{4x}	σ_{zx}	σ_{xy}	S_{4z}^3	$\sigma_{y\bar{z}}$	S_{4y}^3	S_{4z}	σ_{yz}	S_{4y}
σ_y	σ_{yz}	S_{4y}^3	σ_{xy}	$\sigma_{y\bar{z}}$	S_{4y}	$\sigma_{x\bar{y}}$	S_{4z}	S_{4x}	$\sigma_{z\bar{x}}$	S_{4z}^3	S_{4x}^3	σ_{zx}
σ_z	$\sigma_{y\bar{z}}$	σ_{zx}	S_{4z}^3	σ_{yz}	$\sigma_{z\bar{x}}$	S_{4z}	$\sigma_{x\bar{y}}$	S_{4x}^3	S_{4y}	σ_{xy}	S_{4x}	S_{4y}^3
$S_6(1)$	σ_{zx}	σ_{xy}	S_{4x}	S_{4y}	S_{4z}^3	σ_{yz}	S_{4x}^3	S_{4y}^3	S_{4z}	$\sigma_{y\bar{z}}$	$\sigma_{z\bar{x}}$	$\sigma_{x\bar{y}}$
$S_6(2)$	S_{4z}^3	σ_{yz}	S_{4y}	$\sigma_{x\bar{y}}$	S_{4x}^3	$\sigma_{z\bar{x}}$	σ_{zx}	S_{4z}	$\sigma_{y\bar{z}}$	S_{4y}^3	σ_{xy}	S_{4x}
$S_6(3)$	$\sigma_{x\bar{y}}$	S_{4x}	S_{4y}^3	S_{4z}^3	$\sigma_{y\bar{z}}$	σ_{zx}	$\sigma_{z\bar{x}}$	σ_{xy}	S_{4x}^3	S_{4y}	S_{4z}	σ_{yz}
$S_6(4)$	σ_{xy}	$\sigma_{y\bar{z}}$	$\sigma_{z\bar{x}}$	S_{4z}	S_{4x}	S_{4y}	S_{4y}^3	$\sigma_{x\bar{y}}$	σ_{yz}	σ_{zx}	S_{4z}^3	S_{4x}^3
$S_6^5(1)$	S_{4z}	S_{4x}^3	σ_{zx}	σ_{xy}	σ_{yz}	S_{4y}^3	S_{4y}	S_{4z}^3	S_{4x}	$\sigma_{z\bar{x}}$	$\sigma_{x\bar{y}}$	$\sigma_{y\bar{z}}$
$S_6^5(2)$	$\sigma_{z\bar{x}}$	S_{4z}	σ_{yz}	S_{4y}^3	$\sigma_{x\bar{y}}$	S_{4x}	$\sigma_{y\bar{z}}$	S_{4y}	σ_{xy}	S_{4x}^3	σ_{zx}	S_{4z}^3
$S_6^5(3)$	S_{4y}	$\sigma_{x\bar{y}}$	$\sigma_{y\bar{z}}$	σ_{zx}	S_{4z}	S_{4x}^3	σ_{yz}	$\sigma_{z\bar{x}}$	S_{4z}^3	S_{4x}	S_{4y}^3	σ_{xy}
$S_6^5(4)$	S_{4y}^3	S_{4z}^3	S_{4x}^3	$\sigma_{z\bar{x}}$	σ_{xy}	$\sigma_{y\bar{z}}$	S_{4x}	σ_{zx}	$\sigma_{x\bar{y}}$	σ_{yz}	S_{4y}	S_{4z}

Table 32(b)1iv

O_h	S_{4x}	S_{4y}	S_{4z}	S_{4x}^3	S_{4y}^3	S_{4z}^3	σ_{xy}	σ_{yz}	σ_{zx}	$\sigma_{x\bar{y}}$	$\sigma_{y\bar{z}}$	$\sigma_{z\bar{x}}$
E	S_{4x}	S_{4y}	S_{4z}	S_{4x}^3	S_{4y}^3	S_{4z}^3	σ_{xy}	σ_{yz}	σ_{zx}	$\sigma_{x\bar{y}}$	$\sigma_{y\bar{z}}$	$\sigma_{z\bar{x}}$
C_{2x}	S_{4x}^3	$\sigma_{z\bar{x}}$	$\sigma_{x\bar{y}}$	S_{4x}	σ_{zx}	σ_{xy}	S_{4z}^3	$\sigma_{y\bar{z}}$	S_{4y}^3	S_{4z}	σ_{yz}	S_{4y}
C_{2y}	σ_{yz}	S_{4y}^3	σ_{xy}	$\sigma_{y\bar{z}}$	S_{4y}	$\sigma_{x\bar{y}}$	S_{4z}	S_{4x}	$\sigma_{z\bar{x}}$	S_{4z}^3	S_{4x}^3	σ_{zx}
C_{2z}	$\sigma_{y\bar{z}}$	σ_{zx}	S_{4z}^3	σ_{yz}	$\sigma_{z\bar{x}}$	S_{4z}	$\sigma_{x\bar{y}}$	S_{4x}^3	S_{4y}	σ_{xy}	S_{4x}	S_{4y}^3
$C_3^2(1)$	σ_{zx}	σ_{xy}	S_{4x}	S_{4y}	S_{4z}^3	σ_{yz}	S_{4x}^3	S_{4y}^3	S_{4z}	$\sigma_{y\bar{z}}$	$\sigma_{z\bar{x}}$	$\sigma_{x\bar{y}}$
$C_3^2(2)$	S_{4z}^3	σ_{yz}	S_{4y}	$\sigma_{x\bar{y}}$	S_{4x}^3	$\sigma_{z\bar{x}}$	σ_{zx}	S_{4z}	$\sigma_{y\bar{z}}$	S_{4y}^3	σ_{xy}	S_{4x}
$C_3^2(3)$	$\sigma_{x\bar{y}}$	S_{4x}	S_{4y}^3	S_{4z}^3	$\sigma_{y\bar{z}}$	σ_{zx}	$\sigma_{z\bar{x}}$	σ_{xy}	S_{4x}^3	S_{4y}	S_{4z}	σ_{yz}
$C_3^2(4)$	σ_{xy}	$\sigma_{y\bar{z}}$	$\sigma_{z\bar{x}}$	S_{4z}	S_{4x}	S_{4y}	S_{4y}^3	$\sigma_{x\bar{y}}$	σ_{yz}	σ_{zx}	S_{4z}^3	S_{4x}^3
$C_3(1)$	S_{4z}	S_{4x}^3	σ_{zx}	σ_{xy}	σ_{yz}	S_{4y}^3	S_{4y}	S_{4z}^3	S_{4x}	$\sigma_{z\bar{x}}$	$\sigma_{x\bar{y}}$	$\sigma_{y\bar{z}}$
$C_3(2)$	$\sigma_{z\bar{x}}$	S_{4z}	σ_{yz}	S_{4y}^3	$\sigma_{x\bar{y}}$	S_{4x}	$\sigma_{y\bar{z}}$	S_{4y}	σ_{xy}	S_{4x}^3	σ_{zx}	S_{4z}^3
$C_3(3)$	S_{4y}	$\sigma_{x\bar{y}}$	$\sigma_{y\bar{z}}$	σ_{zx}	S_{4z}	S_{4x}^3	σ_{yz}	$\sigma_{z\bar{x}}$	S_{4z}^3	S_{4x}	S_{4y}^3	σ_{xy}
$C_3(4)$	S_{4y}^3	S_{4z}^3	S_{4x}^3	$\sigma_{z\bar{x}}$	σ_{xy}	$\sigma_{y\bar{z}}$	S_{4x}	σ_{zx}	$\sigma_{x\bar{y}}$	σ_{yz}	S_{4y}	S_{4z}
i	C_{4x}	C_{4y}	C_{4z}	C_{4x}^3	C_{4y}^3	C_{4z}^3	C_{2xy}	C_{2yz}	C_{2zx}	$C_{2x\bar{y}}$	$C_{2y\bar{z}}$	$C_{2z\bar{x}}$
σ_x	C_{4x}^3	$C_{2z\bar{x}}$	$C_{2x\bar{y}}$	C_{4x}	C_{2zx}	C_{2xy}	C_{4z}^3	$C_{2y\bar{z}}$	C_{4y}^3	C_{4z}	C_{2yz}	C_{4y}
σ_y	C_{2yz}	C_{4y}^3	C_{2xy}	$C_{2y\bar{z}}$	C_{4y}	$C_{2x\bar{y}}$	C_{4z}	C_{4x}	$C_{2z\bar{x}}$	C_{4z}^3	C_{4x}^3	C_{2zx}
σ_z	$C_{2y\bar{z}}$	C_{2zx}	C_{4z}^3	C_{2yz}	$C_{2z\bar{x}}$	C_{4z}	$C_{2x\bar{y}}$	C_{4x}^3	C_{4y}	C_{2xy}	C_{4x}	C_{4y}^3
$S_6(1)$	C_{2zx}	C_{2xy}	C_{4x}	C_{4y}	C_{4z}^3	C_{2yz}	C_{4x}^3	C_{4y}^3	C_{4z}	$C_{2y\bar{z}}$	$C_{2z\bar{x}}$	$C_{2x\bar{y}}$
$S_6(2)$	C_{4z}^3	C_{2yz}	C_{4y}	$C_{2x\bar{y}}$	C_{4x}^3	$C_{2z\bar{x}}$	C_{2zx}	C_{4z}	$C_{2y\bar{z}}$	C_{4y}^3	C_{2xy}	C_{4x}
$S_6(3)$	$C_{2x\bar{y}}$	C_{4x}	C_{4y}^3	C_{4z}^3	$C_{2y\bar{z}}$	C_{2zx}	$C_{2z\bar{x}}$	C_{2xy}	C_{4x}^3	C_{4y}	C_{4z}	C_{2yz}
$S_6(4)$	C_{2xy}	$C_{2y\bar{z}}$	$C_{2z\bar{x}}$	C_{4z}	C_{4x}	C_{4y}	C_{4y}^3	$C_{2x\bar{y}}$	C_{2yz}	C_{2zx}	C_{4z}^3	C_{4x}^3
$S_6^5(1)$	C_{4z}	C_{4x}^3	C_{2zx}	C_{2xy}	C_{2yz}	C_{4y}^3	C_{4y}	C_{4z}^3	C_{4x}	$C_{2z\bar{x}}$	$C_{2x\bar{y}}$	$C_{2y\bar{z}}$
$S_6^5(2)$	$C_{2z\bar{x}}$	C_{4z}	C_{2yz}	C_{4y}^3	$C_{2x\bar{y}}$	C_{4x}	$C_{2y\bar{z}}$	C_{4y}	C_{2xy}	C_{4x}^3	C_{2zx}	C_{4z}^3
$S_6^5(3)$	C_{4y}	$C_{2x\bar{y}}$	$C_{2y\bar{z}}$	C_{2zx}	C_{4z}	C_{4x}^3	C_{2yz}	$C_{2z\bar{x}}$	C_{4z}^3	C_{4x}	C_{4y}^3	C_{2xy}
$S_6^5(4)$	C_{4y}^3	C_{4z}^3	C_{4x}^3	$C_{2z\bar{x}}$	C_{2xy}	$C_{2y\bar{z}}$	C_{4x}	C_{2zx}	$C_{2x\bar{y}}$	C_{2yz}	C_{4y}	C_{4z}

Table 32(b)2i

O_h	E	C_{2x}	C_{2y}	C_{2z}	$C_3(1)$	$C_3(2)$	$C_3(3)$	$C_3(4)$	$C_3^2(1)$	$C_3^2(2)$	$C_3^2(3)$	$C_3^2(4)$
C_{4x}^3	C_{4x}^3	C_{4x}	C_{2yz}	$C_{2y\bar{z}}$	C_{2zx}	C_{4z}	$C_{2x\bar{y}}$	C_{2xy}	C_{4z}^3	$C_{2z\bar{x}}$	C_{4y}^3	C_{4y}
C_{4y}^3	C_{4y}^3	$C_{2z\bar{x}}$	C_{4y}	C_{2zx}	C_{2xy}	C_{2yz}	C_{4x}^3	$C_{2y\bar{z}}$	C_{4x}	C_{4z}^3	$C_{2x\bar{y}}$	C_{4z}
C_{4z}^3	C_{4z}^3	$C_{2x\bar{y}}$	C_{2xy}	C_{4z}	C_{4x}^3	C_{4y}^3	C_{4y}	$C_{2z\bar{x}}$	C_{2zx}	C_{2yz}	$C_{2y\bar{z}}$	C_{4x}
C_{4x}	C_{4x}	C_{4x}^3	$C_{2y\bar{z}}$	C_{2yz}	C_{4y}^3	$C_{2x\bar{y}}$	C_{4z}	C_{4z}^3	C_{2xy}	C_{4y}	C_{2zx}	$C_{2z\bar{x}}$
C_{4y}	C_{4y}	C_{2zx}	C_{4y}^3	$C_{2z\bar{x}}$	C_{4z}	C_{4x}	$C_{2y\bar{z}}$	C_{4x}^3	C_{2yz}	$C_{2x\bar{y}}$	C_{4z}^3	C_{2xy}
C_{4z}	C_{4z}	C_{2xy}	$C_{2x\bar{y}}$	C_{4z}^3	C_{2yz}	$C_{2z\bar{x}}$	C_{2zx}	C_{4y}^3	C_{4y}	C_{4x}^3	C_{4x}	$C_{2y\bar{z}}$
C_{2xy}	C_{2xy}	C_{4z}	C_{4z}^3	$C_{2x\bar{y}}$	C_{4x}	C_{2zx}	$C_{2z\bar{x}}$	C_{4y}	C_{4y}^3	$C_{2y\bar{z}}$	C_{2yz}	C_{4x}^3
C_{2yz}	C_{2yz}	$C_{2y\bar{z}}$	C_{4x}^3	C_{4x}	C_{4y}	C_{4z}^3	C_{2xy}	$C_{2x\bar{y}}$	C_{4z}	C_{4y}^3	$C_{2z\bar{x}}$	C_{2zx}
C_{2zx}	C_{2zx}	C_{4y}	$C_{2z\bar{x}}$	C_{4y}^3	C_{4z}^3	$C_{2y\bar{z}}$	C_{4x}	C_{2yz}	C_{4x}^3	C_{2xy}	C_{4z}	$C_{2x\bar{y}}$
$C_{2x\bar{y}}$	$C_{2x\bar{y}}$	C_{4z}^3	C_{4z}	C_{2xy}	$C_{2y\bar{z}}$	C_{4y}	C_{4y}^3	C_{2zx}	$C_{2z\bar{x}}$	C_{4x}	C_{4x}^3	C_{2yz}
$C_{2y\bar{z}}$	$C_{2y\bar{z}}$	C_{2yz}	C_{4x}	C_{4x}^3	$C_{2z\bar{x}}$	C_{2xy}	C_{4z}^3	C_{4z}	$C_{2x\bar{y}}$	C_{2zx}	C_{4y}	C_{4y}^3
$C_{2z\bar{x}}$	$C_{2z\bar{x}}$	C_{4y}^3	C_{2zx}	C_{4y}	$C_{2x\bar{y}}$	C_{4x}^3	C_{2yz}	C_{4x}	$C_{2y\bar{z}}$	C_{4z}	C_{2xy}	C_{4z}^3
S_{4x}^3	S_{4x}^3	S_{4x}	σ_{yz}	$\sigma_{y\bar{z}}$	σ_{zx}	S_{4z}	$\sigma_{x\bar{y}}$	σ_{xy}	S_{4z}^3	$\sigma_{z\bar{x}}$	S_{4y}^3	S_{4y}
S_{4y}^3	S_{4y}^3	$\sigma_{z\bar{x}}$	S_{4y}	σ_{zx}	σ_{xy}	σ_{yz}	S_{4x}^3	$\sigma_{y\bar{z}}$	S_{4x}	S_{4z}^3	$\sigma_{x\bar{y}}$	S_{4z}
S_{4z}^3	S_{4z}^3	$\sigma_{x\bar{y}}$	σ_{xy}	S_{4z}	S_{4x}^3	S_{4y}^3	S_{4y}	$\sigma_{z\bar{x}}$	σ_{zx}	σ_{yz}	$\sigma_{y\bar{z}}$	S_{4x}
S_{4x}	S_{4x}	S_{4x}^3	$\sigma_{y\bar{z}}$	σ_{yz}	S_{4y}^3	$\sigma_{x\bar{y}}$	S_{4z}	S_{4z}^3	σ_{xy}	S_{4y}	σ_{zx}	$\sigma_{z\bar{x}}$
S_{4y}	S_{4y}	σ_{zx}	S_{4y}^3	$\sigma_{z\bar{x}}$	S_{4z}	S_{4x}	$\sigma_{y\bar{z}}$	S_{4x}^3	σ_{yz}	$\sigma_{x\bar{y}}$	S_{4z}^3	σ_{xy}
S_{4z}	S_{4z}	σ_{xy}	$\sigma_{x\bar{y}}$	S_{4z}^3	σ_{yz}	$\sigma_{z\bar{x}}$	σ_{zx}	S_{4y}^3	S_{4y}	S_{4x}^3	S_{4x}	$\sigma_{y\bar{z}}$
σ_{xy}	σ_{xy}	S_{4z}	S_{4z}^3	$\sigma_{x\bar{y}}$	S_{4x}	σ_{zx}	$\sigma_{z\bar{x}}$	S_{4y}	S_{4y}^3	$\sigma_{y\bar{z}}$	σ_{yz}	S_{4x}^3
σ_{yz}	σ_{yz}	$\sigma_{y\bar{z}}$	S_{4x}^3	S_{4x}	S_{4y}	S_{4z}^3	σ_{xy}	$\sigma_{x\bar{y}}$	S_{4z}	S_{4y}^3	$\sigma_{z\bar{x}}$	σ_{zx}
σ_{zx}	σ_{zx}	S_{4y}	$\sigma_{z\bar{x}}$	S_{4y}^3	S_{4z}^3	$\sigma_{y\bar{z}}$	S_{4x}	σ_{yz}	S_{4x}^3	σ_{xy}	S_{4z}	$\sigma_{x\bar{y}}$
$\sigma_{x\bar{y}}$	$\sigma_{x\bar{y}}$	S_{4z}^3	S_{4z}	σ_{xy}	$\sigma_{y\bar{z}}$	S_{4y}	S_{4y}^3	σ_{zx}	$\sigma_{z\bar{x}}$	S_{4x}	S_{4x}^3	σ_{yz}
$\sigma_{y\bar{z}}$	$\sigma_{y\bar{z}}$	σ_{yz}	S_{4x}	S_{4x}^3	$\sigma_{z\bar{x}}$	σ_{xy}	S_{4z}^3	S_{4z}	$\sigma_{x\bar{y}}$	σ_{zx}	S_{4y}	S_{4y}^3
$\sigma_{z\bar{x}}$	$\sigma_{z\bar{x}}$	S_{4y}^3	σ_{zx}	S_{4y}	$\sigma_{x\bar{y}}$	S_{4x}^3	σ_{yz}	S_{4x}	$\sigma_{y\bar{z}}$	S_{4z}	σ_{xy}	S_{4z}^3

Table 32(b)2ii

O_h	i	σ_x	σ_y	σ_z	$S_6^5(1)$	$S_6^5(2)$	$S_6^5(3)$	$S_6^5(4)$	$S_6(1)$	$S_6(2)$	$S_6(3)$	$S_6(4)$
C_{4x}^3	S_{4x}^3	S_{4x}	σ_{yz}	$\sigma_{y\bar{z}}$	σ_{zx}	S_{4z}	$\sigma_{x\bar{y}}$	σ_{xy}	S_{4z}^3	$\sigma_{z\bar{x}}$	S_{4y}^3	S_{4y}
C_{4y}^3	S_{4y}^3	$\sigma_{z\bar{x}}$	S_{4y}	σ_{zx}	σ_{xy}	σ_{yz}	S_{4x}^3	$\sigma_{y\bar{z}}$	S_{4x}	S_{4z}^3	$\sigma_{x\bar{y}}$	S_{4z}
C_{4z}^3	S_{4z}^3	$\sigma_{x\bar{y}}$	σ_{xy}	S_{4z}	S_{4x}^3	S_{4y}^3	S_{4y}	$\sigma_{z\bar{x}}$	σ_{zx}	σ_{yz}	$\sigma_{y\bar{z}}$	S_{4x}
C_{4x}	S_{4x}	S_{4x}^3	$\sigma_{y\bar{z}}$	σ_{yz}	S_{4y}^3	$\sigma_{x\bar{y}}$	S_{4z}	S_{4z}^3	σ_{xy}	S_{4y}	σ_{zx}	$\sigma_{z\bar{x}}$
C_{4y}	S_{4y}	σ_{zx}	S_{4y}^3	$\sigma_{z\bar{x}}$	S_{4z}	S_{4x}	$\sigma_{y\bar{z}}$	S_{4x}^3	σ_{yz}	$\sigma_{x\bar{y}}$	S_{4z}^3	σ_{xy}
C_{4z}	S_{4z}	σ_{xy}	$\sigma_{x\bar{y}}$	S_{4z}^3	σ_{yz}	$\sigma_{z\bar{x}}$	σ_{zx}	S_{4y}^3	S_{4y}	S_{4x}^3	S_{4x}	$\sigma_{y\bar{z}}$
C_{2xy}	σ_{xy}	S_{4z}	S_{4z}^3	$\sigma_{x\bar{y}}$	S_{4x}	σ_{zx}	$\sigma_{z\bar{x}}$	S_{4y}	S_{4y}^3	$\sigma_{y\bar{z}}$	σ_{yz}	S_{4x}^3
C_{2yz}	σ_{yz}	$\sigma_{y\bar{z}}$	S_{4x}^3	S_{4x}	S_{4y}	S_{4z}^3	σ_{xy}	$\sigma_{x\bar{y}}$	S_{4z}	S_{4y}^3	$\sigma_{z\bar{x}}$	σ_{zx}
C_{2zx}	σ_{zx}	S_{4y}	$\sigma_{z\bar{x}}$	S_{4y}^3	S_{4z}^3	$\sigma_{y\bar{z}}$	S_{4x}	σ_{yz}	S_{4x}^3	σ_{xy}	S_{4z}	$\sigma_{x\bar{y}}$
$C_{2x\bar{y}}$	$\sigma_{x\bar{y}}$	S_{4z}^3	S_{4z}	σ_{xy}	$\sigma_{y\bar{z}}$	S_{4y}	S_{4y}^3	σ_{zx}	$\sigma_{z\bar{x}}$	S_{4x}	S_{4x}^3	σ_{yz}
$C_{2y\bar{z}}$	$\sigma_{y\bar{z}}$	σ_{yz}	S_{4x}	S_{4x}^3	$\sigma_{z\bar{x}}$	σ_{xy}	S_{4z}^3	S_{4z}	$\sigma_{x\bar{y}}$	σ_{zx}	S_{4y}	S_{4y}^3
$C_{2z\bar{x}}$	$\sigma_{z\bar{x}}$	S_{4y}^3	σ_{zx}	S_{4y}	$\sigma_{x\bar{y}}$	S_{4x}^3	σ_{yz}	S_{4x}	$\sigma_{y\bar{z}}$	S_{4z}	σ_{xy}	S_{4z}^3
S_{4x}^3	C_{4x}^3	C_{4x}	C_{2yz}	$C_{2y\bar{z}}$	C_{2zx}	C_{4z}	$C_{2x\bar{y}}$	C_{2xy}	C_{4z}^3	$C_{2z\bar{x}}$	C_{4y}^3	C_{4y}
S_{4y}^3	C_{4y}^3	$C_{2z\bar{x}}$	C_{4y}	C_{2zx}	C_{2xy}	C_{2yz}	C_{4x}^3	$C_{2y\bar{z}}$	C_{4x}	C_{4z}^3	$C_{2x\bar{y}}$	C_{4z}
S_{4z}^3	C_{4z}^3	$C_{2x\bar{y}}$	C_{2xy}	C_{4z}	C_{4x}^3	C_{4y}^3	C_{4y}	$C_{2z\bar{x}}$	C_{2zx}	C_{2yz}	$C_{2y\bar{z}}$	C_{4x}
S_{4x}	C_{4x}	C_{4x}^3	$C_{2y\bar{z}}$	C_{2yz}	C_{4y}^3	$C_{2x\bar{y}}$	C_{4z}	C_{4z}^3	C_{2xy}	C_{4y}	C_{2zx}	$C_{2z\bar{x}}$
S_{4y}	C_{4y}	C_{2zx}	C_{4y}^3	$C_{2z\bar{x}}$	C_{4z}	C_{4x}	$C_{2y\bar{z}}$	C_{4x}^3	C_{2yz}	$C_{2x\bar{y}}$	C_{4z}^3	C_{2xy}
S_{4z}	C_{4z}	C_{2xy}	$C_{2x\bar{y}}$	C_{4z}^3	C_{2yz}	$C_{2z\bar{x}}$	C_{2zx}	C_{4y}^3	C_{4y}	C_{4x}^3	C_{4x}	$C_{2y\bar{z}}$
σ_{xy}	C_{2xy}	C_{4z}	C_{4z}^3	$C_{2x\bar{y}}$	C_{4x}	C_{2zx}	$C_{2z\bar{x}}$	C_{4y}	C_{4y}^3	$C_{2y\bar{z}}$	C_{2yz}	C_{4x}^3
σ_{yz}	C_{2yz}	$C_{2y\bar{z}}$	C_{4x}^3	C_{4x}	C_{4y}	C_{4z}^3	C_{2xy}	$C_{2x\bar{y}}$	C_{4z}	C_{4y}^3	$C_{2z\bar{x}}$	C_{2zx}
σ_{zx}	C_{2zx}	C_{4y}	$C_{2z\bar{x}}$	C_{4y}^3	C_{4z}^3	$C_{2y\bar{z}}$	C_{4x}	C_{2yz}	C_{4x}^3	C_{2xy}	C_{4z}	$C_{2x\bar{y}}$
$\sigma_{x\bar{y}}$	$C_{2x\bar{y}}$	C_{4z}^3	C_{4z}	C_{2xy}	$C_{2y\bar{z}}$	C_{4y}	C_{4y}^3	C_{2zx}	$C_{2z\bar{x}}$	C_{4x}	C_{4x}^3	C_{2yz}
$\sigma_{y\bar{z}}$	$C_{2y\bar{z}}$	C_{2yz}	C_{4x}	C_{4x}^3	$C_{2z\bar{x}}$	C_{2xy}	C_{4z}^3	C_{4z}	$C_{2x\bar{y}}$	C_{2zx}	C_{4y}	C_{4y}^3
$\sigma_{z\bar{x}}$	$C_{2z\bar{x}}$	C_{4y}^3	C_{2zx}	C_{4y}	$C_{2x\bar{y}}$	C_{4x}^3	C_{2yz}	C_{4x}	$C_{2y\bar{z}}$	C_{4z}	C_{2xy}	C_{4z}^3

Table 32(b)2iii

O_h	C_{4x}	C_{4y}	C_{4z}	C_{4x}^3	C_{4y}^3	C_{4z}^3	C_{2xy}	C_{2yz}	C_{2zx}	$C_{2x\bar{y}}$	$C_{2y\bar{z}}$	$C_{2z\bar{x}}$
C_{4x}^3	E	$C_3^2(2)$	$C_3(3)$	C_{2x}	$C_3(1)$	$C_3(4)$	$C_3^2(1)$	C_{2z}	$C_3^2(3)$	$C_3(2)$	C_{2y}	$C_3^2(4)$
C_{4y}^3	$C_3(2)$	E	$C_3(1)$	$C_3(4)$	C_{2y}	$C_3^2(3)$	$C_3^2(4)$	$C_3^2(1)$	C_{2x}	$C_3^2(2)$	$C_3(3)$	C_{2z}
C_{4z}^3	$C_3^2(3)$	$C_3^2(1)$	E	$C_3^2(2)$	$C_3(4)$	C_{2z}	C_{2x}	$C_3(1)$	$C_3(3)$	C_{2y}	$C_3^2(4)$	$C_3(2)$
C_{4x}	C_{2x}	$C_3^2(4)$	$C_3(2)$	E	$C_3^2(3)$	$C_3^2(1)$	$C_3(4)$	C_{2y}	$C_3(1)$	$C_3(3)$	C_{2z}	$C_3^2(2)$
C_{4y}	$C_3^2(1)$	C_{2y}	$C_3^2(4)$	$C_3(3)$	E	$C_3^2(2)$	$C_3(1)$	$C_3(2)$	C_{2z}	$C_3^2(3)$	$C_3(4)$	C_{2x}
C_{4z}	$C_3^2(4)$	$C_3(3)$	C_{2z}	$C_3(1)$	$C_3(2)$	E	C_{2y}	$C_3^2(2)$	$C_3^2(1)$	C_{2x}	$C_3^2(3)$	$C_3(4)$
C_{2xy}	$C_3(1)$	$C_3(4)$	C_{2x}	$C_3^2(4)$	$C_3^2(1)$	C_{2y}	E	$C_3^2(3)$	$C_3(2)$	C_{2z}	$C_3^2(2)$	$C_3(3)$
C_{2yz}	C_{2z}	$C_3(1)$	$C_3^2(1)$	C_{2y}	$C_3^2(2)$	$C_3(2)$	$C_3(3)$	E	$C_3^2(4)$	$C_3(4)$	C_{2x}	$C_3^2(3)$
C_{2zx}	$C_3(3)$	C_{2x}	$C_3^2(3)$	$C_3^2(1)$	C_{2z}	$C_3(1)$	$C_3^2(2)$	$C_3(4)$	E	$C_3^2(4)$	$C_3(2)$	C_{2y}
$C_{2x\bar{y}}$	$C_3^2(2)$	$C_3(2)$	C_{2y}	$C_3^2(3)$	$C_3(3)$	C_{2x}	C_{2z}	$C_3^2(4)$	$C_3(4)$	E	$C_3(1)$	$C_3^2(1)$
$C_{2y\bar{z}}$	C_{2y}	$C_3^2(3)$	$C_3(4)$	C_{2z}	$C_3^2(4)$	$C_3(3)$	$C_3(2)$	C_{2x}	$C_3^2(2)$	$C_3^2(1)$	E	$C_3(1)$
$C_{2z\bar{x}}$	$C_3(4)$	C_{2z}	$C_3^2(2)$	$C_3(2)$	C_{2x}	$C_3^2(4)$	$C_3^2(3)$	$C_3(3)$	C_{2y}	$C_3(1)$	$C_3^2(1)$	E
S_{4x}^3	i	$S_6(2)$	$S_6^5(3)$	σ_x	$S_6^5(1)$	$S_6^5(4)$	$S_6(1)$	σ_z	$S_6(3)$	$S_6^5(2)$	σ_y	$S_6(4)$
S_{4y}^3	$S_6^5(2)$	i	$S_6^5(1)$	$S_6^5(4)$	σ_y	$S_6(3)$	$S_6(4)$	$S_6(1)$	σ_x	$S_6(2)$	$S_6^5(3)$	σ_z
S_{4z}^3	$S_6(3)$	$S_6(1)$	i	$S_6(2)$	$S_6^5(4)$	σ_z	σ_x	$S_6^5(1)$	$S_6^5(3)$	σ_y	$S_6(4)$	$S_6^5(2)$
S_{4x}	σ_x	$S_6(4)$	$S_6^5(2)$	i	$S_6(3)$	$S_6(1)$	$S_6^5(4)$	σ_y	$S_6^5(1)$	$S_6^5(3)$	σ_z	$S_6(2)$
S_{4y}	$S_6(1)$	σ_y	$S_6(4)$	$S_6^5(3)$	i	$S_6(2)$	$S_6^5(1)$	$S_6^5(2)$	σ_z	$S_6(3)$	$S_6^5(4)$	σ_x
S_{4z}	$S_6(4)$	$S_6^5(3)$	σ_z	$S_6^5(1)$	$S_6^5(2)$	i	σ_y	$S_6(2)$	$S_6(1)$	σ_x	$S_6(3)$	$S_6^5(4)$
σ_{xy}	$S_6^5(1)$	$S_6^5(4)$	σ_x	$S_6(4)$	$S_6(1)$	σ_y	i	$S_6(3)$	$S_6^5(2)$	σ_z	$S_6(2)$	$S_6^5(3)$
σ_{yz}	σ_z	$S_6^5(1)$	$S_6(1)$	σ_y	$S_6(2)$	$S_6^5(2)$	$S_6^5(3)$	i	$S_6(4)$	$S_6^5(4)$	σ_x	$S_6(3)$
σ_{zx}	$S_6^5(3)$	σ_x	$S_6(3)$	$S_6(1)$	σ_z	$S_6^5(1)$	$S_6(2)$	$S_6^5(4)$	i	$S_6(4)$	$S_6^5(2)$	σ_y
$\sigma_{x\bar{y}}$	$S_6(2)$	$S_6^5(2)$	σ_y	$S_6(3)$	$S_6^5(3)$	σ_x	σ_z	$S_6(4)$	$S_6^5(4)$	i	$S_6^5(1)$	$S_6(1)$
$\sigma_{y\bar{z}}$	σ_y	$S_6(3)$	$S_6^5(4)$	σ_z	$S_6(4)$	$S_6^5(3)$	$S_6^5(2)$	σ_x	$S_6(2)$	$S_6(1)$	i	$S_6^5(1)$
$\sigma_{z\bar{x}}$	$S_6^5(4)$	σ_z	$S_6(2)$	$S_6^5(2)$	σ_x	$S_6(4)$	$S_6(3)$	$S_6^5(3)$	σ_y	$S_6^5(1)$	$S_6(1)$	i

Table 32(b)2iv

O_h	S_{4x}	S_{4y}	S_{4z}	S_{4x}^3	S_{4y}^3	S_{4z}^3	σ_{xy}	σ_{yz}	σ_{zx}	$\sigma_{x\bar{y}}$	$\sigma_{y\bar{z}}$	$\sigma_{z\bar{x}}$
C_{4x}^3	i	$S_6(2)$	$S_6^5(3)$	σ_x	$S_6^5(1)$	$S_6^5(4)$	$S_6(1)$	σ_z	$S_6(3)$	$S_6^5(2)$	σ_y	$S_6(4)$
C_{4y}^3	$S_6^5(2)$	i	$S_6^5(1)$	$S_6^5(4)$	σ_y	$S_6(3)$	$S_6(4)$	$S_6(1)$	σ_x	$S_6(2)$	$S_6^5(3)$	σ_z
C_{4z}^3	$S_6(3)$	$S_6(1)$	i	$S_6(2)$	$S_6^5(4)$	σ_z	σ_x	$S_6^5(1)$	$S_6^5(3)$	σ_y	$S_6(4)$	$S_6^5(2)$
C_{4x}	σ_x	$S_6(4)$	$S_6^5(2)$	i	$S_6(3)$	$S_6(1)$	$S_6^5(4)$	σ_y	$S_6^5(1)$	$S_6^5(3)$	σ_z	$S_6(2)$
C_{4y}	$S_6(1)$	σ_y	$S_6(4)$	$S_6^5(3)$	i	$S_6(2)$	$S_6^5(1)$	$S_6^5(2)$	σ_z	$S_6(3)$	$S_6^5(4)$	σ_x
C_{4z}	$S_6(4)$	$S_6^5(3)$	σ_z	$S_6^5(1)$	$S_6^5(2)$	i	σ_y	$S_6(2)$	$S_6(1)$	σ_x	$S_6(3)$	$S_6^5(4)$
C_{2xy}	$S_6^5(1)$	$S_6^5(4)$	σ_x	$S_6(4)$	$S_6(1)$	σ_y	i	$S_6(3)$	$S_6^5(2)$	σ_z	$S_6(2)$	$S_6^5(3)$
C_{2yz}	σ_z	$S_6^5(1)$	$S_6(1)$	σ_y	$S_6(2)$	$S_6^5(2)$	$S_6^5(3)$	i	$S_6(4)$	$S_6^5(4)$	σ_x	$S_6(3)$
C_{2zx}	$S_6^5(3)$	σ_x	$S_6(3)$	$S_6(1)$	σ_z	$S_6^5(1)$	$S_6(2)$	$S_6^5(4)$	i	$S_6(4)$	$S_6^5(2)$	σ_y
$C_{2x\bar{y}}$	$S_6(2)$	$S_6^5(2)$	σ_y	$S_6(3)$	$S_6^5(3)$	σ_x	σ_z	$S_6(4)$	$S_6^5(4)$	i	$S_6^5(1)$	$S_6(1)$
$C_{2y\bar{z}}$	σ_y	$S_6(3)$	$S_6^5(4)$	σ_z	$S_6(4)$	$S_6^5(3)$	$S_6^5(2)$	σ_x	$S_6(2)$	$S_6(1)$	i	$S_6^5(1)$
$C_{2z\bar{x}}$	$S_6^5(4)$	σ_z	$S_6(2)$	$S_6^5(2)$	σ_x	$S_6(4)$	$S_6(3)$	$S_6^5(3)$	σ_y	$S_6^5(1)$	$S_6(1)$	i
S_{4x}^3	E	$C_3^2(2)$	$C_3(3)$	C_{2x}	$C_3(1)$	$C_3(4)$	$C_3^2(1)$	C_{2z}	$C_3^2(3)$	$C_3(2)$	C_{2y}	$C_3^2(4)$
S_{4y}^3	$C_3(2)$	E	$C_3(1)$	$C_3(4)$	C_{2y}	$C_3^2(3)$	$C_3^2(4)$	$C_3^2(1)$	C_{2x}	$C_3^2(2)$	$C_3(3)$	C_{2z}
S_{4z}^3	$C_3^2(3)$	$C_3^2(1)$	E	$C_3^2(2)$	$C_3(4)$	C_{2z}	C_{2x}	$C_3(1)$	$C_3(3)$	C_{2y}	$C_3^2(4)$	$C_3(2)$
S_{4x}	C_{2x}	$C_3^2(4)$	$C_3(2)$	E	$C_3^2(3)$	$C_3^2(1)$	$C_3(4)$	C_{2y}	$C_3(1)$	$C_3(3)$	C_{2z}	$C_3^2(2)$
S_{4y}	$C_3^2(1)$	C_{2y}	$C_3^2(4)$	$C_3(3)$	E	$C_3^2(2)$	$C_3(1)$	$C_3(2)$	C_{2z}	$C_3^2(3)$	$C_3(4)$	C_{2x}
S_{4z}	$C_3^2(4)$	$C_3(3)$	C_{2z}	$C_3(1)$	$C_3(2)$	E	C_{2y}	$C_3^2(2)$	$C_3^2(1)$	C_{2x}	$C_3^2(3)$	$C_3(4)$
σ_{xy}	$C_3(1)$	$C_3(4)$	C_{2x}	$C_3^2(4)$	$C_3^2(1)$	C_{2y}	E	$C_3^2(3)$	$C_3(2)$	C_{2z}	$C_3^2(2)$	$C_3(3)$
σ_{yz}	C_{2z}	$C_3(1)$	$C_3^2(1)$	C_{2y}	$C_3^2(2)$	$C_3(2)$	$C_3(3)$	E	$C_3^2(4)$	$C_3(4)$	C_{2x}	$C_3^2(3)$
σ_{zx}	$C_3(3)$	C_{2x}	$C_3^2(3)$	$C_3^2(1)$	C_{2z}	$C_3(1)$	$C_3^2(2)$	$C_3(4)$	E	$C_3^2(4)$	$C_3(2)$	C_{2y}
$\sigma_{x\bar{y}}$	$C_3^2(2)$	$C_3(2)$	C_{2y}	$C_3^2(3)$	$C_3(3)$	C_{2x}	C_{2z}	$C_3^2(4)$	$C_3(4)$	E	$C_3(1)$	$C_3^2(1)$
$\sigma_{y\bar{z}}$	C_{2y}	$C_3^2(3)$	$C_3(4)$	C_{2z}	$C_3^2(4)$	$C_3(3)$	$C_3(2)$	C_{2x}	$C_3^2(2)$	$C_3^2(1)$	E	$C_3(1)$
$\sigma_{z\bar{x}}$	$C_3(4)$	C_{2z}	$C_3^2(2)$	$C_3(2)$	C_{2x}	$C_3^2(4)$	$C_3^2(3)$	$C_3(3)$	C_{2y}	$C_3(1)$	$C_3^2(1)$	E

Chapter 3
Orthogonality Theorem and Character Tables

3.1 Introduction

In the last two chapters, we studied the basic concepts related to the symmetry elements/operations, matrices, and point groups for both crystallographic and molecular cases. They are required to form the basis to understand the symmetry representations.

In this chapter, we are going to discuss the concepts of reducible/irreducible representations and the basic features of orthogonality theorem. The properties of irreducible representations obtained from the theorem will be used as rules to construct the character tables, in general for non-abelian groups. We will also discuss separately the procedures to construct the character tables for abelian and cyclic groups of finite/infinite order.

3.2 Representations

The sets of matrices for different symmetry operations of different point groups, obeying the laws of group multiplication table, are said to form representations. Further, based on the character of the matrix, a representation is said to be either reducible or irreducible. A representation is called reducible if it is possible to find a similarity transformation matrix which is able to convert all the matrices in the representation such that the resulting matrices can be blocked into smaller matrices in the form as given by Eq. 3.1.

$$\Gamma(A) = \begin{pmatrix} \Gamma_1(A) & & & & \\ & \Gamma_2(A) & & & \\ & & \cdot & & \\ & & & \cdot & \\ & & & & \Gamma_n(A) \end{pmatrix} a \qquad (3.1)$$

M. A. Wahab, *Symmetry Representations of Molecular Vibrations*, Springer Series in Chemical Physics 126, https://doi.org/10.1007/978-981-19-2802-4_3

In other words, a reducible representation $\Gamma(R)$ is a linear combination of a number of irreducible representations $\Gamma(IR)$ of the point group under consideration. That is

$$\Gamma(R) = a_1\Gamma_1(IR) + a_2\Gamma_2(IR) + \ldots a_n\Gamma_n(IR)$$

where they are numbered in increasing order of their dimensions and the a_n are 1, 2, 3 ... etc., or zero.

These irreducible representations are either a set of 1×1 matrices or a set of matrices which cannot be reduced further by similarity transformations.

On the other hand, if it is not possible to find a similarity transformation matrix which can reduce the matrices of representation under consideration, then the representation is said to be irreducible. All one-dimensional representations are always irreducible.

Example 3.1 Show that the similarity transformation changes the reducible representation matrix A of larger dimensions into a matrix B which is the sum of irreducible representation matrices of smaller dimensions.

Solution: Let us consider the reducible representation matrix A of dimensions 3×3. That is,

$$A = \begin{pmatrix} a_{11} & a_{12} & a_{13} \\ a_{21} & a_{22} & a_{23} \\ a_{31} & a_{32} & a_{33} \end{pmatrix}$$

Further, let the similarity transformation matrix is Q and its inverse as Q^{-1}, then

$$Q^{-1}AQ = B$$

$$= Q^{-1} \begin{pmatrix} a_{11} & a_{12} & a_{13} \\ a_{21} & a_{22} & a_{23} \\ a_{31} & a_{32} & a_{33} \end{pmatrix} Q$$

$$= \begin{pmatrix} b_{11} & b_{12} & 0 \\ b_{21} & b_{22} & 0 \\ 0 & 0 & b_{33} \end{pmatrix}$$

$$= B$$

$\Longrightarrow$ The matrix B is regarded to be built up from matrices B_1 (of dimensions 2) and B_2 (of dimension 1), where

$$B_1 = \begin{pmatrix} b_{11} & b_{12} \\ b_{21} & b_{22} \end{pmatrix} \text{ and } B_2 = (b_{33}).$$

Thus, the matrix B is the sum of two irreducible representation matrices B_1 and B_2.

3.3 Orthogonality Theorem

The orthogonality theorem deals with the elements of matrices constituting the irreducible representations of a point group. A complete proof of this theorem may be found in any textbook on group theory. However, here we are interested in its properties which can be taken as rules to construct the character table of a point group.

For the purpose of a qualitative idea of the theorem, let us consider two irreducible representations i and j of a point group. Let l_i and l_j be the dimensions of these representations, h be the order (total number of independent symmetry elements) of the point group, and R a particular symmetry operation in the group, respectively. Further, suppose that $[\Gamma_i(R)]_{mn}$ is an element in the mth row and nth column of a matrix in the ith irreducible representation. Similarity, its complex conjugate element $[\Gamma_j(R)]^*_{m'n'}$ is in the m'th row and n'th column of a matrix in the jth irreducible representation. The two matrix elements are related to h, l_i, and l_j through the following orthogonality relation

$$\sum_R [\Gamma_i(R)]_{mn}[\Gamma_j(R)]^*_{m'n'} = \frac{h}{\sqrt{l_i}\sqrt{l_j}}\delta_{ij}\delta_{mm'}\delta_{nn'} \tag{3.2}$$

where δ_{ij}, $\delta_{mm'}$ and $\delta_{nn'}$ are Kronecker delta symbols, and

$$\delta_{ij} = 0 \text{ for i} \neq \text{j}$$
$$= 1 \text{ for i} = \text{j}$$

Therefore, depending on the values of i, j, m, m', n, and n', the following three special cases of this theorem arise:

Case I: i $\neq$ j, m $=$ m', and n $=$ n'

In this case, $[\Gamma_i(R)]_{mn}$ and $[\Gamma_j(R)]_{mn}$ represent two real elements in the mth row and nth column of the matrix for the operation R in the irreducible representations i and j. This consideration modifies Eq. 3.2 as

$$\sum_R [\Gamma_i(R)]_{mn}[\Gamma_j(R)]_{m'n'} = \frac{h}{\sqrt{l_i}\sqrt{l_j}}\delta_{ij} = 0 \tag{3.3}$$

To understand its meaning, Eq. 3.3 can be applied to two irreducible representations of a group containing the operations E and A such that

$$\text{(a)} \quad E = \begin{pmatrix} E_{11} & E_{12} \\ E_{21} & E_{22} \end{pmatrix} \quad A = \begin{pmatrix} A_{11} & A_{12} \\ A_{21} & A_{22} \end{pmatrix}$$

$$\text{(b)} \quad E = \begin{pmatrix} E_{xx} & E_{xy} \\ E_{yx} & E_{yy} \end{pmatrix} \quad A = \begin{pmatrix} A_{xx} & A_{xy} \\ A_{yx} & A_{yy} \end{pmatrix}$$

they provide an important result given by the equation

$$E_{11}E_{xx} + A_{11}A_{xx} = 0 \tag{3.4}$$

Case II: $i = j$, $m \neq m'$ $n \neq n'$

In this case, the elements $[\Gamma_i(R)]_{mn}$ are in the mth row and nth column and $\left[\Gamma_j(R)\right]_{m'n'}$ in the m'th row and n'th column of a matrix for operation R, both in the ith irreducible representation. They modify Eq. 3.2 as

$$\sum_R [\Gamma_i(R)]_{mn} \left[\Gamma_j(R)\right]_{m'n'} = \frac{h}{l_i} \delta_{mm'} \delta_{nn'} \tag{3.5}$$

Case III: $i = j$, $m = m'$ and $n = n'$

In this case, the element $[\Gamma_i(R)]_{mn}$ is in the mth row and nth column of a matrix for operation in the ith irreducible representation. This modifies Eq. 3.2 as

$$\sum_R [\Gamma_i(R)]^2_{mn} = \frac{h}{l_i} \tag{3.6}$$

3.4 Properties of Irreducible Representation

Orthogonality theorem provides a number of important relationships related to irreducible representations of point groups and character of their representations. Some of these are very important from crystallographic point of view. They can be used as rules for constructing character tables for crystallographic and molecular point groups. We present them with simple proof.

1. The number of irreducible representations of a group is equal to the number of conjugate classes (say k) in the group.

Proof According to Orthogonality theorem, we have

$$\sum_{p=1}^{k} g_p \chi_i(R_p) \chi_j(R_p) = h\delta_{ij}$$

$$= 0 \text{ for } i \neq j$$
$$= h \text{ for } i = j \qquad (3.7)$$

where k represents the number of classes in the point group, g_p refers to the number of symmetry operations in a class, R_p is the symmetry operation in the pth class, $\chi_i(R_p)$ and $\chi_j(R_p)$ are the characters of the matrix for operation R_p in the ith and jth irreducible representations, respectively.

Equation 3.7 implies that the k quantities, $\chi_l(R_p)$, in each representation Γ_l behave like the components of k-dimensional vector and these k vectors are mutually orthogonal. Since only k number of k-dimensional vectors can be mutually orthogonal, therefore there can be no more than k irreducible representations in the group which has k number of classes.

2. The sum of the squares of the dimensions of all the irreducible representations of a group is equal to the order of the group, i.e.,

$$\sum_{i=1}^{k} l_i^2 = h \qquad (3.8)$$

where l_i can assume only positive integer values other than zero.

Proof A complete proof of this is quite lengthy; however, it is easy to show that

$$\sum l_i^2 \leq h$$

We know that in a matrix of order l, the number of matrix elements is l^2. Thus, each irreducible representation Γ_i will provide l^2 h-dimensional vectors. The basic theorem requires the set of $l_1^2 + l_2^2 + l_3^2 + \cdots$ vectors to be mutually orthogonal. Since, there can be no more than h orthogonal h-dimensional vectors, the sum $l_1^2 + l_2^2 + l_3^2 + \cdots$ may not exceed h. This implies that

$$\sum l_i^2 \leq h$$

Further, we know that the character representation of E in the ith irreducible representation, $\chi_i(E)$ is equal to the order of representation. Therefore, the above rule can also be written as

$$\sum_{i=1}^{k} [\chi_i(E)]^2 = h \qquad (3.9)$$

3. In a given representation (reducible or irreducible), the characters of all matrices (symmetry operations) belonging to the same class are identical.

Proof Since, all symmetry elements (operations) in a class are conjugate to one another, therefore all matrices corresponding to these operations in the class in any representation (reducible or irreducible) must also be conjugate. But, we have shown in Sect. 1.5 that the conjugate matrices will have identical characters.

4. The sum of the squares of the characters in any irreducible representation is equal to the order of the group, i.e.,

$$\sum_R [\chi_i(R)]^2 = h \tag{3.10}$$

where R is any operation of the group and the summation is taken over all operations.

Proof We can simplify Eq. 3.10 by using the fact that all operations in a class have the same character (from point 3, mentioned above). Therefore, considering Eq. 3.7 again for $i = j$, we can write

$$\sum_{p=1}^{k} g_p [\chi_i(R_p)]^2 = h \quad \text{for } i = j \tag{3.11}$$

The summation in Eq. 3.11 is taken over all classes of operations.

5. The character of the distinct irreducible representations must be orthogonal, i.e.,

$$\sum_{p=1}^{k} g_p \chi_i(R_p) \chi_j(R_p) = h\delta_{ij}$$
$$= 0 \text{ for } i \neq j$$
$$= h \text{ for } i = j \tag{3.12}$$

Proof Equation 3.12 can be written separately for $i \neq j$ as

$$\sum_{p=1}^{k} g_p \chi_i(R_p) \chi_j(R_p) = 0 \tag{3.13}$$

In this case too, the summation in Eq. 3.13 is taken over all classes of operations.

Verify the rules obtained from orthogonality theorem using the character table for the point group $C_{2h}(2/m)$.

Solution: *Given*: Point group is $C_{2h}(2/m)$, it contains four symmetry operations: E, C_2, i and σ_h order of the group, $g = 4$.

Since all the symmetry operations of this point group are their own inverses, therefore they belong to separate classes. This implies that four symmetry operations form four different classes. The character table of the point group C_{2h} (2/m) is

C_{2h}	E	C_2	i	σ_h
A_g	1	1	1	1
B_g	1	-1	1	-1
A_u	1	1	-1	-1
B_u	1	-1	-1	1

Now, let us verify the five rules one by one.

1. Since there are four classes, therefore there will be four irreducible representations in the point group $C_{2h}(2/m)$. Rule 1 is verified to be true.
2. All the four irreducible representations are one-dimensional. This indicates that there are four positive integers which satisfy the relation

$$\sum_{i=1}^{4} l_i^2 = l_1^2 + l_2^2 + l_3^2 + l_4^2 = 4$$

This can be only if $l_1 = l_2 = l_3 = l_4 = 1$.

Similarly, one can easily verify that

$$\sum_{i=1}^{4}[\chi_i(E)]^2 = 1^2 + 1^2 + 1^2 + 1^2 = 4$$

The rule 2 is verified to be true.

3. All the symmetry operations in the point group C_{2h} belong to separate classes; therefore, their corresponding characters are expected to be different. In this case, we have

$$\chi(E) = 3, \chi(C_2) = -1, \chi(i) = -3 \chi(\sigma_h) = 1$$

4. To verify rule 4, let us check the equation

$$\sum_{p=1}^{4} g_p[\chi_i(R_p)]^2 = h = 4$$

In the point group C_{2h}, each class has only one member in it, therefore

$$g_1 = g_2 = g_3 = g_4 = 1$$

Taking this into account, the above equation for A_u representation can be written as

$$\sum_{p=1}^{4} g_p[\chi_{\Gamma_2}(R_p)]^2 = 1(1)^2 + 1(1)^2 + 1(-1)^2 + 1(-1)^2 = 4 = h \text{ (order of the group)}$$

Thus rule 4 is verified to be true.

5. To check the orthogonality condition of any two representations, let us use the equation

$$\sum_{p=1}^{4} g_p \chi_i(R_p) \chi_j(R_p) = 0$$

To get the product of B_g and A_u representations, we have

$$\sum_{p=1}^{4} g_p \chi_{\Gamma_2}(R_p) \chi_{\Gamma_3}(R_p) = 1(1 \times 1) + 1(-1 \times 1) + 1(1 \times -1) + 1(-1 \times -1)$$

$$= 1 - 1 - 1 + 1 = 0$$

Similarly, the product of other pairs can also be verified.

Example 3.3 Verify the rules obtained from orthogonality theorem using the character table for the point group $C_{3v}(3m)$.

Solution: *Given*: Point group is $C_{3v}(3m)$, it contains six symmetry operations: E, C_3, C_3^2, σ_x, σ_y, and σ_{xy} $\Rightarrow$ order of the group, $g = 6$.

Here, C_3 or C_3^2 are inverses of each other and similarity transform on C_3 or C_3^2 will show that they belong to the same class. All other operations are their own inverses. However, again similarity transform σ_x, σ_y, or σ_{xy} will show that they belong to the same class. This implies that six operations form three classes. They are E, $2C_3$, $3\sigma_v$.

The character table of the point group $C_{3v}(3m)$ is

C_{3v}	E	$2C_3$	$3\sigma_v$
A_1	1	1	1
A_2	1	1	-1
E	2	-1	0

Now, let us verify the five rules one by one.

Since there are three classes, therefore there will be three irreducible representations in the point group C_{3v}. Rule 1 is verified to be true.

1. Out of the three irreducible representations, two are one-dimensional and one is two-dimensional. This indicates that there are three integers which satisfy the relation

$$\sum_{i=1}^{3} l_i^2 = l_1^2 + l_2^2 + l_3^2 = 6$$

This can be only if $l_1 = l_2 = 1$ and $l_3 = 2$.
Similarly, one can easily verify that

$$\sum_{i=1}^{3} [\chi_i(E)]^2 = 1^2 + 1^2 + 2^2 = 6$$

Rule 2 is verified to be true.

2. In the group C_{3v}, there are three classes; E, $2C_3$ and $3\sigma_v$ (where within the same class all operation matrices will have same character). In this case, we have

$$\chi(E) = 3, \quad \chi(C_3) = 0, \quad \chi(\sigma_v) = 1$$

3. To verify rule 4, let us check the equation

$$\sum_{p=1}^{3} g_p [\chi_i(R_p)]^2 = h = 6$$

In the group $C_{3v} : g_1 = 1, g_2 = 2, g_3 = 3$
Taking this into account, the above equation for A_2 representation can be written as

$$\sum_{p=1}^{3} g_p [\chi_{\Gamma_2}(R_p)]^2 = 1(1)^2 + 2(1)^2 + 3(-1)^2$$

$$= 1 + 2 + 3 = 6 = \text{order of the group}$$

Similarly, one can verify for other representations, let us do it for E representation

$$\sum_{p=1}^{3} g_p [\chi_{\Gamma_3}(R_p)]^2 = 1(2)^2 + 2(-1)^2 + 3(0)^2$$

$$= 4 + 2 + 0 = 6 = \text{order of the group}$$

The rule 4 is verified to be true.

4. To check the orthogonality condition of any two representations, let us use the equation

$$\sum_{p=1}^{3} g_p \chi_i(R_p) \chi_j(R_p) = 0$$

To get the product of A_2 and E representations, we have

$$\sum_{p=1}^{3} g_p \chi_{\Gamma_2}(R_p) \chi_{\Gamma_3}(R_p) = 1(1 \times 2) + 2(1 \times -1) + 3(-1 \times 0)$$

$$= 2 - 2 + 0 = 0$$

Similarly, the product of other pairs can also be verified.

Example 3.4 Verify the rules derived from orthogonality theorem using the character table for the point group $T_d\left(\overline{4}3m\right)$.

Solution: *Given*: point group is $T_d\left(\overline{4}3m\right)$, contains twenty four symmetry operations:
E, C_{2x}, C_{2y}, C_{2z}, $C_3(1)$, $C_3(2)$, $C_3(3)$, $C_3(4)$, $C_3^2(1)$, $C_3^2(2)$, $C_3^2(3)$, $C_3^2(4)$, S_{4x}, S_{4y}, S_{4z}, S_{4x}^3, S_{4y}^3, S_{4z}^3, σ_{xy}, σ_{yz}, σ_{zx}, $\sigma_{x\overline{y}}$, $\sigma_{y\overline{z}}$, $\sigma_{z\overline{x}}$.
$\Longrightarrow$ Order of the group, $g = 24$.

However, $C_3(1)$, $C_3(2)$, $C_3(3)$, $C_3(4)$, $C_3^2(1)$, $C_3^2(2)$, $C_3^2(3)$, $C_3^2(4)$ form a class.
Similarly, C_{2x}, C_{2y}, C_{2z}; S_{4x}, S_{4y}, S_{4z}, S_{4x}^3, S_{4y}^3, S_{4z}^3; σ_{xy}, σ_{yz}, σ_{zx}, $\sigma_{x\overline{y}}$, $\sigma_{y\overline{z}}$, $\sigma_{z\overline{x}}$ form separate classes. Therefore, in all there are five classes. They are E, $8C_3$, $3C_2$, $6S_4$, $6\sigma_d$. The character table of the point group $T_d\left(\overline{4}3m\right)$ is

T_d	E	$8C_3$	$3C_2$	$6S_4$	$6\sigma_d$
A_1	1	1	1	1	1
A_2	1	1	1	-1	-1
E	2	-1	2	0	0
T_1	3	0	-1	1	-1
T_2	3	0	-1	-1	1

Now, let us verify the five rules, one by one.

1. Since there are five classes, therefore there will be five irreducible representations in the point group $T_d(\overline{4}3m)$. Rule 1 is verified to be true.
2. Out of the five irreducible representations, two are of one-dimensional, one is two-dimensional and the remaining two are three-dimensional. This indicates that there are five integers which satisfy the relation

$$\sum_{i=1}^{5} l_i^2 = l_1^2 + l_2^2 + l_3^2 + l_4^2 + l_5^2 = 24$$

Solution of this is $l_1 = l_2 = 1$, $l_3 = 2$, $l_4 = l_5 = 3$.
Similarly, one can easily verify that

$$\sum_{i=1}^{5} [\chi_i(E)]^2 = 1^2 + 1^2 + 2^2 + 3^2 + 3^2$$

$$= 1 + 1 + 4 + 9 + 9 = 24$$

Rule 2 is verified to be true.

3. In the point group T_d, there are five classes; E, $8C_3$, $3C_2$, $6S_4$, $6\sigma_d$ (where within the same class all operation matrices will have same character). In this case, we have

$$\chi(E) = 3, \ \chi(C_3) = 0, \ \chi(C_2) = -1, \ \chi(S_4) = -1, \ \chi(\sigma_d) = 1$$

4. To verify rule 4, let us check the equation

$$\sum_{p=1}^{5} g_p[\chi_i(R_p)]^2 h = 24$$

In the group T_d: $g_1 = 1$, $g_2 = 1$, $g_3 = 2$, $g_4 = 3$, $g_5 = 3$.
Taking this into account, the above equation for E representation can be written as

$$\sum_{p=1}^{5} g_p[\chi_i(R_p)]^2 = 1(2)^2 + 8(-1)^2 + 3(2)^2 + 6(0)^2 + 6(0)^2$$

$$= 4 + 8 + 12 = 24 = \text{order of the group}$$

Similarly for T_2 representations, we have

$$\sum_{p=1}^{5} g_p[\chi_i(R_p)]^2 = 1(3)^2 + 8(0)^2 + 3(-1)^2 + 6(-1)^2 + 6(1)^2$$

$$= 9 + 0 + 3 + 6 + 6 = 24 = h$$

5. To check the orthogonality condition of any two representations, let us use the equation

$$\sum_{p=1}^{5} g_p \chi_i(R_p) \chi_j(R_p) = 0$$

To get the product of A_2 and E representations, we have

$$\sum_{p=1}^{5} g_p \chi_{\Gamma_2}(R_p) \chi_{\Gamma_3}(R_p) = 1(1 \times 2) + 8(1 \times -1) + 3(1 \times 2) + 6(1 \times 0) + 6(-1 \times 0)$$

$$= 2 - 8 + 6 + 0 + 0 = 0$$

Similarly, the product $T_1 \times T_2$ will give us

$$\sum_{p=1}^{5} g_p \chi_{\Gamma_4}(R_p) \chi_{\Gamma_5}(R_p) = 1(3 \times 3) + 8(0 \times 0) + 3(-1 \times -1) + 6(1 \times -1) + 6(-1 \times 1)$$

$$= 9 + 0 + 3 - 6 - 6 = 0$$

Hence, rule 5 is verified to be true.

3.5 Parts of a Character Table

A character table can be conveniently divided into five parts (I–V) on the basis of the type of information they contain.

Top corner of part I contains a single point group (or isomorphic point groups) generally in Schoenflies notation whose character table is to be prepared. Below this, the Mulliken symbols of different irreducible representations are written, the number of which is equal to the number of classes (consisting of the symmetry elements/operations) in the given point group.

On the top row under part II, the symmetry elements/operations related to the given point group are listed in terms of their classes. Below this, the character values for all elements/operations against different irreducible representations are written. This is the main part of the character table.

Parts III, IV, and V are basically dealing with the different transformation properties that are relevant for crystals/molecules. They are of great interest while applying group theory to physical or chemical problems. Therefore, it is necessary to list them properly in the character table.

Part III gives the detail of the transformation properties about the Cartesian coordinates x, y, z and the rotation along the x, y, z axes, etc., correspond to infrared (IR) activity of the particular species. On the other hand, part IV gives the detail of the transformation properties about the binary products of coordinate, (related to d-orbitals), such as $x^2 - y^2$, $x^2 + y^2 + z^2$ and squares of coordinates such as x^2, y^2, and z^2 correspond to Raman activity of the particular species.

Similarly, part V gives the detail of the transformation properties about the triple products of coordinates (of f-orbitals) such as z^3, xyz, xz^2, yz^2, $x(x^2 - 3y^2)$, $y(3x^2 - y^2)$, and $z(x^2 - y^2)$.

All, the five parts of the character table corresponding to a given point group can be presented in the following tabular form as shown in Table 3.1.

Table 3.1 Various parts of a character table

I	II	III	IV	V
Point group	List of symmetry elements/ operations in terms of number of classes	Transformation Properties		
Irreducible representation $\Gamma_2\Gamma_1$ Γ_n	Character values	Position coordinates (x, y, z) Translation vectors (T_x, T_y, T_z) Rotation vectors (R_x, R_y, R_z) Orbitals P (P_x, P_y, P_z) (s and p-orbitals basis operations)	Binary product of coordinates s-orbital $x^2 + y^2$ (2D) planar molecule $x^2 + y^2 + z^2$ (3D) d-orbitals $x^2 - y^2$, xy xz and yz Polarizability α_{xx}, α_{yy}, α_{zz}, α_{xy}, etc, (s and p-orbitals basis function)	Cubic or triple product of coordinates of (f-orbitals) such as z^3, xyz, xz^2, yz^2, $x(x^2 - 3y^2)$, $y(3x^2 - y^2)$, $z(x^2 - y^2)$ (s and f-orbitals basis function)

3.6 Characters of Representations in Point Groups

In Sect. 2.3, we observed that the point groups can broadly be classified as abelian or non-abelian, where abelian includes the cyclic groups. Let us discuss the nature of character representations in them separately in brief. This will be helpful in constructing the character tables for different point groups under each category.

Abelian Group

We know that in an abelian group, all symmetry elements/operations satisfy the condition AB = BA (i.e., they all commute with each other). Since each symmetry element/operation represents a class in itself (hence belongs to an irreducible representation), the number of classes in an abelian group is equal to the order of the group. Hence, all the irreducible representations must be one-dimensional. This makes the problem of finding the irreducible representations in an abelian group simpler. Further, we know that the order of any symmetry element/operation of a group is finite and the characters of all symmetry elements/operations are roots of unity.

Based on the above, one can easily conclude that the matrix representations corresponding to the symmetry element/operation A of a given (point) group is simply the character of the group, i.e.,

$$D(A) = \chi(A)$$

For crystallographic (point) groups, the order of the group (or the order of the symmetry element/operation) is always finite, then

$$A^h = E$$

Therefore,

$$D(A) = \chi(A) = \exp\left(\frac{2\pi il}{h}\right), \quad (l = 1, 2, \ldots h) \tag{3.14}$$

Thus, by knowing the order of the group and taking different values of l, one can easily construct the complete character table of the required group. However, the character table of higher order (point) groups can also be obtained by taking direct product of two of its subgroups.

Cyclic Groups

We know that all the elements/operations of a cyclic group are generated/by one element/operation such as A, A^2, A^3, ... $A^n(= E)$, where n is the order of the group. Also all cyclic groups are abelian, but the converse is not true. Thus, a cyclic group of order n will contain n number of classes and hence there will be n number of irreducible representations.

In order to obtain the character table of a cyclic group, let us start with the identity condition using Schoenflies notation, i.e.,

$$C_n^n = E$$
$$\text{So that } \chi(C_n)^n = \chi(E) = 1$$

This indicates that the characters of all irreducible representations in the first column are 1. Further, it follows that other characters $\chi(C_n)$ may be one of the nth roots of unity. Such numbers are in general complex of the form

$$\chi(C_n)^n = \left(e^{i\theta}\right)^n = 1 = e^{(2\pi il)}$$

where $l = 0, 1, 2, 3.......(n - 1)$ is the number of irreducible representations in the cyclic group and $\theta = \frac{2\pi l}{n}$, n is the order of the cyclic group.

Therefore, the character for a general rotation is given by

$$\chi(C_n)^n = e^{\left(\frac{2\pi il}{n}\right)} \tag{3.15}$$

Here, if one complex number is defined as

$$\varepsilon^l = e^{\left(\frac{2\pi i l}{n}\right)} = \cos\left(\frac{2\pi i l}{n}\right) + i \sin\left(\frac{2\pi i l}{n}\right)$$

Then the other number is

$$\varepsilon^{-l} = e^{-\left(\frac{2\pi i l}{n}\right)} = \cos\left(\frac{2\pi i l}{n}\right) - i \sin\left(\frac{2\pi i l}{n}\right)$$

$$\text{So that } \varepsilon^l + \varepsilon^{-l} = 2 \cos\left(\frac{2\pi i l}{n}\right)$$

For a totally symmetric representation, Γ_1 is obtained by substituting $l = 0$ in Eq. 3.15 this gives us

$$\chi(C_n) = e^0 = 1$$

$\Longrightarrow$ All characters in the first row will be equal to 1. Other characters can be determined by taking powers of $\chi(C_n)$. For example,

$$\chi(C_n^m) = e^{\left(\frac{2\pi i l m}{n}\right)} \tag{3.16}$$

where m represents the number of operations carried out (the power on the generating symmetry elements), i.e.,

$$m = 1, 2, 3, \ldots n - 1$$

Based on the information obtained after substituting the values of l, m, and n, a general form of the character table for a cyclic group C_n can be written as shown in Table 3.2.

The character table for a given cyclic group can be obtained either directly by using the general table mentioned above or by using ab-initio method.

Table 3.2 General form of character table for cyclic groups

C_n	$E(= C_n^n)$	C_n^1	C_n^2	C_n^3	...	C_n^{n-1}
Γ_1	1	1	1	1	...	1
Γ_2	1	ε	ε^2	ε^3	...	ε^{n-1}
...	...	...	...	...	...	...
...	...	...	...	...	...	...
Γ_m	1	ε^{m-1}	$\varepsilon^{2(m-1)}$	$\varepsilon^{3(m-1)}$	...	$\varepsilon^{(n-1)(m-1)}$
Γ_n	1	ε^{n-1}	$\varepsilon^{2(n-1)}$	$\varepsilon^{3(n-1)}$	...	$\varepsilon^{(n-1)^2}$

Non-abelian Groups

Non-abelian groups are treated in general way and their character tables can be constructed by using the rules derived from orthogonality theorem. However, a number of character tables under this category can also be obtained from the direct product of two smaller groups, which are the subgroups of a super group.

Direct Product Group Method

A group G is said to be the direct product of two (or more) of its subgroups, say $G_a = E, A_2, \ldots A_{ha}$ and $G_b = E, B_2, \ldots B_{hb}$, if

(i)   All elements of G_a commute with all elements of G_b.
(ii) Every element of G is expressible in one and only one way, i.e., as a product of an element of G_a by an element of G_b. Therefore,

$$G = G_a \times G_b = E, A_2, \ldots A_{ha}, B_2, A_2 B_2, \ldots A_{ha} B_2, \ldots, A_{ha} B_{hb}$$

forms a group of $h_a h_b$ elements, assuming that the only common element in the group is the identity element and h_a and h_b are the orders of the subgroups G_a and G_b, respectively.

Following the above-mentioned criteria, the character table of a point group can be obtained by taking direct product of its two subgroups. The possible point groups and their product subgroups are given below.

n—odd	n—even
$C_{nh} = C_n \times C_s$	$C_{nh} = C_n \times C_i$
$C_{ni} = C_n \times C_i$	$D_{nh} = D_n \times C_i$
$D_{nh} = D_n \times C_s$	$T_h = T \times C_i$
$C_{nv} = C_n \times C_s$	$O_h = O \times C_i$

3.7 Construction of Character Tables

We know that there are 32 crystallographic point groups that are possible in 3D, so that there should be in principle the same number of representations. However, it has been found that there are some sets of these representations that exhibit identical group multiplication tables; that is they are isomorphic. Taking this into account, we need to construct only 18 distinct crystallographic character tables. This number can further be reduced by using the direct product group method.

In the following, we shall discuss about the construction of character tables of all representative cases both from crystallographic and molecular point groups without assigning the Mulliken symbols and transformation properties of different irreducible representations. However, the character tables of all crystallographic point groups and a number of important molecular point groups complete in all respects will be provided at the end of this chapter as an appendix.

1. **The point group $C_1(1)$**

This is a trivial group consisting of only the identity elements E. Therefore,

$$D(E) = \chi(E) = \exp(2\pi i l) = 1 \quad (l = 0)$$

Thus, it has only one irreducible representation. The character table is.

E	E
Γ	1

2. **The Point Groups $C_2(2)$, $C_s(m)$ and $C_i(\overline{1})$**

These three point groups exhibit identical group multiplication table and hence are isomorphic. Therefore, they must have identical character table. They are cyclic (abelian) groups of order 2 and hence will have two irreducible representations (each of one-dimension). Hence,

$$D(A) = \chi(A) = exp\left(\frac{2\pi i l}{h}\right) = \pm 1, (l = 0, 1)$$

The character table is

C_i C_s C_2	E	i σ C_2
Γ_1	1	1
Γ_2	1	-1

3. **The Point Groups $C_{2v}(mm2)$, $C_{2h}\left(\frac{2}{m}\right)$, and D_2 (222)**

These three point groups also exhibit identical group multiplication table and hence are isomorphic. Further, they are abelian groups of order 4, hence they will have four irreducible representations (each of one-dimension). Looking at three non-identity elements in each of these point groups, we observe that the product of any two yields the third one, hence all characters are either $+1$, or two are $+1$ and other two are -1.

Following these criteria, the required character table is

C_{2v}		C_2	$\sigma(xz)$	$\sigma(yz)$
C_{2h}	E	C_2	i	σ_h
D_2		C_{2z}	C_{2y}	C_{2x}
Γ_1	1	1	1	1
Γ_2	1	1	-1	-1
Γ_3	1	-1	1	-1
Γ_4	1	-1	-1	1

Alternative Method

We can also obtain the character table of the point group C_{2h} (and its other two isomorphic partners) by taking the direct product of its two subgroups C_2 and C_i. Both the subgroups are of order 2, so that the order of the resulting C_{2h} point group is $2 \times 2 = 4$. The symmetry elements of the C_{2h} group will be

$$C_2 \times C_i = (E, C_2) \times (E, i) = E, C_2, i, \sigma_h$$

Therefore, the character table of C_{2h} group will be

C_2	E	C_2
Γ_1	1	1
Γ_2	1	-1

$\times$

C_i	E	i
Γ_1	1	1
Γ_2	1	-1

$=$

$C_2 * C_i = C_{2h}$	E	C_2	i	σ_h
Γ_1	1	1	1	1
Γ_2	1	-1	1	-1
Γ_3	1	1	-1	-1
Γ_4	1	-1	-1	1

Interchanging the position of Γ_2 and Γ_3, it represents the same character table as above. Following a similar procedure, the character tables of $C_{4h}\left(\frac{4}{m}\right) = C_4 \times C_i$ and $C_{6h}\left(\frac{6}{m}\right) = C_6 \times C_i$ can be obtained easily.

4. The Point Group $C_3(3)$

This is a cyclic (abelian) group of order 3, hence it will have three irreducible representations (each of one-dimension). Further, as we know that a cyclic group of order ≥ 3 involves complex numbers in its irreducible representation, the tentative form of the character table obtained from the general table (Table 3.2) can be written as.

C_3	E	C_3	C_3^2
Γ_1	1	1	1
Γ_2	1	ε	ε^2
Γ_3	1	ε^2	ε

$$\text{where } \varepsilon = \exp\left(\frac{2\pi i}{3}\right) \text{ and } \varepsilon^* = \varepsilon^2$$

In this case, the orthogonality relation reduces to $1 + \varepsilon + \varepsilon^2 = 0$, a special case of the orthogonality theorem that the sum of the nth roots of unity is zero.

Substituting the value of ε and ε^2, the character table of C_3 becomes

C_3	E	C_3	C_3^2
Γ_1	1	1	1
Γ_2	1	$-\frac{1}{2}+\frac{\sqrt{3}i}{2}$	$-\frac{1}{2}-\frac{\sqrt{3}i}{2}$
Γ_3	1	$-\frac{1}{2}-\frac{\sqrt{3}i}{2}$	$-\frac{1}{2}+\frac{\sqrt{3}i}{2}$

Either form of the character table can be used.

Alternative Method

Character table of $C_3(3)$ can also be obtained by using ab-initio method. For this purpose, let us start with the Eq. 3.16, i.e.,

$$\chi(C_n^m) = e^{\left(\frac{2\pi i l m}{n}\right)}, \quad \text{for } C_3,\ n = 3$$

The members of $C_3(3)$ are C_3, C_3^2 and $C_3^3 = E$, they belong to separate classes.
$\implies$ There are three irreducible representations.

Let us consider three different representations, Γ_1, Γ_2, and Γ_3. Further, the values of l and m are $l = 0, 1, 2$ and m $= 1, 2, 3$.

For Γ_1: $l = 0$ and m $= 1, 2, 3$
Since $l = 0$, $\chi(C_3) = e^0 = 1$, $\chi(C_3^2) = e^0 = 1$ and $\chi(C_3^3) = e^0 = 1$.
So, for Γ_1, the characters are 1, 1, 1
For Γ_2 : $l = 1$ and m $= 1, 2, 3$

$$\chi(C_3) = \exp\left(\frac{2\pi i}{3}\right) = \varepsilon$$
$$\chi(C_3^2) = \exp\left(\frac{2.2\pi i}{3}\right) = \exp\left(\frac{4\pi i}{3}\right)$$
$$\chi(C_3^3) = \exp\left(\frac{3.2\pi i}{3}\right) = \exp\left(\frac{6\pi i}{3}\right)$$

Now, we have

$$\chi(C_3) = \exp\left(\frac{2\pi i}{3}\right) = \cos\left(\frac{2\pi}{3}\right) + i \sin\left(\frac{2\pi}{3}\right) = \varepsilon$$

$$\chi(C_3^2) = \exp\left(\frac{4\pi i}{3}\right) = \cos\left(\frac{4\pi}{3}\right) + i \sin\left(\frac{4\pi}{3}\right)$$

$$= \cos\left(2\pi - \frac{2\pi}{3}\right) + i \sin\left(2\pi - \frac{2\pi}{3}\right)$$

$$= \cos\left(\frac{2\pi}{3}\right) - i \sin\left(\frac{2\pi}{3}\right) = \varepsilon^*$$

$$\chi(C_3^3) = \exp\left(\frac{6\pi i}{3}\right) = \cos\left(\frac{6\pi}{3}\right) + i \sin\left(\frac{6\pi}{3}\right)$$

$$= \cos(2\pi) + i \sin(2\pi)$$

$$= \cos(2\pi - 0) + i \sin(2\pi - 0)$$

$$= \cos(0) + i \sin(0) = 1$$

So, for Γ_2, the characters are 1, ε, ε^*
Now, for Γ_3: $l = 2$ and m = 1, 2, 3

$$\chi(C_3) = \exp\left(\frac{2\pi i.2.1}{3}\right) = \exp\left(\frac{4\pi i}{3}\right)$$

$$\chi(C_3^2) = \exp\left(\frac{2\pi i.2.2}{3}\right) = \exp\left(\frac{8\pi i}{3}\right)$$

$$\chi(C_3^3) = \exp\left(\frac{2\pi i.2.3}{3}\right) = \exp(4\pi i)$$

$$\Rightarrow \chi(C_3) = \exp\left(\frac{4\pi i}{3}\right) = \cos\left(2\pi - \frac{2\pi}{3}\right) + i \sin\left(2\pi - \frac{2\pi}{3}\right)$$

$$= \cos\left(\frac{2\pi}{3}\right) - i \sin\left(\frac{2\pi}{3}\right) = \varepsilon^*$$

$$\chi(C_3^2) = \exp\left(\frac{8\pi i}{3}\right) = \cos\left(2\pi + \frac{2\pi}{3}\right) + i \sin\left(2\pi + \frac{2\pi}{3}\right)$$

$$= \cos\left(\frac{2\pi}{3}\right) + i \sin\left(\frac{2\pi}{3}\right) = \varepsilon$$

and $$\chi(C_3^3) = \exp(4\pi) = \cos(2\pi + 2\pi) + i \sin(2\pi + 2\pi)$$

$$= \cos(2\pi) + i \sin(2\pi)$$

$$= \cos(2\pi - 0) + i \sin(2\pi - 0)$$

$$= \cos 0 + i \sin 0 = 1$$

So, for Γ_3 the characters are 1, ε^*, ε
Therefore, the character table is

C_3	E	C_3	C_3^2
Γ_1	1	1	1
Γ_2	1	e	e^*
Γ_3	1	e^*	e

5. The point groups $C_4(4)$ and $S_4(\overline{4})$

Both C_4 and S_4 are cyclic (abelian) groups of order 4, hence they will have four irreducible representations (each of one-dimension). They are also isomorphic and will have identical character table, which can be obtained by using Table 3.2 prepared for a cyclic group. Therefore, the tentative character table of C_4/S_4 is

S_4/C_4	E	C_4	C_2	C_4^3
Γ_1	1	1	1	1
Γ_2	1	ε	ε^2	ε^3
Γ_3	1	ε^2	ε^4	ε^6
Γ_4	1	ε^3	ε^6	ε^9

Taking $\varepsilon = i$, we have $\varepsilon^2 = -1$, $\varepsilon^3 = -i$, $\varepsilon^4 = 1$, $\varepsilon^6 = -1$, and $\varepsilon^9 = i$, respectively. Substituting these values in the above table and interchanging the characters of Γ_2 and Γ_3 (to obtain the group representations involving complex number together), the table becomes

S_4/C_4	E	C_4	C_2	C_4^3
Γ_1	1	1	1	1
Γ_2	1	-1	1	-1
Γ_3	1	i	-1	$-i$
Γ_4	1	$-i$	-1	i

Alternative Method

Character table of $C_4(4)$ can also be obtained by using ab-initio method. For this purpose, let us start with the Eq. 3.16, i. e.,

$$\chi(C_n^m) = e^{\left(\frac{2\pi i l m}{n}\right)}, \quad \text{for } C_4, \quad n = 4$$

The members of $C_4(4)$ are C_4, $C_4^2 = C_2$, C_4^3 and $C_4^4 = E$, they belong to separate classes.

$\implies$ There are four irreducible representations. Let us suppose, they are Γ_1, Γ_2, Γ_3, and Γ_4. Further, the values of l and m are $l = 0, 1, 2, 3$ and m = 1, 2, 3, 4.

For Γ_1: $l = 0$ and m = 1, 2, 3, 4. The resulting four characters are

$$\chi(C_4) = e^0 = 1, \ \chi(C_2) = e^0 = 1, \ \chi(C_4^3) = e^0 = 1 \text{ and } \chi(E) = e^0 = 1$$

Thus, for Γ_1 the four characters are 1, 1, 1, 1
For Γ_2: $l = 1$ and m = 1, 2, 3, 4

$$\chi(C_4) = \exp\left(\frac{2\pi i}{4}\right) = \exp\left(\frac{\pi i}{2}\right) = \cos\left(\frac{\pi}{2}\right) + i \ \sin\left(\frac{\pi}{2}\right) = i$$

$$\chi(C_2) = \exp\left(\frac{2.2\pi i}{4}\right) = \exp(\pi i) = \cos(\pi) + i \ \sin(\pi) = -1$$

$$\chi(C_4^3) = \exp\left(\frac{3.2\pi i}{4}\right) = \exp\left(\frac{3\pi i}{2}\right)$$

$$= \cos\left(\frac{3\pi}{2}\right) + i \ \sin\left(\frac{3\pi}{2}\right)$$

$$= \cos\left(2\pi - \frac{\pi}{2}\right) + i \ \sin\left(2\pi - \frac{\pi}{2}\right)$$

$$= \cos\left(\frac{\pi}{2}\right) - i \ \sin\left(\frac{\pi}{2}\right) = -i$$

and $\quad \chi(C_4^4) = \exp\left(\frac{4.2\pi i}{4}\right) = \exp\left(\frac{4\pi i}{2}\right)$

$$= \exp(2\pi i) = \cos(2\pi) + i \ \sin(2\pi) = 1$$

Thus, for Γ_2 the four characters are i, −1, −i, 1
For Γ_3: $l = 2$ and m = 1, 2, 3, 4

$$\chi(C_4) = \exp\left(\frac{2.2\pi i}{4}\right) = \exp(\pi i) = \cos(\pi) + i \ \sin(\pi) = -1$$

$$\chi(C_2) = \exp\left(\frac{2.2\pi i \times 2}{4}\right) = \exp(2\pi i) = \cos(2\pi) + i \ \sin(2\pi) = 1$$

$$\chi(C_4^3)) = \exp\left(\frac{2.2\pi i \times 3}{4}\right) = \exp(3\pi i) = \cos(3\pi) + i \ \sin(3\pi)$$

$$= \cos(2\pi + \pi) + i \ \sin(2\pi + \pi)$$

$$= \cos(\pi) + i \ \sin(\pi) = -1$$

$$\chi(C_4^4) = \exp\left(\frac{2.2\pi i \times 4}{4}\right) = \exp(4\pi i) = \cos(4\pi) + i \ \sin(4\pi)$$

$$= \cos(2\pi + 2\pi) + i \ \sin(2\pi + 2\pi)$$

$$= \cos(2\pi) + i \ \sin(2\pi) = 1$$

Thus, for Γ_3 the four characters are −1, 1, −1, 1
For Γ_4: $l = 3$ and m = 1, 2, 3, 4

$$\chi(C_4) = \exp\left(\frac{3.2\pi i}{4}\right) = \exp\left(\frac{3\pi i}{2}\right) = \cos\left(\frac{3\pi}{2}\right) + i \sin\left(\frac{3\pi}{2}\right)$$
$$= \cos\left(2\pi - \frac{\pi}{2}\right) + i \sin\left(2\pi - \frac{\pi}{2}\right)$$
$$= \cos\left(\frac{\pi}{2}\right) - i \sin\left(\frac{\pi}{2}\right) = -i$$

$$\chi(C_2) = \exp\left(\frac{3.2\pi i \times 2}{4}\right) = \exp(3\pi i) = \cos(3\pi) + i \sin(3\pi)$$
$$= \cos(2\pi + \pi) + i \sin(2\pi + \pi)$$
$$= \cos(\pi) + i \sin(\pi) = -1$$

$$\chi(C_4^3) = \exp\left(\frac{3.2\pi i \times 3}{4}\right) = \exp\left(\frac{9\pi i}{2}\right) = \cos\left(\frac{9\pi}{2}\right) + i \sin\left(\frac{9\pi}{2}\right)$$
$$= \cos\left(4\pi + \frac{\pi}{2}\right) + i \sin\left(4\pi + \frac{\pi}{2}\right)$$
$$= \cos\left(\frac{\pi}{2}\right) + i \sin\left(\frac{\pi}{2}\right) = -i$$

$$\chi(C_4^4)) = \exp\left(\frac{3.2\pi i \times 4}{4}\right) = \exp(6\pi i) = \cos(6\pi) + i \sin(6\pi)$$
$$= \cos(4\pi + 2\pi) + i \sin(4\pi + 2\pi)$$
$$= \cos(2\pi) + i \sin(2\pi) = 1$$

Thus, for Γ_4 the four characters are $-1, -i, i, 1$

Rearranging the character values, we can obtain the same character table as above.

6. **The Point Groups $C_6(6)$, C_{3i} ($\overline{3}$) and $C_{3h}(\overline{6})$**

The point groups C_6, $C_{3i}(\equiv S_6)$ and C_{3h} are cyclic (abelian) groups of order 6, hence they will have six irreducible representations (each of one-dimension). They are also isomorphic and will have identical character table which can be obtained in three different ways: (i) Directly by using the Table 3.2, (ii) By ab-initio method prepared for cyclic group, and (iii) By using the direct product subgroup method.

Let us use the third method. Any one of the following three products can be used:

$$C_{3i} = S_6 = C_3 \times C_i$$
$$C_6 = C_3 \times C_2$$
$$C_{3h} = C_3 \times C_s$$

For example, the character table of C_6 can be obtained as

C_3	E	C_3	C_3^2
Γ_1	1	1	1
Γ_2	1	1	ε^*
Γ_3	1	ε^*	ε

×

C_2	E	C_2
Γ_1	1	1
Γ_2	1	-1

=

$C_{3h}/C_{3i}/C_6$	E	C_3	C_3^2	C_2	C_6	C_6^5
Γ_1	1	1	1	1	1	1
Γ_2	1	ε	ε^*	1	ε	ε^*
Γ_3	1	ε^*	ε	1	ε^*	ε
Γ_4	1	1	1	-1	-1	-1
Γ_5	1	ε	ε^*	-1	$-\varepsilon$	$-\varepsilon^*$
Γ_6	1	ε^*	ε	-1	$-\varepsilon^*$	$-\varepsilon$

7. **The point groups $D_3(32)$ and $C_{3v}(3m)$**

The point groups D_3 and C_{3v} are non-abelian, and we use the general rules derived from the orthogonal theorem to determine their character table. Since the two point groups are isomorphic, they will have identical character table. Let us consider the point group D_3 whose symmetry elements are E, C_3, C_3^2, C_{2x}, C_{2y}, and C_{2xy}, i.e., the order of the group is 6. However, they can be put into three classes: E, $2C_3$, $3C_2$.

Thus according to rule 1, the number of irreducible representations = 3.

According to rule 2, we have

$$\sum_{i=1}^{3} l_i^2 = 6 \quad or \quad l_1^2 + l_2^2 + l_3^2 = 6$$

Since l can assume only positive integer values other than zero. Therefore, positive values of l_1, l_2, and l_3 are 1, 1, and 2.

Further, according to rule 3, we have

$$\sum_{i=1}^{3}[\chi(E)]^2 = h \quad or \quad [\chi_1(E)]^2 + [\chi_2(E)]^2 + [\chi_3(E)]^2 = 6$$

Again, the possible values of $\chi_1(E)$, $\chi_2(E)$ and $\chi_3(E)$ are 1, 1, and 2. On the basis of obtained information, the tentative character table of the point group D_3 and its isomorph can be written as

C_{3v}/D_3	E	$2C_3$	$3C_2$
Γ_1	1	1	1
Γ_2	1	L	M
Γ_3	2	P	Q

Let us apply rule 4 to obtain the character values L, M, P, and Q in the representations Γ_2 and Γ_3. Orthogonality condition for Γ_1 and Γ_2 representations give us

$$\sum_{i=1}^{3} g_p \chi_i(R_p) \, \chi_j(R_p) = 0$$

or $g_1 \chi_1(E)\chi_2(E) + g_2 \chi_1(C_3)\chi_2(C_3) + g_3 \chi_1(C_2)\chi_2(C_2) = 0$

where $g_1 = 1$, $g_2 = 2$, and $g_3 = 3$

Hence, for Γ_1 and Γ_2 representations, we get

$$1 + 2L + 3M = 0 \quad \text{(i)}$$

Further, the orthogonality condition for Γ_2 representation gives us

$$\sum_{i=1}^{3} g_p [X_i(R_p)]^2 = h$$

or $1 + 2L^2 + 3M^2 = 6$ (ii)

After solving equations (i) and (ii), we obtain $L = 1$ and $M = -1$. Now, applying the orthogonality condition (taking the product) for first two classes, we can obtain two other character values P and Q. This will give us

$$1 \times 1 + 1 + 2 \times \chi(C_3) = 0$$
$$\Rightarrow P = -1$$

Similarly, the product of first and third classes gives us

$$1 \times 1 + 1 \times (-1) + 2 \times \chi(C_2) = 0$$
$$or \ 1 - 1 + 2Q = 0$$
$$\Rightarrow Q = 0$$

Therefore, substituting the obtained character values, the final form of the character table is

C_{3v}/D_3	E	$2C_3$	$3C_2$
Γ_1	1	1	1
Γ_2	1	1	-1
Γ_3	2	-1	0

We know that Γ_3 representation is two-dimensional (or it is doubly degenerate), therefore the character value -1, corresponding to class C_3 is a sum of two complex numbers, i.e.,

$$\varepsilon + \varepsilon^* = -1 = 2\cos\left(\frac{2\pi}{3}\right) = 2\cos 120°$$

Accordingly, this table can also be written in a slightly different form as

C_{3v}/D_3	E	$2\,C_3$	$3\,C_2$
Γ_1	1	1	1
Γ_2	1	1	-1
Γ_3	2	$2\cos\frac{2\pi}{3}$	0

8. The point group D_∞ and $C_{\infty v}$

The two point groups D_∞ and $C_{\infty v}$ contain E, C_∞^Φ, $C_\infty^{-\Phi}$, $(\infty\ \sigma_v)$ and an infinite number of perpendicular C_2 symmetry elements. However, they can be put into only three classes, i.e., E, $2\,C_\infty^\Phi$, and ∞C_2. Since these point groups contain infinite operations, a general method cannot be used to construct their character table. However, because of their resemblance with the $D_3/(C_{3v})$ character table, they can be obtained by extending D_3/C_{3v} group to $D_\infty/C_{\infty v}$ groups. In doing so, the first two representations are same as in D_3/C_{3v} group while other representations are denoted as Γ_3, Γ_4, ... , Γ_k and so on as shown below. The character values corresponding to C_3 operation in these representations are written as $2\cos\varphi$, $2\cos 2\varphi$, $2\cos 3\varphi$ ..., $2\cos k\varphi$ and so on. Therefore, the character table of $D_\infty/C_{\infty v}$ obtained as (the abbreviation of $\cos\varphi$ is written as $c\varphi$).

$C_{\infty v}/D_\infty$	E	$2C_\infty^\varphi$	...	$C_2/(\infty\sigma_v)$
Γ_1	1	1	...	1
Γ_2	1	1	...	-1
Γ_3	2	$2c\varphi$	...	0
Γ_4	2	$2c2\varphi$	...	0
...	...	...	...	...
Γ_k	2	$2ck\varphi$	...	0
...	...	...	...	...

9. The point group $D_{\infty h}$

On a similar line as discussed above, the character table of $D_{\infty h}$ cannot be obtained by any general method. However, the same can be obtained by generalization of the direct product subgroup method. In this case, the identity used is

$$D_{\infty h} = D_\infty \times C_i$$

Therefore, the character table of $D_{\infty h}$ is given by

$D_\infty \times C_i =$

D_∞	E	$2C_\infty^\varphi$	$\cdots$	∞C_2
Γ_1	1	1	$\cdots$	1
Γ_2	1	1	$\cdots$	-1
Γ_3	2	$2c\varphi$	$\cdots$	0
Γ_4	2	$2c2\varphi$	$\cdots$	0
$\cdots$	$\cdots$	$\cdots$	$\cdots$	$\cdots$

$\times$

C_i	E	i
Γ_1	1	1
Γ_2	1	-1

=

$D_{\infty h}$	E	$2C_\infty^\varphi$	$\cdots$	∞C_2	i	$2S_\infty^\varphi$	$\cdots$	∞C_v
Γ_1	1	1	$\cdots$	1	1	1	$\cdots$	1
Γ_2	1	1	$\cdots$	-1	1	1	$\cdots$	-1
Γ_3	2	$2c\varphi$	$\cdots$	0	2	$-2c\varphi$	$\cdots$	0
Γ_4	2	$2c2\varphi$	$\cdots$	0	2	$2c2\varphi$	$\cdots$	0
$\cdots$	$\cdots$	$\cdots$	$\cdots$	$\cdots$	$\cdots$	$\cdots$	$\cdots$	$\cdots$
Γ_a	1	1	$\cdots$	1	-1	-1	$\cdots$	-1
Γ_b	1	1	$\cdots$	-1	-1	-1	$\cdots$	-1
Γ_c	2	$2c\varphi$	$\cdots$	0	-2	$2c\varphi$	$\cdots$	0
Γ_d	2	$2c2\varphi$	$\cdots$	0	-2	$-2c2\varphi$	$\cdots$	0
$\cdots$	$\cdots$	$\cdots$	$\cdots$	$\cdots$	$\cdots$	$\cdots$	$\cdots$	$\cdots$

10. **The point group T(23)**

This is a non-cyclic/non-abelian point group belonging to cubic crystal system. Because of the presence of C_3 symmetry axes, some of the characters in irreducible representation involve complex numbers as in cyclic or abelian group, such as ε and ε^*, where

$$\varepsilon = \exp\left(\frac{2\pi i}{3}\right) = \cos\left(\frac{2\pi}{3}\right) + i \ \sin\left(\frac{2\pi i}{3}\right)$$

The point group T consists of 12 symmetry elements which can be put into four classes E, $4C_3$, $4C_3^2$, and $3C_2$. Therefore, according to rule 1, there are four irreducible representations. According to rule 2, we have

$$\sum_{i=1}^{4} l_i^2 = 12 \ \text{ or } \ l_1^2 + l_2^2 + l_3^2 + l_4^2 = 12$$

$$\Rightarrow l_1 = 1, \ l_2 = 1, l_3 = 1, \ \text{and } l_4 = 3$$

That is, three representations are of one-dimension each and the fourth representation is triply degenerate. Further, according to rule 3, we obtain $l_i = \chi_i(E)$. However, we know that the Γ_1 representation is always totally symmetric. Therefore, let the characters corresponding to C_3, C_3^2 and C_2 in Γ_2 and Γ_3 representations involve complex numbers, ε and ε^*. Orthogonal condition reduces to $1 + \varepsilon + \varepsilon^* = 0$. Thus, on the basis of obtained information, the tentative character table of the point group T can be written as

T	E	$4C_3$	$4C_3^2$	$3C_2$
Γ_1	1	1	1	1
Γ_2	1	ε	ε^*	1
Γ_3	1	ε^*	ε	1
Γ_4	3	A	B	C

Let us now obtain the character values of A, B, and C of Γ_4 representation. The product $\Gamma_1 \times \Gamma_4$ gives us

$$3 + 4A + 4B + 3C = 0 \qquad\qquad \text{(i)}$$

Now, let us consider the orthogonal condition of different classes. The product of classes I and II will give us

$$1 \times 1 + 1 \times \varepsilon + 1 \times \varepsilon^* + 3 \times \chi(C_3) = 0$$
$$\text{or } 1 + \varepsilon + \varepsilon^* + 3A = 0 \qquad .$$

Since $1 + \varepsilon + \varepsilon^* = 0, \Rightarrow A = 0$
Similarly, the product of classes I and III will give us

$$1 \times 1 + 1 \times \varepsilon^* + 1 \times \varepsilon + 3 \times \chi(C_3^2) = 0$$
$$\text{or } 1 + \varepsilon^* + \varepsilon + 3B = 0$$
$$\text{Since } 1 + \varepsilon + \varepsilon^* = 0, \Rightarrow B = 0$$

Substituting the values of A and B in Eq. (i), we obtain $3C = -3$ or $C = -1$. Therefore, the final form of the character table of the point group T is given as

T	E	$4\,C_3$	$4\,C_3^2$	$3\,C_2$
Γ_1	1	1	1	1
Γ_2	1	ε	ε^*	1
Γ_3	1	ε^*	ε	1
Γ_4	3	0	0	-1

11. The point groups O (432) and T_d $\left(\overline{4}3m\right)$

The two point groups, O and T_d are of order 24. Their symmetry elements can be divided into five different classes; therefore there are five irreducible representations. The symmetry elements for the point group O in five different classes are E, $8C_3$, $3C_2$, $6C_4$, and $6C_2'$. The point groups of O and T_d are isomorphic, hence they will have identical character table. Let us work out the character table for the point group O (and its isomorph T_d).

According to rule 2, we have

$$\sum_{i=1}^{5} l_i^2 = 24 \ \ or \ \ l_1^2 + l_2^2 + l_3^2 + l_4^2 + l_5^2 = 24$$

$$\Rightarrow l_1 = 1, \ \ l_2 = 1, \ \ l_3 = 2, \ \ l_4 = 3, \ \ l_5 = 3$$

On the basis of obtained information, the tentative character table of the point group O (and its isomorph T_d) can be written as

T_d/O	E	$8C_3$	$3C_2$	$6C_4$	$6C_2'$
Γ_1	1	1	1	1	1
Γ_2	1	A	B	C	D
Γ_3	2	E	F	G	H
Γ_4	3	K	L	M	N
Γ_5	3	P	Q	R	S

Let us now obtain the character values of $\Gamma_2 - \Gamma_5$ representations. The product $\Gamma_1 \times \Gamma_2$ will give us

$$1 + 8A + 3B + 6C + 6D = 0, \ \Rightarrow A = 1, \ B = 1, \ C = -1, D = -1$$

Similarly, the product $\Gamma_1 \times \Gamma_3$ will give us

$$2 + 8E + 3F + 6G + 6H = 0, \ \Rightarrow E = -1, \ F = 2, \ G = 0, \ H = 0$$

$\Gamma_1 \times \Gamma_4$ will give us

$$3 + 8K + 3L + 6M + 6N = 0, \ \Rightarrow K = 0, \ L = -1, \ M = 1, \ N = -1$$

and $\Gamma_1 \times \Gamma_5$ will give us

$$3 + 8P + 3Q + 6R + 6S = 0 \ \Rightarrow P = 0, \ Q = -1, \ R = -1, \ S = 1$$

Substituting all the character values, the final form of the character table of the point group O (and its isomorph T_d) is as follows:

T_d/O	E	$8\,C_3$	$3\,C_2$	$6\,C_4$	$6C_2'$
Γ_1	1	1	1	1	-1
Γ_2	1	1	1	-1	-1
Γ_3	2	-1	2	0	0
Γ_4	3	0	-1	1	-1
Γ_5	3	0	-1	-1	1

All important character tables for crystallographic/molecular point groups are provided at the end of this chapter as an appendix.

3.8 Mulliken Symbols

While constructing character tables, it is necessary to use some suitable symbols (notations) to represent various symmetry operations and distinct irreducible representations. The letters A and B are used to denote one-dimensional non-degenerate representations (sometimes with subscript numbers to distinguish the similar ones), while the letter E usually denotes the two-dimensional degenerate representations. However, cyclic groups of order ≥ 3 have one-dimensional representations generated by the complex nth roots of unity. The conjugate pairs of these are often linked with each other and also denoted by E. Three, four, and five-dimensional degenerate representations are denoted by letters F or T, G, and H, respectively, where G and H appear only in I and I_h point groups. Table 3.3 provides the use of symbols that are based on the dimensions of the representations.

In Mulliken symbols, the letter A stands for symmetric with respect to the principal axis C_n of the highest order in the system. Accordingly in the character table, it always has $+1$ character. On the other hand, the letter B stands for anti-symmetric with respect to the axis C_n, and always has -1 character.

In the infinite order point groups such as $C_{\infty v}$ and $D_{\infty h}$ corresponding to linear molecules, the Greek letter notations are used. Non-degenerate representations are denoted by the symbol Σ, while doubly degenerate representations are denoted by the symbols π, Δ, and Φ.

Table 3.3 Symbols based on dimension

Dimension of representation	Symbol
1	A, B
2	E
3	F or T
4	G
5	H

Table 3.4 Character values and symbols

Character value	C_n axes	Symbol C_2 axes $\perp$ to C_s or $\sigma_v(\sigma_d)$	σ_h	i
1	A	1	$'$	g
-1	B	2	$''$	u

Any of these primary symbols may be modified by adding subscript or superscript notations to indicate symmetry or anti-symmetry with respect to some symmetry operations other than the principal axis.

In any centrosymmetric groups with any degeneracy, a subscript g (for German gerade = even) indicates symmetry and subscript u (for German ungerade = odd) indicates anti-symmetry with respect to inversion (i).

For finite order non-degenerate representation and non-linear groups, the subscripts 1 and 2, respectively, indicate symmetry and anti-symmetry with respect to a non-principal rotation (lower than the principal axis) or to a vertical mirror plane (σ_v).

For non-degenerate representations of the infinite order linear groups $C_{\infty v}$ and $D_{\infty h}$, the subscripts + and − have the same meaning as 1 and 2 in the finite order groups. On the other hand, with any degeneracy in non-centrosymmetric non-linear groups, the addition of a prime ($'$) or double prime ($''$) to a primary symbol indicates symmetry or anti-symmetry with respect to horizontal mirror plane (σ_h). Table 3.4 provides a summary of such symbols. The first listed irreducible representation in the character table is always a totally symmetric representation, irrespective of the group. It is composed of only +1 character for all symmetry operations of the group. Depending on the group, the Mulliken symbol for the representation may be selected from A, A_1, A$'$, A_g, A_{1g}, Σ^+ Σ_g^+.

3.9 Transformation Properties

When different symmetry operations of a point group are applied, different kinds of effects on the position coordinates and rotation axes are observed in general and on their unit vectors associated with the atoms of a molecule in particular. The resulting effects on the coordinates (or their unit vectors) are associated with the corresponding irreducible representations (also called species), different types of transformation properties observed in the character tables are briefly summarized.

As pointed out in the text (Table 3.1), third and fourth columns of each character table list the infrared (IR) and Raman activities of the particular species. If one or more of the translation components T_x, T_y, and T_z is listed in the third column, the species is infrared active. Further, both the components of change in the dipole moment (μ_x, μ_y, μ_z) as well as the components of translation (T_x, T_y, T_z) should be listed in third column. However, since they always occur together (as both are vectors and transform in the same way if the symmetry operation is carried out similarly), the dipole moment part is omitted from the character tables (also in appendix). Also listed in the third column are the components for the rotational coordinates R_x, R_y, R_z, where the subscripts x, y, z indicate the direction in which the translation or dipole moment change occurs for a particular vibrational species, that species is IR active. Similarly, if one or more of the components of the polarizability α_{xx}, α_{yy}, α_{zz}, α_{xy}, etc., is listed in the row for a certain species in the fourth column, that species is Raman active. When the components of translation, change in dipole moment, or polarizability are degenerate, they are enclosed in parentheses. Thus, the transformation properties provide us the infrared and Raman active species by mere inspection. In some more explicit form, we have

(a) Third column of a character table lists the transformation properties of Cartesian coordinates x, y, z and rotation about x, y, and z axes (or linear and rotational vectors).

    1. The coordinate z and the rotation R_z are in general transform as A and B representations (species) in all non-cubic point groups non-degenerately. They correspond to the infrared activity of the particular species.

    2. The coordinates x, y (as in C_s, C_2,C_{2h}) or x, y, z (as in C_i) transform as non-degenerate species to which they are associated. However, the coordinates (x, y) and rotations (R_x, R_y) transform as doubly degenerate E representations in all C_n and D_n $(n \geq 3)$ point groups. Similarity, the coordinates (x, y, z) and rotations (R_x, R_y, R_z) transform as triply degenerate T representations in all cubic point groups.

The symbol (x, y) indicates that there is no difference in symmetry between the x and y directions hence they may be treated as equivalent and indistinguishable. Orbitals such as P_x and P_y in such situations are said to have same energy.

(b) Fourth column of the character table lists the transformation properties of the binary products of linear vectors. They correspond to Raman activity of the particular species.

    1. Among other things, the binary products of linear vectors correspond to the transformation of d-orbitals. The notations such as z^2, $x^2 - y^2$, xy, yz, or zx indicate the species by which the d-orbital of the same designation transforms.

A notation such as $2z^2 - x^2 - y^2$ listed for E_g in the point group O_h indicates the transformation property of d_{z^2} orbital. On the other hand, a notation such as $x^2+y^2+z^2$ (as in O_h) or $x^2 + y^2$ (as in C_{3v}) do not correspond to d-orbital transformations. They occur because of the direct product of certain pairs of vectors spread across two representations in the point groups belonging to cubic system.

(iii) Fifth column of the character table lists the transformation properties of triple (cubic) product of linear vectors.

    1. In this case, the notations such as

$$z^3, \ xyz, \ xz^2, \ yz^2, \ x(x^2 - 3y^2), \ y(3x^2 - y^2), \ z(x^2 - y^2)$$

correspond to different non-degenerate representations depending on symmetry. On the other hand, the notations within the parentheses and square brackets such as

$$\left(xz^2, \ yz^2\right), \ [xyz, \ z(x^2 - y^2)], \ [x(x^2 - 3y^2), \ y(3x^2 - y^2)]$$

in general correspond to degenerate E representations in various point groups.

2. The notations such as

$$(x^3, \ y^3, \ z^3), \ [x(z^2 - x^2), \ y(z^2 - x^2), \ z(x^2 - y^2)]$$

correspond to triply degenerate T representation in cubic point groups.

3.10 Summary

1. The set of matrices for different symmetry operations of a point group (which obey the laws of group multiplication table) are said to form representations and based on the character of the matrix, a representation is said to be either reducible or irreducible.
2. A representation is called reducible if it is possible to find a similarity transformation matrix which is able to convert all the matrices in the representation such that the resulting matrices can be blocked into smaller matrices in the form of a diagonal matrix.
3. In a representation, if it is not possible to find a similarity transformation matrix which can reduce the matrices of representation under consideration, then the representation is said to be irreducible. All one-dimensional representations are always irreducible.
4. The orthogonality theorem deals with the elements of matrices constituting the irreducible representation of a point group. It provides a number of important relationships related to the character of their representations. Some of these are very important from crystallographic point of view. They can be used as rules for constructing character tables for crystallographic and molecular point groups. Its important properties are

   (i)   The number of irreducible representations of a group is equal to the number of conjugate classes (say k) in the group.
   (ii) The sum of the squares of the dimensions of all the irreducible representations of a group is equal to the order of the group.
   (iii) In a given representation (reducible or irreducible), the characters of all matrices (symmetry operations) belonging to the same class are identical.
   (iv) The sum of the squares of the characters in any irreducible representation is equal to the order of the group.
   (v)   The character of the distinct irreducible representations must be orthogonal.

5. A character table can be conveniently divided into five parts (I - V) on the basis of the type of information they contain.
6. While constructing character tables, it is necessary to use some suitable notations (known as Mulliken symbols) to represent various symmetry operations and distinct irreducible representations.

Appendix: Character Tables

Character Tables for Some Important Point Groups using Schoenflies Notation

1. The Nonaxial Groups

C_1	E
A	1

C_s	E	σ_h	Transformation properties		
A'	1	1	x, y, R_z	x^2, y^2, z^2, xy	xz^2, yz^2, $x(x^2-3y^2)$, $y(3x^2-y^2)$
A''	1	−1	z, R_x, R_y	yz, xz	z^3, xyz, $z(x^2-y^2)$
$\chi_s(R)$	3	1			

C_i	E	i	Transformation properties		
A_g	1	1	R_x, R_y, R_z	x^2, y^2, z^2, xy, xz, yz	
A_u	1	−1	x, y, z		(xz^2, yz^2), $[xyz, z(x^2-y^2)]$ z^3, xyz, z (x^2-y^2), xz^2, yz^2, x(x^2-3y^2), y ($3x^2-y^2$)
χ_s (R)	3	−3			

2. The C_n Groups

C_2	E	C_2	Transformation properties		
A	1	1	z, R_z	x^2, y^2, z^2, xy	$z^3, xyz, z(x^2 - y^2)$
B	1	-1	x, y, R_x, R_y	yz, xz	$xz^2, yz^2, x(x^2 - 3y^2),$ $y(3x^2 - y^2)$
$\chi_s(R)$	3	-1			

C_3	E	C_3	C_3^2	Transformation properties		
A	1	1	1	z, R_z	$x^2 + y^2, z^2$	$z^3, x(x^2 - 3y^2), y(3x^2 - y^2)$
E	1 1	ε ε^*	ε^* ε	$(x,y),$ (R_x, R_y)	$(x^2 - y^2, xy), (yz, xz)$	$(xz^2, yz^2), [xyz, z(x^2 - y^2)]$
$\chi_s(R)$	3	0	0	$\varepsilon = \exp\left(\frac{2\pi i}{3}\right)$		

C_4	E	C_4	C_2	C_4^3	Transformation properties		
A	1	1	1	1	z, R_z	$x^2 + y^2, z^2$	z^3
B	1	-1	1	-1		$x^2 - y^2, xy$	$xyz, z(x^2 - y^2)$
E	1 1	i $-i$	-1 -1	$-i$ i	$(x,y),$ (R_x, R_y)	(yz, xz)	$(xz^2, yz^2)[x(x^2 - 3y^2),$ $y(3x^2 - y^2)$
$\chi_s(R)$	3	1	-1	1			

C_5	E	C_5	C_5^2	C_5^3	C_5^4	Transformation properties		
						z, R_z	$x^2+y^2,\ z^2$	z^3
A	1	1	1	1	1			
E_1	1	ε	ε^2	ε^{2*}	ε^*	$(x,y),$	$(xz,\ yz)$	$(xz^2,\ yz^2)$
	1	ε^*	ε^{2*}	ε^2	ε	$(R_x,\ R_y)$		
E_2	1	ε^2	ε^*	ε	ε^{2*}		$(x^2-y^2,\ xy)$	$[xyz,\ z(x^2-y^2)],\ [x(x^2-3y^2),\ y(3x^2-y^2)]$
	1	ε^{2*}	ε	ε^*	ε^2			
$\chi_s(R)$	3	$1+2c\frac{2\pi}{5}$	$1+2c\frac{4\pi}{5}$	$1+2c\frac{4\pi}{5}$	$1+2c\frac{2\pi}{5}$	$\varepsilon=\exp\left(\frac{2\pi i}{5}\right)$		

C_6	E	C_6	C_3	C_2	C_3^2	C_6^5	Transformation properties		
A	1	1	1	1	1	1	z, R_z	$x^2+y^2,\ z^2$	z^3
B	1	-1	1	-1	1	-1			$x(x^2-3y^2),\ y(3x^2-y^2)$
E_1	1 1	$-\varepsilon$ ε^*	$-\varepsilon^*$ $-\varepsilon$	-1 -1	ε ε^*	ε^* ε	$(x,y),$ (R_x, R_y)	(xz, yz)	$(xz^2,\ yz^2)$
E_2	1 1	$-\varepsilon$ $-\varepsilon^*$ $-\varepsilon$	$-\varepsilon$ $-\varepsilon^*$	1 1	ε^* $-\varepsilon$	$-\varepsilon$ $-\varepsilon^*$		$(x^2-y^2,\ xy)$	$[xyz,\ z(x^2-y^2)]$
$\chi_s(R)$	3	2	0	-1	0	2	$\varepsilon = \exp\left(\frac{2\pi i}{6}\right)$		

3. The D_n Groups

D_2	E	$C_2(z)$	$C_2(y)$	$C_2(x)$	Transformation properties		
A	1	1	1	1		$x^2,\ y^2,\ z^2$	xyz
B_1	1	1	-1	-1	z, R_z	xy	$z^3,\ z(x^2-y^2)$
B_2	1	-1	1	-1	y, R_y	xz	$yz^2,\ y(3x^2-y^2)$
B_3	1	-1	-1	1	x, R_x	yz	$xz^2,\ x(x^2-3y^2)$
$\chi_s(R)$	3	-1	-1	-1			

D_3	E	$2C_3$	$3C_2$	Transformation properties		
A_1	1	1	1		$x^2+y^2,\ z^2$	$x(x^2-3y^2)$
A_2	1	1	-1	z, R_z		$z^3,\ y(3x^2-y^2)$
E	2	-1	0	$(x,y),$ (R_x, R_y)	$(x^2-y^2,\ xy),\ (xz, yz)$	$(xz^2, yz^2),\ [xyz,\ z(x^2-y^2)]$
$\chi_s(R)$	3	0	-1			

D_4	E	$2C_4$	$C_2(=C_4^2)$	$2C_2'$	$2C_2''$	Transformation properties		
A_1	1	1	1	1	1		$x^2+y^2,\ z^2$	
A_2	1	1	1	-1	-1	z, R_z		z^3
B_1	1	-1	1	1	-1		x^2-y^2	xyz
B_2	1	-1	1	-1	1		xy	$z(x^2-y^2)$
E	2	0	-2	0	0	$(x,y),$ (R_x, R_y)	(xz, yz)	$(xz^2,\ yz^2),\ [x(x^2-3y^2), y(3x^2-y^2)]$
$\chi_s(R)$	3	1	-1	-1	-1			

D_5	E	$2C_5$	$2C_5^2$	$5C_2$	Transformation properties		
A_1	1	1	1	1		$x^2 + y^2,\ z^2$	z^3
A_2	1	1	1	-1	z, R_z		$xyz,\ z(x^2 - y^2)$
E_1	2	$2c\frac{2\pi}{5}$	$2c\frac{4\pi}{5}$	0	$(x,y),$ (R_x, R_y)	(xz, yz)	$(xz^2,\ yz^2)$
E_2	2	$2c\frac{4\pi}{5}$	$2c\frac{2\pi}{5}$	0		$(x^2 - y^2, xy)$	$[xyz,\ z(x^2 - y^2)],\ [x(x^2 - 3y^2),\ y(3x^2 - y^2)]$
$\chi_s(R)$	3	$1 + 2c\frac{2\pi}{5}$	$1 + 2c\frac{4\pi}{5}$	-1			

D_6	E	$2C_6$	$2C_3$	$C_2(=$ $C_6^3)$	$3C_2'$	$3C_2''$	Transformation properties		
A_1	1	1	1	1	1	1		$x^2 + y^2,$ z^2	
A_2	1	1	1	1	-1	-1	z, R_z		z^3
B_1	1	-1	1	-1	1	-1			$x(x^2 - 3y^2)$
B_2	1	-1	1	-1	-1	1			$y(3x^2 - y^2)$
E_1	2	1	-1	-2	0	0	$(x,y),$ (R_x, R_y)	(xz, yz)	(xz^2, yz^2)
E_2	2	-1	-1	2	0	0		$(x^2 - y^2,$ $xy)$	$[xyz, z(x^2 - y^2)]$
$\chi_s(R)$	3	2	0	-1	-1	-1			

4. The C_{nv} Groups

C_{2v}	E	C_2	$\sigma_v(xz)$	$\sigma_v'(yz)$	Transformation properties		
A_1	1	1	1	1	z	x^2, y^2, z^2	$z^3, z(x^2 - y^2)$
A_2	1	1	-1	-1	R_z	xy	xyz
B_1	1	-1	1	-1	x, R_y	xz	$xz^2, x(x^2 - 3y^2)$
B_2	1	-1	-1	1	y, R_x	yz	$yz^2, y(3x^2 - y^2)$
$\chi_s(R)$	3	-1	1	1			

C_{3v}	E	$2C_3$	$3\sigma_v$	Transformation properties		
A_1	1	1	1	z	$x^2 + y^2, z^2$	$z^3, x(x^2 - 3y^2)$
A_2	1	1	-1	R_z		$y(3x^2 - y^2)$
E	2	-1	0	$(x,y),$ (R_x, R_y)	$(x^2 - y^2, xy)$ (xz, yz)	$(xz^2, yz^2), [xyz, z(x^2 - y^2)]$
$\chi_s(R)$	3	0	1			

C_{4v}	E	$2C_4$	C_2	$2\sigma_v$	$2\sigma_d$	Transformation properties		
A_1	1	1	1	1	1	z	$x^2+y^2,\ z^2$	z^3
A_2	1	1	1	-1	-1	R_z		
B_1	1	-1	1	1	-1		x^2-y^2	$z(x^2-y^2)$
B_2	1	-1	1	-1	1		xy	xyz
E	2	0	-2	0	0	$(x,y),$ $(R_x,\ R_y)$	$(xz,\ yz)$	$(xz^2,\ yz^2),\ \left[x(x^2-3y^2),\ y(3x^2-y^2)\right]$
$\chi_s(R)$	3	1	-1	1	1			

C_{5v}	E	$2C_5$	$2C_5^2$	$5\sigma_v$	Transformation Properties		
A_1	1	1	1	1	z	$x^2 + y^2,\, z^2$	z^3
A_2	1	1	1	-1	R_z		
E_1	2	$2c\frac{2\pi}{5}$	$2c\frac{4\pi}{5}$	0	$(x,y),$ (R_x, R_y)	(xz, yz)	$(xz^2,\, yz^2)$
E_2	2	$2c\frac{4\pi}{5}$	$2c\frac{2\pi}{5}$	0		$(x^2 - y^2,\, xy)$	$[xyz,\, z(x^2 - y^2)],\, [x(x^2 - 3y^2),\, y(3x^2 - y^2)]$
$\chi_s(R)$	3	$1 + 2c\frac{2\pi}{5}$	$1 + 2c\frac{4\pi}{5}$	1			

C_{6v}	E	$2C_6$	$2C_3$	C_2	$3\sigma_v$	$3\sigma_d$	Transformation properties		
A_1	1	1	1	1	1	1	z	$x^2 + y^2, z^2$	z^3
A_2	1	1	1	1	-1	-1	R_z		
B_1	1	-1	1	-1	1	-1			$x(x^2 - 3y^2)$
B_2	1	-1	1	-1	-1	1			$y(3x^2 - y^2)$
E_1	2	1	-1	-2	0	0	$(x,y), (R_x, R_y)$	(xz, yz)	(xz^2, yz^2)
E_2	2	-1	-1	2	0	0		$(x^2 - y^2, xy)$	$[xyz, z(x^2 - y^2)]$
$\chi_s(R)$	3	2	0	-1	1	1			

5. The C_{nh} Groups

C_{2h}	E	C_2	i	σ_h	Transformation properties		
A_g	1	1	1	1	R_z	x^2, y^2, z^2, xy	
B_g	1	-1	1	-1	R_x, R_y	xz, yz	
A_u	1	1	-1	-1	z		$z^3, xyz, z(x^2 - y^2)$
B_u	1	-1	-1	1	x, y		$xz^2, yz^2, x(x^2 - 3y^2), y(3x^2 - y^2)$
$\chi_s(R)$	3	-1	-3	1			

C_{3h}	E	C_3	C_3^2	σ_h	S_3	S_3^5	Transformation properties		
A'	1	1	1	1	1	1	R_z	$x^2+y^2,\ z^2$	$x(x^2-3y^2),\ y(3x^2-y^2)$
E'	1	ε	ε^*	1	ε	ε^*	(x,y)	$(x^2-y^2,\ xy)$	$(xz^2,\ yz^2)$
	1	ε^*	ε	1	ε^*	ε			
A''	1	1	1	-1	-1	-1	z		z^3
E''	1	ε	ε^*	-1	$-\varepsilon$	$-\varepsilon^*$	$(R_x,\ R_y)$	$(xz,\ yz)$	$[xyz,\ z(x^2-y^2)]$
	1	ε^*	ε	-1	$-\varepsilon^*$	$-\varepsilon$			
$\chi_s(\mathbf{R})$	3	0	0	1	-2	-2			

C_{4h}	E	C_4	C_2	C_4^3	i	S_4^3	σ_h	S_4		
A_g	1	1	1	1	1	1	1	1	R_z	$x^2+y^2,\ z^2$
B_g	1	-1	1	-1	1	-1	1	-1		$(x^2-y^2,\ xy)$
E_g	1	i	-1	$-i$	1	i	-1	$-i$	$(R_x,\ R_y)$	$(xz,\ yz)$
	1	$-i$	-1	i	1	$-i$	-1	i		
A_u	1	1	1	1	-1	-1	-1	-1	z	z^3
B_u	1	-1	1	-1	-1	1	-1	1		$xyz, z(x^2-y^2)$
E_u	1	i	-1	$-i$	-1	$-i$	1	i	(x,y)	$(xz^2,\ yz^2),\ [x(x^2-3y^2),\ y(3x^2-y^2)]$
	1	$-i$	-1	i	-1	i	1	$-i$		
$\chi_s(R)$	3	1	-1	1	-3	-1	1	-1		

Transformation properties

C_{5h}	E	C_5	C_5^2	C_5^3	C_5^4	σ_h	S_5	S_5^7	S_5^3	S_5^9	Transformation properties		
A'	1	1	1	1	1	1	1	1	1	1	R_z	x^2+y^2, z^2	
E_1'	1	ε	ε^2	ε^{2*}	ε^*	1	ε	ε^2	ε^{2*}	ε^*	(x, y)		$(xz^2,\ yz^2)$
	1	ε^*	ε^{2*}	ε^2	ε	1	ε^*	ε^{2*}	ε^2	ε			
E_2'	1	ε^2	ε^*	ε	ε^{2*}	1	ε^2	ε^*	ε	ε^{2*}		$(x^2-y^2,\ xy)$	$[x(x^2-3y^2),\ y(3x^2-y^2)]$
	1	ε^{2*}	ε	ε^*	ε^2	1	ε^{2*}	ε	ε^*	ε^2			
A''	1	1	1	1	1	-1	-1	-1	-1	-1	z		z^3
E_1''	1	ε	ε^2	ε^{2*}	ε^*	-1	$-\varepsilon$	$-\varepsilon^2$	$-\varepsilon^{2*}$	$-\varepsilon^*$	$(R_x,\ R_y)$	$(xz,\ yz)$	
	1	ε^*	ε^{2*}	ε^2	ε	-1	$-\varepsilon^*$	$-\varepsilon^{2*}$	$-\varepsilon^2$	$-\varepsilon$			
E_2''	1	ε^2	ε^*	ε	ε^{2*}	-1	$-\varepsilon^2$	$-\varepsilon^*$	$-\varepsilon$	$-\varepsilon^{2*}$			$xyz,\ z(x^2-y^2)$
	1	ε^{2*}	ε	ε^*	ε^2	-1	$-\varepsilon^{2*}$	$-\varepsilon$	$-\varepsilon^*$	$-\varepsilon^2$			
$\chi_s(R)$	3	$1+2c\frac{2\pi}{5}$	$1+2c\frac{4\pi}{5}$	$1+2c\frac{4\pi}{5}$	$1+2c\frac{4\pi}{5}$	1	$-1+2c\frac{2\pi}{5}$	$-1+2c\frac{4\pi}{5}$	$-1+2c\frac{4\pi}{5}$	$-1+2c\frac{2\pi}{5}$	$\varepsilon = \exp\left(\frac{2\pi i}{5}\right)$		

C_{6h}	E	C_6	C_3	C_2	C_3^2	C_6^5	i	S_3^5	S_6^5	σ_h	S_6	S_3	Transformation properties	
A_g	1	1	1	1	1	1	1	1	1	1	1	1	R_z	x^2+y^2, z^2
B_g	1	-1	1	-1	1	-1	1	-1	1	-1	1	-1		
E_{1g}	1	ε	$-\varepsilon^*$	-1	$-\varepsilon$	ε^*	1	ε	$-\varepsilon^*$	-1	$-\varepsilon$	ε^*	(R_x, R_y)	(xz, yz)
	1	ε^*	$-\varepsilon$	-1	$-\varepsilon^*$	ε	1	ε^*	$-\varepsilon$	-1	$-\varepsilon^*$	ε		
E_{2g}	1	$-\varepsilon^*$	$-\varepsilon$	1	$-\varepsilon^*$	$-\varepsilon$	1	$-\varepsilon^*$	$-\varepsilon$	1	$-\varepsilon^*$	$-\varepsilon$		(x^2-y^2, xy)
	1	$-\varepsilon$	$-\varepsilon^*$	1	$-\varepsilon$	$-\varepsilon^*$	1	$-\varepsilon$	$-\varepsilon^*$	1	$-\varepsilon$	$-\varepsilon^*$		
A_u	1	1	1	1	1	1	-1	-1	-1	-1	-1	-1	z	z^3
B_u	1	-1	1	-1	1	-1	-1	1	-1	1	-1	1		$[x(x^2-3y^2), y(3x^2-y^2)]$
E_{1u}	1	ε	$-\varepsilon^*$	-1	$-\varepsilon$	ε^*	-1	$-\varepsilon$	ε^*	1	ε	$-\varepsilon^*$	(x, y)	(xz^2, yz^2)
	1	ε^*	$-\varepsilon$	-1	$-\varepsilon^*$	ε	-1	$-\varepsilon^*$	ε	1	ε^*	$-\varepsilon$		
E_{2u}	1	$-\varepsilon^*$	$-\varepsilon$	1	$-\varepsilon^*$	$-\varepsilon$	-1	ε^*	ε	-1	ε^*	ε		$[xyz, z(x^2-y^2)]$
	1	$-\varepsilon$	$-\varepsilon^*$	1	$-\varepsilon$	$-\varepsilon^*$	-1	ε	ε^*	-1	ε	ε^*		
$\chi_s(R)$	3	2	0	-1	0	2	-3	-2	0	1	0	-2	$\varepsilon = \exp\left(\frac{2\pi i}{6}\right)$	

6. The $\mathbf{D}_{nh}$ *Groups*

D_{2h}	E	$C_2(z)$	$C_2(y)$	$C_2(x)$	i	$\sigma(xy)$	$\sigma(xz)$	$\sigma(yz)$	Transformation properties		
A_g	1	1	1	1	1	1	1	1		x^2, y^2, z^2	
B_{1g}	1	1	-1	-1	1	1	-1	-1	R_z	xy	
B_{2g}	1	-1	1	-1	1	-1	1	-1	R_y	xz	
B_{3g}	1	-1	-1	1	1	-1	-1	1	R_x	yz	
A_u	1	1	1	1	-1	-1	-1	-1			xyz
B_{1u}	1	1	-1	-1	-1	-1	1	1	z		$z^3,\ z(x^2-y^2)$
B_{2u}	1	-1	1	-1	-1	1	-1	1	y		$yz^2,\ y(3x^2-y^2)$
B_{3u}	1	-1	-1	1	-1	1	1	-1	x		$xz^2,\ x(x^2-3y^2)$
$\chi_s(R)$	3	-1	-1	-1	-3	1	1	1			

D_{3h}	E	$2C_3$	$3C_2$	σ_h	$2S_3$	$3\sigma_v$	Transformation properties		
A_1'	1	1	1	1	1	1		$x^2 + y^2,$ z^2	$x(x^2 - 3y^2)$
A_2'	1	1	-1	1	1	-1	R_z		$y(3x^2 - y^2)$
E'	2	-1	0	2	-1	0	(x, y)	$(x^2 - y^2,$ $xy)$	(xz^2, yz^2)
A_1''	1	1	1	-1	-1	-1			
A_2''	1	1	-1	-1	-1	1	z		z^3
E''	2	-1	0	-2	1	0	(R_x, R_y)	(xz, yz)	$\left[xyz, z(x^2 - y^2)\right]$
$\chi_s(R)$	3	0	-1	1	-2	1			

D_{4h}	E	$2C_4$	C_2	$2C_2'$	$2C_2''$	i	$2S_4$	σ_h	$2\sigma_v$	$2\sigma_d$	Transformation properties	
A_{1g}	1	1	1	1	1	1	1	1	1	1		x^2+y^2, z^2
A_{2g}	1	1	1	-1	-1	1	1	1	-1	-1	R_z	
B_{1g}	1	-1	1	1	-1	1	-1	1	1	-1		x^2-y^2
B_{2g}	1	-1	1	-1	1	1	-1	1	-1	1		xy
E_g	2	0	-2	0	0	2	0	-2	0	0	(R_x, R_y)	(xz, yz)
A_{1u}	1	1	1	1	1	-1	-1	-1	-1	-1		
A_{2u}	1	1	1	-1	-1	-1	-1	-1	1	1	z	z^3
B_{1u}	1	-1	1	1	-1	-1	1	-1	-1	-1		xyz
B_{2u}	1	-1	1	-1	1	-1	1	-1	1	-1		$z(x^2-y^2)$
E_u	2	0	-2	0	0	-2	0	2	0	0	(x, y)	$(xz^2, yz^2), [x(x^2-3y^2),\ y(3x^2-y^2)]$
$\chi_s(R)$	3	1	-1	-1	-1	-3	-1	1	1	1		

D_{5h}	E	$2C_5$	$2C_5^2$	$5C_2$	σ_h	$2S_5$	$2S_5^3$	$5\sigma_v$	Transformation properties		
A_1'	1	1	1	1	1	1	1	1		x^2+y^2, z^2	
A_2'	1	1	1	-1	1	1	1	-1	R_z		
E_1'	2	$2c\frac{2\pi}{5}$	$2c\frac{4\pi}{5}$	0	2	$2c\frac{2\pi}{5}$	$2c\frac{4\pi}{5}$	0	(x, y)		(xz^2, yz^2)
E_2'	2	$2c\frac{4\pi}{5}$	$2c\frac{2\pi}{5}$	0	2	$2c\frac{4\pi}{5}$	$2c\frac{2\pi}{5}$	0		(x^2-y^2, xy)	$[x(x^2-3y^2), y(3x^2-y^2)]$
A_1''	1	1	1	1	-1	-1	-1	-1			
A_2''	1	1	1	-1	-1	-1	-1	1	z		z^3
E_1''	2	$2c\frac{2\pi}{5}$	$2c\frac{4\pi}{5}$	0	-2	$-2c\frac{2\pi}{5}$	$-2c\frac{4\pi}{5}$	0	(R_x, R_y)	(xz, yz)	
E_2''	2	$2c\frac{4\pi}{5}$	$2c\frac{2\pi}{5}$	0	-2	$-2c\frac{4\pi}{5}$	$-2c\frac{2\pi}{5}$	0			$[xyz, z(x^2-y^2)]$
$\chi_s(R)$	3	$1+2c\frac{2\pi}{5}$	$1+2c\frac{4\pi}{5}$	-1	1	$-1+2c\frac{2\pi}{5}$	$-1+2c\frac{4\pi}{5}$	1			

D_{6h}	E	$2C_6$	$2C_3$	C_2	$3C_2'$	$3C_2''$	i	$2S_3$	$2S_6$	σ_h	$3\sigma_d$	$3\sigma_v$	Transformation properties		
A_{1g}	1	1	1	1	1	1	1	1	1	1	1	1		$x^2 + y^2,\ z^2$	
A_{2g}	1	1	1	1	-1	-1	1	1	1	1	-1	-1	R_z		
B_{1g}	1	-1	1	-1	1	-1	1	-1	1	-1	1	1			
B_{2g}	1	-1	1	-1	-1	1	1	-1	1	-1	1	-1			
E_{1g}	2	1	-1	-2	0	0	2	-1	-1	-2	0	0	(R_x, R_y)	(xy, yz)	
E_{2g}	2	-1	-1	2	0	0	2	-1	-1	2	0	0		$(x^2 - y^2, xy)$	
A_{1u}	1	1	1	1	1	1	-1	-1	-1	-1	-1	-1			
A_{2u}	1	1	1	1	-1	-1	-1	-1	-1	-1	1	1	z		z^3
B_{1u}	1	-1	1	-1	1	-1	-1	1	-1	1	-1	1			$x(x^2 - 3y^2)$
B_{2u}	1	-1	1	-1	-1	1	-1	1	-1	1	1	-1			$y(3x^2 - y^2)$
E_{1u}	2	1	-1	-2	0	0	-2	-1	1	2	0	0	(x, y)		$(xz^2,\ yz^2)$
E_{2u}	2	-1	-1	2	0	0	-2	1	1	-2	0	0			$[xyz,\ z(x^2 - y^2)]$
$\chi_s(R)$	3	2	0	-1	-1	-1	-3	-2	0	1	1	1			

7. The D_{nd} Groups

D_{2d}	E	$2S_4$	C_2	$2C_2'$	$2\sigma_d$	Transformation properties		
A_1	1	1	1	1	1		$x^2 + y^2, z^2$	xyz
A_2	1	1	1	-1	-1	R_z		$z(x^2 - y^2)$
B_1	1	-1	1	1	-1		$x^2 - y^2$	
B_2	1	-1	1	-1	1	z	xy	z^3
E	2	0	-2	0	0	$(x, y), (R_x, R_y)$	(xz, yz)	$(xz^2, yz^2), [x(x^2 - 3y^2), y(3x^2 - y^2)]$
$\chi_s(R)$	3	-1	-1	-1	1			

D_{3d}	E	$2C_3$	$3C_2$	i	$2S_6$	$3\sigma_d$	Transformation properties	
A_{1g}	1	1	1	1	1	1		$x^2 + y^2,\ z^2$
A_{2g}	1	1	-1	1	1	-1	R_z	
E_g	2	-1	0	2	-1	0	(R_x, R_y)	$(x^2 - y^2,\ xy),\ (xz, yz)$
A_{1u}	1	1	1	-1	-1	-1		$x(x^2 - 3y^2),\ z^3$
A_{2u}	1	1	-1	-1	-1	1	z	$y(3x^2 - y^2)$
E_u	2	-1	0	-2	1	0	(x, y)	$(xz^2,\ yz^2),\ [xyz,\ z(x^2 - y^2)]$
$\chi_s(R)$	3	0	-1	-3	0	1		

D_{4d}	E	$2S_8$	$2C_4$	$2S_8^3$	C_2	$4C_2'$	$4\sigma_d$	Transformation properties		
									$x^2 + y^2$, z^2	
A_1	1	1	1	1	1	1	1		$x^2 + y^2$, z^2	
A_2	1	1	1	1	1	-1	-1	R_z		
B_1	1	-1	1	-1	1	1	-1			
B_2	1	-1	1	-1	1	-1	1	z		z^3
E_1	2	$\sqrt{2}$	0	$-\sqrt{2}$	-2	0	0	(x, y)		$(xz^2,\ yz^2)$
E_2	2	0	-2	0	2	0	0		$(x^2 - y^2,\ xy)$	$[xyz,\ z(x^2 - y^2)]$
E_3	2	$-\sqrt{2}$	0	$\sqrt{2}$	-2	0	0	$(R_x,\ R_y)$	$(yz,\ xz)$	$[x(x^2 - 3y^2),\ y(3x^2 - y^2)]$
$\chi_s(R)$	3	$-1 + \sqrt{2}$	1	$-1 - \sqrt{2}$	-1	-1	1			

D_{5d}	E	$2C_5$	$2C_5^2$	$5C_2$	i	$2S_{10}^3$	$2S_{10}$	$5\sigma_d$	Transformation properties		
A_{1g}	1	1	1	1	1	1	1	1		$x^2+y^2,\ z^2$	
A_{2g}	1	1	1	-1	1	1	1	-1	R_z		
E_{1g}	2	$2c\frac{2\pi}{5}$	$2c\frac{4\pi}{5}$	0	2	$2c\frac{2\pi}{5}$	$2c\frac{4\pi}{5}$	0	(R_x, R_y)	$(xz,\ yz)$	
E_{2g}	2	$2c\frac{4\pi}{5}$	$2c\frac{2\pi}{5}$	0	2	$2c\frac{4\pi}{5}$	$2c\frac{2\pi}{5}$	0		$(x^2-y^2,\ xy)$	
A_{1u}	1	1	1	1	-1	-1	-1	-1			
A_{2u}	1	1	1	-1	-1	-1	-1	1	z		z^3
E_{1u}	2	$2c\frac{2\pi}{5}$	$2c\frac{4\pi}{5}$	0	-2	$-2c\frac{2\pi}{5}$	$-2c\frac{4\pi}{5}$	0	(x, y)		$(xz^2,\ yz^2)$
E_{2u}	2	$2c\frac{4\pi}{5}$	$2c\frac{2\pi}{5}$	0	-2	$-2c\frac{4\pi}{5}$	$-2c\frac{2\pi}{5}$	0			$[xyz,\ z(x^2-y^2)],\ [x(x^2-3y^2),\ y(3x^2-y^2)]$
$\chi_s(R)$	3	$1+2c\frac{2\pi}{5}$	$1+2c\frac{4\pi}{5}$	-1	-3	$-1-2c\frac{2\pi}{5}$	$-1-2c\frac{4\pi}{5}$	1			

D_{6d}	E	$2S_{12}$	$2C_6$	$2S_4$	$2C_3$	$2S_{12}^5$	C_2	$6C_2'$	$6\sigma_d$	Transformation properties		
A_1	1	1	1	1	1	1	1	1	1		$x^2+y^2,\ z^2$	
A_2	1	1	1	1	1	1	1	-1	-1	R_z		
B_1	1	-1	1	-1	1	-1	1	1	-1			
B_2	1	-1	1	-1	1	-1	1	-1	1	z		z^3
E_1	2	$\sqrt{3}$	1	0	-1	$-\sqrt{3}$	-2	0	0	(x,y)		$(xz^2,\ yz^2)$
E_2	2	1	-1	-2	-1	1	2	0	0		$(x^2-y^2,\ xy)$	
E_3	2	0	-2	0	2	0	-2	0	0			$[x(x^2-3y^2),\ y(3x^2-y^2)]$
E_4	2	-1	-1	2	-1	-1	2	0	0			$[xyz,\ z(x^2-y^2)]$
E_5	2	$-\sqrt{3}$	1	0	-1	$\sqrt{3}$	-2	0	0	$(R_x,\ R_y)$	$(xz,\ yz)$	
$\chi_s(R)$	3	$-1+\sqrt{3}$	2	-1	0	$-1-\sqrt{3}$	-1	-1	1			

8. The S_n Groups

S_4	E	S_4	C_2	S_4^3	Transformation properties		
A	1	1	1	1	R_z	$x^2 + y^2, z^2$	$xyz, z(x^2 - y^2)$
B	1	−1	1	−1	x	$x^2 - y^2, xy$	z^3
E	1	i	−1	−i	(x,	(xz, yz)	$(xz^2, yz^2)[x(x^2 - 3y^2), y(3x^2 - y^2)]$
	1	−i	−1	i	y), (R_x, R_y)		
$\chi_s(R)$	3	−1	−1	−1			

S_6	E	C_3	C_3^2	i	S_6^5	S_6	Transformation properties	
							R_z	$x^2 + y^2, z^2$
A_g	1	1	1	1	1	1	R_z	$x^2 + y^2, z^2$
E_g	1	ε	ε^*	1	ε	ε^*	(R_x, R_y)	$(x^2 - y^2, xy), (xz, yz)$
	1	ε^*	ε	1	ε^*	ε		
A_u	1	1	1	-1	-1	-1	z	$z^3, x(x^2 - 3y^2), y(3x^2 - y^2)$
E_u	1	ε	ε^*	-1	$-\varepsilon$	$-\varepsilon^*$	(x, y)	$(xz^2, yz^2), [xyz, z(x^2 - y^2)]$
	1	ε^*	ε	-1	$-\varepsilon^*$	$-\varepsilon$		
$\chi_s(R)$	3	0	0	-3	0	0		

9. The Cubic Groups

T	E	$4C_3$	$4C_3^2$	$3C_2$	Transformation properties		
A	1	1	1	1		$x^2 + y^2,\ z^2$	xyz
E	1	ε	ε^*	1		$(2z^2 - x^2 - y^2,\ x^2 - y^2)$	
	1	ε^*	ε	1			
T	3	0	0	-1	$(R_x, R_y, R_z),\ (x, y, z)$	(xy, xz, yz)	$(x^3, y^3, z^3),\ [x(z^2 - y^2),$ $y(z^2 - x^2),\ z(x^2 - y^2)]$
$\chi_s(R)$	3	0	0	-1	$\varepsilon = \exp\left(\frac{2\pi i}{3}\right)$		

T_d	E	$8C_3$	$3C_2$	$6S_4$	$6\sigma_d$	Transformation properties		
A_1	1	1	1	1	1		$x^2 + y^2, z^2$	xyz
A_2	1	1	1	-1	-1			
E	2	-1	2	0	0		$\left(2z^2 - x^2 - y^2,\ x^2 - y^2\right)$	
T_1	3	0	-1	1	-1	$\left(R_x, R_y, R_z\right)$		$[x(z^2 - y^2), y(z^2 - x^2),\ z(x^2 - y^2)]$
T_2	3	0	-1	-1	1	(x, y, z)	(xy, xz, yz)	$\left(x^3, y^3, z^3\right)$
$\chi_s(R)$	3	0	-1	-1	1			

O	E	$6C_4$	$3C_2$	$8C_3$	$6C_2'$	Transformation properties		
A_1	1	1	1	1	1		$x^2+y^2,\ z^2$	
A_2	1	-1	1	1	-1			xyz
E	2	0	2	1	0		$(2z^2-x^2-y^2,\ x^2-y^2)$	
T_1	3	1	-1	1	-1	$(R_x,\ R_y,\ R_z),\ (x,y,z)$		(x^3,y^3,z^3)
T_2	3	1	-1	-1	1		$(xy,\ xz,\ yz)$	$[x(z^2-y^2),\ y(z^2-x^2),\ z(x^2-y^2)]$
$\chi_s(R)$	3	1	-1	-1	1			

T_h	E	$4C_3$	$4C_3^2$	$3C_2$	i	$4S_6$	$4S_6^5$	$3\sigma_h$	Transformation properties	
A_g	1	1	1	1	1	1	1	1	$x^2 + y^2 + z^2$	
A_u	1	1	1	1	-1	-1	-1	-1		xyz
E_g	1	ε	ε^*	1	1	ε	ε^*	1	$(2z^2 - x^2 - y^2,$	
	1	ε^*	ε	1	1	ε^*	ε	1	$x^2 - y^2)$	
E_u	1	ε	ε^*	1	-1	$-\varepsilon$	$-\varepsilon^*$	-1		
	1	ε^*	ε	1	-1	$-\varepsilon^*$	$-\varepsilon$	-1		
T_g	3	0	0	-1	3	0	0	-1	$(R_x,\ R_y,\ R_z)$ (xz, yz, xy)	
T_u	3	0	0	-1	-3	0	0	1	(x, y, z)	$(x^3, y^3, z^3),\ [x(z^2 - y^2),\ y(z^2 - x^2),\ z(x^2 - y^2)]$
$\chi_s(R)$	3	0	0	-1	-3	0	0	1	$\varepsilon = \exp\!\left(\frac{2\pi i}{3}\right)$	

O_h	E	$8C_3$	$6C_2$	$6C_4$	$3C_2(=C_4^2)$	i	$6S_4$	$8S_6$	$3\sigma_h$	$6\sigma_d$	Transformation properties	
A_{1g}	1	1	1	1	1	1	1	1	1	1		$x^2 + y^2 + z^2$
A_{2g}	1	1	-1	-1	1	1	-1	1	1	-1		
E_g	2	-1	0	0	2	2	0	-1	2	0		$(2z^2 - x^2 - y^2,\ x^2 - y^2)$
T_{1g}	3	0	-1	1	-1	3	1	0	-1	-1	$(R_x,\ R_y,\ R_z)$	
T_{2g}	3	0	1	-1	-1	3	-1	0	-1	1		$(xz,\ yz,\ xy)$
A_{1u}	1	1	1	1	1	-1	-1	-1	-1	-1		
A_{2u}	1	1	-1	-1	1	-1	1	-1	-1	1		$x\,y\,z$
E_u	2	-1	0	0	2	-2	0	1	-2	0		
T_{1u}	3	0	-1	1	-1	-3	-1	0	1	1	(x, y, z)	$(x^3,\ y^3,\ z^3)$
T_{2u}	3	0	1	-1	-1	-3	1	0	1	-1		$[x(z^2 - y^2),\ y(z^2 - x^2),\ z(x^2 - y^2)]$
$\chi_s(R)$	3	0	-1	1	-1	-3	-1	0	1	1		

10. Icosahedral Group

I_h	E	$12C_5$	$12C_5^2$	$20C_3$	$15C_2$	i	$12S_{10}$	$12S_{10}^3$	$20S_6$	$15\sigma_d$	Transformation properties
A_g	1	1	1	1	1	1	1	1	1	1	$x^2+y^2+z^2$
T_{1g}	3	$2c\frac{\pi}{5}$	$2c\frac{3\pi}{5}$	0	-1	3	$2c\frac{3\pi}{5}$	$2c\frac{\pi}{5}$	0	-1	(R_x, R_y, R_z)
T_{2g}	3	$2c\frac{3\pi}{5}$	$2c\frac{\pi}{5}$	0	-1	3	$2c\frac{\pi}{5}$	$2c\frac{3\pi}{5}$	0	-1	
G_g	4	-1	-1	1	0	4	-1	-1	1	0	
H_g	5	0	0	-1	1	5	0	0	-1	1	$(2z^2-x^2-y^2,$ $x^2-y^2,\ xy,\ yz,\ xz)$
A_u	1	1	1	1	1	-1	-1	-1	-1	-1	
T_{1u}	3	$2c\frac{\pi}{5}$	$2c\frac{3\pi}{5}$	0	-1	-3	$-2c\frac{3\pi}{5}$	$-2c\frac{\pi}{5}$	0	1	(x, y, z)
T_{2u}	3	$2c\frac{3\pi}{5}$	$2c\frac{\pi}{5}$	0	-1	-3	$-2c\frac{\pi}{5}$	$-2c\frac{3\pi}{5}$	0	1	
G_u	4	-1	-1	1	0	-4	1	1	-1	0	
H_u	5	0	0	-1	1	-5	0	0	1	-1	
$\chi_s(R)$	3	$2c\frac{\pi}{5}$	$2c\frac{3\pi}{5}$	0	-1	-3	$-2c\frac{3\pi}{5}$	$-2c\frac{\pi}{5}$	0	1	

Character Tables for Linear Molecules

11. *The C$_{\infty v}$ Group*

C$_{\infty v}$	E	2C$_\infty^\varphi$	2C$_\infty^{2\varphi}$	2C$_\infty^{3\varphi}$	...	$\infty\sigma_v$	Transformation properties		
A$_1 \equiv \Sigma^+$	1	1	1	1	...	1	z	x^2+y^2, z^2	z^3
A$_2 \equiv \Sigma^-$	1	1	1	1	...	-1	R$_z$		
E$_1 \equiv \pi$	2	$2c\varphi$	$2c2\varphi$	$2c3\varphi$	...	0	(x, y), (R$_x$, R$_y$)	(xz, yz)	(xz^2, yz^2)
E$_2 \equiv \Delta$	2	$2c2\varphi$	$2c4\varphi$	$2c6\varphi$	...	0		(x^2-y^2, xy)	$[xyz, z(x^2-y^2)]$
E$_3 \equiv \Phi$	2	$2c3\varphi$	$2c6\varphi$	$2c9\varphi$	...	0			$[x(x^2-3y^2), y(3x^2-y^2)]$
E$_4 \equiv \Gamma$	2	$2c4\varphi$	$2c6\varphi$	$2c12\varphi$	...	0			
:									
$\chi_s(R)$	3	$1+2c\varphi$	$1+2c2\varphi$	$1+2c3\varphi$	...	1			

12. The D∞h Group

$D_{\infty h}$	E	$2C_\infty^\varphi$	$2C_\infty^{2\varphi}$	$2C_\infty^{3\varphi}$	$\cdots$	$\infty\sigma_v$	i	$2S_\infty^\varphi$	$\cdots$	∞C_2	Transformation properties	
$\sum_g^+$	1	1	1	1	$\cdots$	1	1	1	$\cdots$	1		$x^2+y^2,\ z^2$
$\sum_g^-$	1	1	1	1	$\cdots$	-1	1	1	$\cdots$	-1	R_z	
π_g	2	$2c\varphi$	$2c2\varphi$	$2c3\varphi$	$\cdots$	0	2	$-2c\varphi$	$\cdots$	0	$(R_x,\ R_y)$	$(xz,\ yz)$
Δ_g	2	$2c2\varphi$	$2c4\varphi$	$2c6\varphi$	$\cdots$	0	2	$2c2\varphi$	$\cdots$	0		$(x^2-y^2,\ xy)$
Φ_g	2	$2c3\varphi$	$2c6\varphi$	$2c9\varphi$	$\cdots$	0	2	$-2c3\varphi$	$\cdots$	0		
Γ_g	2	$2c4\varphi$	$2c8\varphi$	$2c12\varphi$	$\cdots$	0	2	$2c4\varphi$	$\cdots$	0		
$\cdots$					$\cdots$				$\cdots$			
$\sum_u^+$	1	1	1	1	$\cdots$	1	-1	-1	$\cdots$	-1	z	z^3
$\sum_u^-$	1	1	1	1	$\cdots$	-1	-1	-1	$\cdots$	-1		

(continued)

(continued)

$D_{\infty h}$	E	$2C_\infty^\varphi$	$2C_\infty^{2\varphi}$	$2C_\infty^{3\varphi}$	...	$\infty\sigma_v$	i	$2S_\infty^\varphi$	...	∞C_2	Transformation properties	
π_u	2	$2c\varphi$	$2c2\varphi$	$2c3\varphi$	...	0	-2	$2c\varphi$	...	0	(x, y)	$(xz^2,\ yz^2)$
Δ_u	2	$2c2\varphi$	$2c4\varphi$	$2c6\varphi$	...	0	-2	$-2c2\varphi$	...	0		$[xyz,\ z(x^2 - y^2)]$
Φ_u	2	$2c3\varphi$	$2c6\varphi$	$2c9\varphi$	...	0	-2	$2c3\varphi$	...	0		$[x(x^2 - 3y^2),\ y(3x^2 - y^2)]$
$\Gamma_{u^\bullet}$	2	$2c4\varphi$	$2c8\varphi$	$2c12\varphi$	...	0	-2	$-2c4\varphi$	...	0		
$\chi_s(R)$	3	$1 + 2c\varphi$	$1 + 2c2\varphi$	$1 + 2c3\varphi$	...	1	-3	$1 + 2c\varphi$	...	-1		

Chapter 4
Normal Modes of Molecular Vibrations

4.1 Introduction

In the first three chapters, we made a comprehensive study about the fundamentals of symmetry operations and their matrix representations, molecular and crystallographic point groups, construction of group multiplication tables, character tables, and their related aspects.

In this chapter, we are going to derive the normal modes of vibration with the application of all the above-studied concepts using different techniques for some selected molecules as examples. In the process, we shall also learn about the reducible and irreducible representations and the relationships between them, the character representations of matrices of fundamental symmetry operations for some nonlinear and linear molecules, and the presence of vibrational modes in them.

Theoretically derived normal modes of vibration for the selected molecules will help us not only to know the physics involved in different cases but also to determine the types of vibrational modes that are present in them. Theoretically obtained results can be easily compared with the experimental results obtained from IR and Raman-related spectra, for a better understanding of the problem.

4.2 Molecular Motions

We know that atoms or molecules are constantly in motion in solids, liquids, and gases at temperatures above absolute zero. Such motions include translational motion, rotational motion, and vibrational motion. Let us discuss these motions in diatomic and some simple triatomic (polyatomic) molecules.

© The Author(s), under exclusive license to Springer Nature Singapore Pte Ltd. 2022   191
M. A. Wahab, *Symmetry Representations of Molecular Vibrations*, Springer Series
in Chemical Physics 126, https://doi.org/10.1007/978-981-19-2802-4_4

Diatomic Molecules

1. *Translational Motion*

Let us suppose that two atoms A and B of a diatomic molecule move parallel in the same direction, resulting in a translation of the molecule in space. As this is not a periodic motion, no interaction with the electromagnetic radiations occurs and hence there is no possibility for the translation motion to be detected by IR or Raman techniques. Since any motion in space can be resolved along the x, y, and z axes of a Cartesian coordinate system, therefore for every molecule there exist three translational degrees of freedom as shown in Fig. 4.1.

2. *Rotational Motion*

In this case, let us suppose that the two atoms A and B of the molecule rotate as shown in Fig. 4.2. However, unlike translation, rotations can be detected through spectroscopic techniques because of their periodic natures. Frequencies of such rotations will lie in the microwave region only as low frequency bands.

   For a diatomic molecule, there exist only two rotational degrees of freedom, that is along the x and y axes as the rotation along the molecular (z) axis makes no change in the atomic coordinates. This result is also true for any linear molecule irrespective of the number of atoms comprising it.

Fig. 4.1 Three translational degrees of freedom

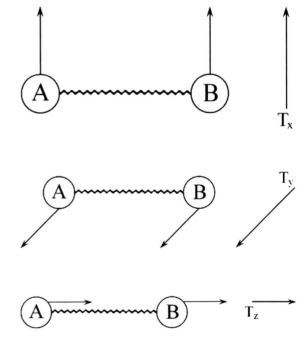

Fig. 4.2 Rotational degrees
of freedom

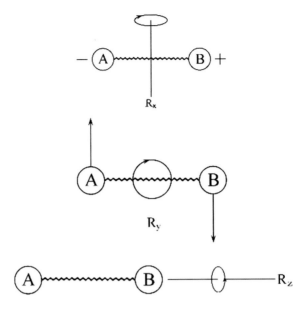

3.   *Vibrational Motion*

Finally, let us consider a simple model for the vibration of diatomic AB molecule
(Fig. 4.3) having masses m_1 and m_2. Suppose that the two atoms are connected
through an elastic spring whose spring constant (or force constant) is k and obeys
Hooke's law. Let, initially, they be at an equilibrium intermolecular distance r_e (called
bond length) and free from both ends (Fig. 4.3a). Now, suppose a force F is required
to stretch the spring by a certain distance q from the equilibrium position w.r.t. an
arbitrary origin. Both the atoms will experience a restoring force of magnitude kq
directed opposite to their motions. Therefore, according to Newton's second law of
motion, the restoring force on the mass m_1 (displacement is along the direction of
force) is

$$\frac{m_1 d^2 q_1}{dt^2} = kq \tag{4.1}$$

Similarly, the restoring force on the mass m_2 (displacement is opposite to the
direction of force) is

$$\frac{m_2 d^2 q_2}{dt^2} = -kq \tag{4.2}$$

where q_1 and q_2 are the displacements made by m_1 and m_2 w.r.t. an arbitrary origin
during vibration.

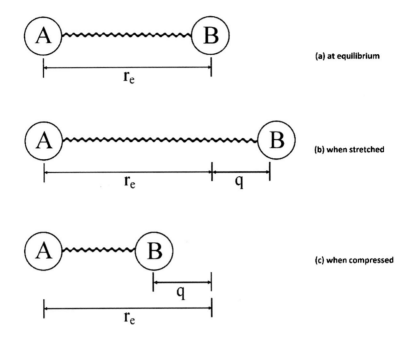

Fig. 4.3 Vibrational motion of diatomic AB molecule

Now multiplying Eqs. 4.1 and 4.2, respectively, by m_2 and m_1, we obtain

$$m_1 m_2 \frac{d^2 q_1}{dt^2} = m_2 kq \qquad \text{(i)}$$

$$\text{and} \quad m_1 m_2 \frac{d^2 q_2}{dt^2} = -m_1 kq \qquad \text{(ii)}$$

Subtracting Eq. (i) from Eq. (ii), we get

$$m_1 m_2 \left(\frac{d^2 q_2}{dt^2} - \frac{d^2 q_1}{dt^2} \right) = -(m_1 + m_2)kq$$

$$\text{or} \quad \frac{d^2 (q_2 - q_1)}{dt^2} = - \left(\frac{m_1 + m_2}{m_1 m_2} \right) kq$$

But $q_2 - q_1 = r_e + q$, where r_e is constant, and

$$\frac{d^2 r_e}{dt^2} = 0$$

Therefore, we are left with

$$\frac{d^2q}{dt^2} = -\left(\frac{m_1 + m_2}{m_1 m_2}\right)kq = -\frac{k}{\mu} q \tag{4.3}$$

where $\mu = \frac{m_1 m_2}{(m_1 + m_2)}$, known as "reduced mass" of the system. Now, let the solution of Eq. 4.3 be given by

$$q = q_0 \cos(2\pi \nu t + \varphi) \tag{4.4}$$

Differentiating Eq. 4.4 twice with respect to "t", we obtain

$$\frac{d^2q}{dt^2} = -4\pi^2 \nu^2 q_0 \cos(2\pi \nu t + \varphi) = -4\pi^2 \nu^2 q \tag{4.5}$$

From Eqs. 4.3 and 4.5, we obtain

$$4\pi^2 \nu^2 = \frac{k}{\mu} \quad \text{or} \quad \nu = \frac{1}{2p}\sqrt{\frac{k}{\mu}} \tag{4.6}$$

Equation 4.3 has exactly the same form as that of a single body simple harmonic oscillator, except that here q is the relative displacement of two bodies from their equilibrium separation. Hence, the relative motion of the two masses m_1 and m_2 is just the same as the motion of a simple harmonic oscillator of mass, $\mu = \frac{m_1 m_2}{(m_1 + m_2)}$. Similarly, Eq. 4.6 has exactly the same form as that of a single body oscillator of mass μ connected with one end of the same spring, the other end of which is fixed in an inertial frame of reference.

The potential energy of such a system is given by

$$dV = -Fdq = kq\, dq$$

$$\text{or} \quad V = k\int q\, dq$$

$$= \frac{1}{2}kq^2 \tag{4.7}$$

Since the energy of a real vibrating molecule is subject to quantum mechanical vibrations, therefore substituting the value of potential energy V in the Schrodinger equation, the vibrational energy of such a harmonic oscillator can be obtained as

$$E_V = \left(v + \frac{1}{2}\right)h\nu \tag{4.8}$$

where $v = 0, 1, 2, \ldots$ is called the vibrational quantum number, ν is the vibrational frequency (given by Eq. 4.6) in Hertz; and h is Planck's constant.

Fig. 4.4 Potential energy
curve of diatomic molecule

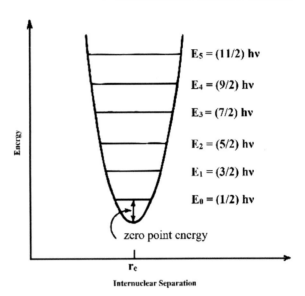

$E_5 = (11/2)\ \mathbf{hv}$

$E_4 = (9/2)\ \mathbf{hv}$

$E_3 = (7/2)\ \mathbf{hv}$

$E_2 = (5/2)\ \mathbf{hv}$

$E_1 = (3/2)\ \mathbf{hv}$

$E_0 = (1/2)\ \mathbf{hv}$

zero point energy

r_e

Internuclear Separation

Expressing the energy in terms of wavenumbers (which is the inverse of wavelength, i.e., in cm^{-1}, called "term value"), we have

$$\varepsilon_v = \frac{m_1 m_2}{(m_1 + m_2)} = \frac{E_v}{hc} = \left(v + \frac{1}{2}\right)h\bar{v} \tag{4.9}$$

Equation 4.8 (or 4.9) gives us a series of equally spaced energy levels as shown in Fig. 4.4. The minimum energy of the system corresponding to $v = 0$ is called the vibrational ground state,

$$E_0 = \frac{1}{2}hv \qquad\qquad \text{(from Eq. 4.8)}$$

$$\varepsilon_0 = \frac{1}{2}h\bar{v} \qquad\qquad \text{(from Eq. 4.9)}$$

It is to be noted that this (minimum) does not coincide with the minimum of the parabola defined by Eq. 4.7. The difference between the two is called the "zero point energy" as indicated in Fig. 4.4. Nonzero value of the lowest energy is a consequence of quantum mechanics.

The energy involved in transition from the ground state ($v = 0$) to higher energy states under harmonic approximation is given by

$$\Delta E = \varepsilon_v - \varepsilon_0 = \left(v + \frac{1}{2}\right)\bar{v} - \left(\frac{1}{2}\right)\bar{v} = v\bar{v} \tag{4.10}$$

Fig. 4.5 Morse potential curve

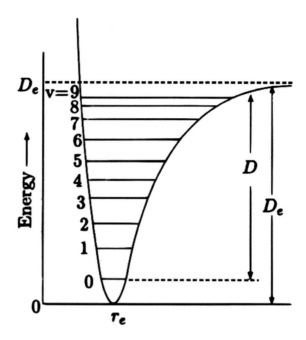

The parabolic potential well shown above is based on the classical harmonic oscillator. However, real molecules are not perfect harmonic oscillators. Consequently, the vibration of the potential energy curve is not a symmetric parabola but shows a skewed appearance and describes the behavior of an anharmonic oscillator. A purely analytical function with this shape in the region of interest for real molecules is known as Morse potential (Fig. 4.5). It is of the form

$$V = D_e \left(1 - e^{\beta q}\right)^2 \tag{4.11}$$

where D_e is the dissociation energy, β is a measure of the curvature at the bottom of the potential well, and $q = r - r_e$, so that $V(r)_{r\,=\,re} = 0$. On the basis of this potential, the solution of the Schrodinger equation gives energy levels of the system as

$$E_V = \left(v + \frac{1}{2}\right)h\nu_e - \left(v + \frac{1}{2}\right)^2 \chi_e h\nu_e + \left(v + \frac{1}{2}\right)^3 \Upsilon_e h\nu_e + \dots$$

where ν_e is the frequency, χ_e and Υ_e are anharmonicity constants which are very small (~0.01) but positive. The higher power terms are usually small and are omitted. Therefore, retaining only the first two terms, we have

$$E_V = \left(v + \frac{1}{2}\right)h\nu_e - \left(v + \frac{1}{2}\right)^2 \chi_e h\nu_e \tag{4.12}$$

From Eq. 4.12, the zero point energy corresponding to the Morse function is given as

$$E_0 = \left(1 - \frac{\chi_e}{2}\right)\frac{h\nu_e}{2} \tag{4.13}$$

Now, Eq. 4.12 can be rewritten as

$$E_V = \left[1 - \left(v + \frac{1}{2}\right)\chi_e\right]\left(v + \frac{1}{2}\right)h\nu_e \tag{4.14}$$

Comparing Eqs. 4.8 and 4.14, we obtain

$$\nu = \nu_e\left[1 - \left(v + \frac{1}{2}\right)\chi_e\right] \tag{4.15}$$

This shows that an anharmonic oscillator behaves like a harmonic oscillator except that its frequency decreases with the increase of vibrational quantum number v.

Writing Eqs. 4.12 and 4.13 in terms of wavenumbers, then their difference will provide the energy involved in transitions from ground state ($v = 0$) to higher energy states, i.e.,

$$\Delta E = \varepsilon_V - \varepsilon_0 = \left(v + \frac{1}{2}\right)\bar{\nu}_e - \left(v + \frac{1}{2}\right)^2 \chi_e\bar{\nu}_e - \left(1 - \frac{\chi_e}{2}\right)\frac{\bar{\nu}_e}{2}$$
$$= \bar{\nu}_e v - \bar{\nu}_e \chi_e v(v + 1) \tag{4.16}$$

where $\bar{\nu}_e$ may be defined as the (hypothetical) equilibrium oscillation frequency of the anharmonic system.

Now, using Eqs. 4.10 and 4.16, the energy required for transitions from the ground state to a few higher energy states for both harmonic and anharmonic cases can be easily calculated. They are provided in Table 4.1.

From Table 4.1, we observe that according to a harmonic oscillator model, a diatomic molecule has equally spaced vibrational energy levels, while an anharmonic model suggests a systematic decrease in energy with the increase of vibrational quantum number. Some of these allowed transitions are shown in Fig. 4.6.

Table 4.1 Energy involved for various allowed transitions

Translation	Energy involved		Modes of vibration
	harmonic oscillator	Anharmonic oscillator	
$v = 0 \rightarrow v = 1$	$\bar{\nu}$	$\bar{\nu}_e(1 - 2\chi_e)$	Fundamental absorption
$v = 0 \rightarrow v = 2$	$2\bar{\nu}$	$2\bar{\nu}_e(1 - 3\chi_e)$	First overtone
$v = 0 \rightarrow v = 3$	$3\bar{\nu}$	$3\bar{\nu}_e(1 - 4\chi_e)$	Second overtone
$v = 0 \rightarrow v = 4$	$4\bar{\nu}$	$4\bar{\nu}_e(1 - 5\chi_e)$	Third overtone

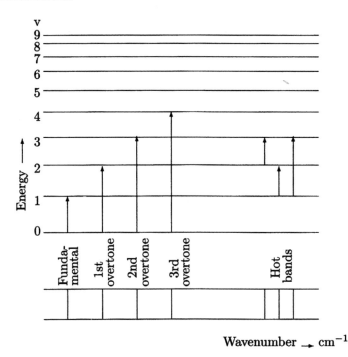

Fig. 4.6 Allowed energy transitions

On a similar line, the energy involved in transitions from the first level ($v = 1$) to other higher energy levels can be obtained from Eq. 4.12 as

$$\Delta E = \varepsilon_{V+1} - \varepsilon_V = \bar{v}_e - 2\bar{v}_e \chi_e(v + 1) \tag{4.17}$$

A few of these transitions are

$$\Delta E_{1 \to 2} = \bar{v}_e(1 - 4\chi_e)$$

and $\quad \Delta E_{2 \to 3} = \bar{v}_e(1 - 6\chi_e)$

So that $\quad \Delta E_{1 \to 3} = \Delta E_{1 \to 2} + \Delta E_{2 \to 3}$

$$= \bar{v}_e(1 - 4\chi_e) + \bar{v}_e(1 - 6\chi_e)$$

$$= 2\bar{v}_e(1 - 5\chi_e)$$

They are also shown in Fig. 4.6. These absorptions are weak and generally known as "Hot bands", and they require high temperature for their generation.

Example 4.1 The spectrum of HCl shows a very intense absorption at 2886 cm^{-1}, comparatively a weaker one at 5668 cm^{-1} and a very weak one at 8347 cm^{-1}, respectively. Determine the equilibrium frequency of the molecule and anharmonicity constant χ_e.

Solution: *Given*: HCl spectrum containing three absorption lines:
Intense absorption line at 2886 cm^{-1} (fundamental frequency).
Weak absorption line at 5668 cm^{-1} (first overtone).
Very weak absorption line at 8347 cm^{-1} (second overtone).
$\bar{v}_e = ?$ $\chi_e = ?$
According to the question, three given lines are fundamental absorption, first overtone, and second overtone. Therefore, from Table 4.1, using the corresponding expressions, we obtain

$$\bar{v}_e = (1 - 2\chi_e) = 2886 \tag{i}$$

$$2\bar{v}_e = (1 - 3\chi_e) = 5668 \tag{ii}$$

$$3\bar{v}_e = (1 - 4\chi_e) = 8347 \tag{iii}$$

Now, dividing Eq. (ii) by Eq. (i), we have

$$\frac{2(1 - 3\chi_e)}{(1 - 2\chi_e)} = \frac{5688}{2886}$$

Solving this equation, we obtain $\chi_e = 0.0174$. Substituting the value of χ_e in Eq. (i), we obtain

$$\bar{v}_e(1 - 2 \times 0.0174) = 2886$$

This gives us $\bar{v}_e = 2990$ cm^{-1}.
This shows that for a real molecule, the observed fundamental frequency and equilibrium frequency differ considerably.

Example 4.2 The fundamental absorption and first overtone transitions of NO molecule are observed at 1876.06 cm^{-1} and 3724.20 cm^{-1}, respectively. Determine the equilibrium frequency, harmonicity constant, and zero point energy.

Solution: *Given*: NO molecule containing two absorption lines:
Line 1 at 1876.06 cm^{-1} (fundamental frequency) and line 2 at 3724.20 cm^{-1} (first overtone). $\bar{v}_e = ?$ $\chi_e = ?$ and $\varepsilon_0 = ?$
For fundamental transition, we have

$$(\Delta E)_{V = 0 \to 1} = \bar{v}_e(1 - 2\chi_e) = 1876.06 \text{ cm}^{-1} \tag{i}$$

and for first overtone, we have

$$(\Delta E)_{V=0\to 2} = \bar{v}_e(1 - 3\chi_e) = 3724.20 \text{ cm}^{-1} \tag{ii}$$

Now, dividing Eq. (ii) by Eq. (i), we have

$$\frac{2(1 - 3\chi_e)}{(1 - 2\chi_e)} = \frac{3724.20}{1876.06}$$

Solving this equation, we obtain $\chi_e = 7.332 \times 10^{-3}$. Further, substituting the value of χ_e in Eq. (i), we obtain

$$\bar{v}_e\left(1 - 2 \times 7.332 \times 10^{-3}\right) = 1876.06$$

Again solving this, we obtain

$$\bar{v}_e = 1903.98 \text{ cm}^{-1}$$

The zero point energy in terms of wavenumbers is given by

$$\begin{aligned}
\varepsilon_0 &= \left(1 - \frac{\chi_e}{2}\right)\frac{\bar{v}_e}{2} \\
&= \left(1 - \frac{7.332 \times 10^{-3}}{2}\right)\frac{1903.98}{2} \\
&= \left(1 - 3.666 \times 10^{-3}\right) \times 951.99 \\
&= 948.5 \text{ cm}^{-1}
\end{aligned}$$

Polyatomic Molecules

According to the theory of molecular vibrations, a small disturbance of a molecule from its equilibrium position gives rise to vibrations of N atoms about their mean positions. Each atom in the molecule has three degrees of translational freedom and hence the molecule containing N number of atoms can be considered as a system of 3N number of harmonic oscillators with 3N degrees of freedom.

These degrees of freedom correspond to translational, rotational, and vibrational motions of the molecule. A nonlinear (angular) molecule has six degrees of freedom, three each as translational and rotational, and hence the number of vibrational degrees of freedom associated with this is given by

$$v_{vib} = 3N - 6 \tag{4.18}$$

On the other hand, a linear molecule has five degrees of freedom, three translational and two rotational (because the rotation about the molecular axis does not bring any change in the atomic coordinates), and hence the number of vibrational degrees of freedom associated with this is given by

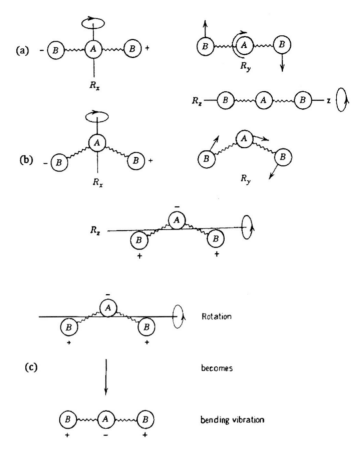

Fig. 4.7 Rotation motions in triatomic linear and nonlinear molecules

$$\nu_{vib} = 3N - 5 \tag{4.19}$$

Prima facie, it appears that a linear molecule has one more vibrational mode than its nonlinear counterpart with the same number of atoms. However, the fact is when one arbitrary rotation of nonlinear molecule makes it look linear during its rotation, then such a rotation changes into a bending vibration. Figure 4.7 shows the rotation motions in triatomic linear and nonlinear molecules.

Since a given molecule with N atoms will have (N – 1) bonds, therefore out of 3N – 6 (or 3N – 5) vibrations, (N – 1) bonds will be stretching type and the rest 2N – 5 (or 2N – 4) will be bending or torsional type.

For a diatomic molecule (N = 2), which is perfectly linear, the mode of vibration is (3 × 2) – 5 = 1. This rule simply tells us that there can be only one mode of vibration possible in a diatomic molecule but it neither says anything about the type of vibration nor provides any other information. This is actually governed by simply anharmonicity.

4.3 Relationship Between Reducible and Irreducible Representations

From Sect. 1.5, we know that a similarity transformation leaves the trace (T)/character (χ) of a matrix unchanged. Based on this principle, we can write the character of a matrix in the reducible representation in terms of the characters of the matrices in the irreducible representations as

$$\chi(R) = \sum_j n_j \chi_j(R) \tag{4.20}$$

where

$\chi(R)$ the character of the matrix corresponding to an operation R in the reducible representation,

n_j the number of times the block constituting the irreducible representation repeats itself along diagonal of the matrix, and

$\chi_j(R)$ the character of the matrix for an operation R in the jth irreducible representation.

Now, let us multiply both sides of Eq. 4.20 by $\chi_i(R)$, i.e., the character of the matrix for an operation R in ith irreducible representation. Taking the summation over all the operations of the group, we obtain

$$\sum_R \chi(R)\chi_i(R) = \sum_j n_j \chi_j(R)\chi_i(R) \tag{4.21}$$

But, from the orthogonality theorem, we know that

$$\sum_j \chi_j(R)\chi_i(R) = h\delta_{ij}$$

$$= h \text{ for } i = j$$
$$= 0 \text{ for } i \neq j$$

Therefore, Eq. 4.21 reduces to

$$\sum_R \chi(R)\chi_i(R) = \sum_j n_i h \, \delta_{ij}$$

$$= n_i h \quad (\text{for } i = j)$$

$$\text{or} \quad n_i = \frac{1}{h}\sum_R \chi(R)\chi_i(R) \tag{4.22}$$

If g_p refers to the number of operations in the pth class of the group, then Eq. 4.22 becomes

$$n_i = \frac{1}{h} \sum_{R_p} g_p \, \chi\left(R_p\right) \chi_i\left(R_p\right) \tag{4.23}$$

where R_p = a symmetry operation in the pth class. This equation is used to determine the number of times the ith irreducible representation occurs in the corresponding reducible representation.

4.4   Characters of Matrices of Some Fundamental Symmetry Operations

We know that nonlinear and linear molecules exhibit slightly different point group symmetries. Therefore, it is expected that the characters of the matrices of their fundamental symmetry operations may also differ. Let us analyze them separately.

Nonlinear (Angular) Molecules

From the initial sections of Chap. 1, we know that a nonlinear (angular) molecule (or the external shape of a crystal) can have the following fundamental symmetry elements/operations:

1	Identity element/operation	E
2	Rotation (proper)	C_n (where n = 1, 2, 3, 4 and 6)
3	Reflection (mirror)	σ
4	Inversion	i
5	Rotation (improper)	S_n

where each symmetry element/operation has its own matrix representation. Let us determine the character value (refer to Table 1.4) per un-shifted (stationary) atom/vector for all above-mentioned symmetry operations, one by one. Here, it is to be made clear that only the un-shifted or stationary atoms (vectors) after the symmetry operation will contribute to the character of the matrix.

1.   *Identity Operation* (**E**)

The identity operation E leaves all the three vectors (or the corresponding atom) unmoved as shown in Fig. 1.11. Therefore, we can write

$$x \rightarrow x, y \rightarrow y, z \rightarrow z$$

In the matrix notation, this is written as

$$E. \begin{pmatrix} x \\ y \\ z \end{pmatrix} = \begin{pmatrix} 1 & 0 & 0 \\ 0 & 1 & 0 \\ 0 & 0 & 1 \end{pmatrix} \begin{pmatrix} x \\ y \\ z \end{pmatrix} = \begin{pmatrix} x \\ y \\ z \end{pmatrix}$$

This gives $\chi(E) = 3$.

2. *Proper Rotation* (C_n)

Proper rotation axis in general is taken as the principal axis (i.e., the z-axis). The matrix corresponding to an anticlockwise rotation through an angle α is given by (Eq. 1.19)

$$C_n = \begin{pmatrix} \cos\alpha & -\sin\alpha & 0 \\ \sin\alpha & \cos\alpha & 0 \\ 0 & 0 & 1 \end{pmatrix} \tag{4.24}$$

where $\alpha = \frac{2\pi}{n}$ and n = 1, 2, 3, 4, and 6. The character of this matrix is

$$\chi(C_n) = 1 + 2\cos\alpha$$

Therefore, the character values per un-shifted (stationary) atom for different rotational symmetry operations are:

$\chi(C_1) = 1 + 2\cos 360° = 1 + 2 \times (1) = 3$ (same as identity operation)

$\chi(C_2) = 1 + 2\cos 180° = 1 + 2 \times (-1) = -1$

$\chi(C_3) = 1 + 2\cos 120° = 1 + 2 \times \left(-\frac{1}{2}\right) = 0$

$\chi(C_4) = 1 + 2\cos 90° = 1 + 2 \times (0) = 1$

$\chi(C_6) = 1 + 2\cos 60° = 1 + 2 \times \left(\frac{1}{2}\right) = 2$, etc.

The character values of various proper rotational operations can also be obtained directly from their matrices by substituting the value of α for different n in Eq. 4.24. They are

$$C_1 = E = \begin{pmatrix} 1 & 0 & 0 \\ 0 & 1 & 0 \\ 0 & 0 & 1 \end{pmatrix}$$

$$\Rightarrow \chi(C_1) = \chi(E) = 3$$

$$C_{2z} = 2(z) = 2[001] = \begin{pmatrix} -1 & 0 & 0 \\ 0 & -1 & 0 \\ 0 & 0 & 1 \end{pmatrix}, \quad \chi(C_{2z}) = -1$$

$$C_{2y} = 2(y) = 2[010] = \begin{pmatrix} -1 & 0 & 0 \\ 0 & 1 & 0 \\ 0 & 0 & -1 \end{pmatrix}, \quad \chi(C_{2y}) = -1$$

$$C_{2x} = 2(x) = 2[100] = \begin{pmatrix} 1 & 0 & 0 \\ 0 & -1 & 0 \\ 0 & 0 & -1 \end{pmatrix}, \quad \chi(C_{2x}) = -1$$

$$\Rightarrow \chi(C_2) = -1$$

$$C_{3z} = 3(z) = 3[001] = \begin{pmatrix} -1/2 & -\sqrt{3}/2 & 0 \\ \sqrt{3}/2 & -1/2 & 0 \\ 0 & 0 & 1 \end{pmatrix}, \quad \chi(C_{3z}) = 0$$

$$3[111] = \begin{pmatrix} 0 & 0 & 1 \\ 1 & 0 & 0 \\ 0 & 1 & 0 \end{pmatrix}, \quad \chi(3[111]) = 0$$

$$\Rightarrow \chi(C_3) = 0$$

$$C_{4z} = 4[001] = \begin{pmatrix} 0 & -1 & 0 \\ 1 & 0 & 0 \\ 0 & 0 & 1 \end{pmatrix}$$

$$\Rightarrow \chi(C_4) = 1$$

$$C_{6z} = 6[001] = \begin{pmatrix} 1/2 & -\sqrt{3}/2 & 0 \\ \sqrt{3}/2 & 1/2 & 0 \\ 0 & 0 & 1 \end{pmatrix}$$

$$\Rightarrow \chi(C_6) = 2$$

For improper rotational symmetry operations, we have

$$S_n = C_n \sigma_h$$

$$= \begin{pmatrix} \cos\alpha & -\sin\alpha & 0 \\ \sin\alpha & \cos\alpha & 0 \\ 0 & 0 & 1 \end{pmatrix} \begin{pmatrix} 1 & 0 & 0 \\ 0 & 1 & 0 \\ 0 & 0 & -1 \end{pmatrix}$$

$$= \begin{pmatrix} \cos\alpha & \sin\alpha & 0 \\ -\sin\alpha & \cos\alpha & 0 \\ 0 & 0 & -1 \end{pmatrix} \tag{4.25}$$

The character of the matrix is

$$\chi(S_n) = 2\cos\alpha - 1$$

Therefore, the character values per un-shifted (stationary) atom for different improper rotational symmetry operations are

$$\chi(S_1) = 2\cos 360° - 1 = 2 \times (1) - 1 = 1$$
$$\chi(S_2) = 2\cos 180° - 1 = 2 \times (-1) - 1 = -3 \text{ (same as identity operation)}$$
$$\chi(S_3) = 2\cos 120° - 1 = 2 \times \left(-\frac{1}{2}\right) - 1 = -2$$
$$\chi(S_4) = 2\cos 90° - 1 = 2 \times (0) - 1 = -1$$
$$\chi(S_6) = 2\cos 60° - 1 = 2 \times \left(\frac{1}{2}\right) - 1 = 0, \text{ etc.}$$

Similar to the above, the character values of various improper rotational operations can also be obtained directly from their matrices by substituting the value of α for different n in Eq. 4.25. They are

$$S_1 = \sigma_h = \begin{pmatrix} 1 & 0 & 0 \\ 0 & 1 & 0 \\ 0 & 0 & -1 \end{pmatrix}$$

$$\Rightarrow \chi(S_1) = \chi(s_h) = 1$$

$$S_2 = \begin{pmatrix} -1 & 0 & 0 \\ 0 & -1 & 0 \\ 0 & 0 & -1 \end{pmatrix}$$

$$\Rightarrow \chi(S_2) = -3 = \chi(i)$$

$$S_3 = \begin{pmatrix} -1/2 & \sqrt{3}/2 & 0 \\ -\sqrt{3}/2 & -1/2 & 0 \\ 0 & 0 & -1 \end{pmatrix}$$

$$\Rightarrow \chi(S_3) = -2$$

$$S_4 = \begin{pmatrix} 0 & 1 & 0 \\ -1 & 0 & 0 \\ 0 & 0 & -1 \end{pmatrix}$$

$$\Rightarrow \chi(S_4) = -1$$

$$S_6 = \begin{pmatrix} 1/2 & \sqrt{3}/2 & 0 \\ -\sqrt{3}/2 & 1/2 & 0 \\ 0 & 0 & -1 \end{pmatrix}$$

$$\Rightarrow \chi(S_6) = \chi(C_3) = 0$$

3. **Reflection (Mirror)** $\sigma_{(xz)}$

The effect of $\sigma_{(xz)}$ on an un-shifted atom is shown in Fig. 1.14. It only reverses the sense of the y vector, while x and z vectors remain unmoved, i.e.,

$$x \to x \quad y \to -y \quad z \to z$$

In the matrix notation, this is written as

$$\sigma_{(xz)} \cdot \begin{pmatrix} x \\ y \\ z \end{pmatrix} = \begin{pmatrix} 1 & 0 & 0 \\ 0 & -1 & 0 \\ 0 & 0 & 1 \end{pmatrix} \begin{pmatrix} x \\ y \\ z \end{pmatrix} = \begin{pmatrix} x \\ -y \\ z \end{pmatrix}$$

Therefore,

$$\chi\left(s_{(xz)}\right) = 1$$

Similarly, we can obtain

$$\chi\left(s_{(yz)}\right) = 1$$

4. **Inversion**

The effect of an inversion operation on an un-shifted atom is shown in Fig. 1.12. This operation simply reverses the sense of all three vectors, i.e.,

$$x \to -x \quad y \to -y \quad z \to -z$$

In the matrix notation, this is written as

$$i \cdot \begin{pmatrix} x \\ y \\ z \end{pmatrix} = \begin{pmatrix} -1 & 0 & 0 \\ 0 & -1 & 0 \\ 0 & 0 & -1 \end{pmatrix} \begin{pmatrix} x \\ y \\ z \end{pmatrix} = \begin{pmatrix} -x \\ -y \\ -z \end{pmatrix}$$

Therefore,

$$\chi(i) = -3$$

Table 4.2 Symmetry operations and the formulae of corresponding character values in reducible representation

Symmetry operation (R)	Character value $\chi(R)$ in reducible representation
E	$3N^* - 5$
C_∞^φ	$(N-2)(1+2\cos\varphi)+1$
σ_v	$N-1$
i	$1; N = 2, 4, 6, \ldots$ $-2; N = 3, 5, 7, \ldots$
S_∞^φ	$1; N = 2, 4, 6, \ldots$ $2\cos\varphi; \quad N = 3, 5, 7, \ldots$
C_2	$1; N = 2, 4, 6, \ldots$ $0; N = 3, 5, 7, \ldots$

N* refers to the number of atoms in the linear molecules

Linear Molecules

Linear molecules with or without a center of inversion can have infinite symmetry elements/operations. This makes the treatment of linear molecules different from the treatment of nonlinear molecules. The formulae of character values for symmetry elements/operations (of linear molecules) in reducible representation are provided in Table 4.2.

4.5 Determination of Overall Reducible Representation of Nonlinear Molecules

The overall reducible representation of nonlinear molecules can be determined by using two different methods. They are the First principle method and un-shifted (or stationary) atom method.

First Principle Method

In this method, the following procedure is adopted:

1. Determine the point group of the nonlinear molecule, if not known. Write its symmetries, number of classes in it, order of the point group, and the number of atoms (N) in the molecule.
2. Construct a 3N × 3N matrix in the form of reducible representation for each symmetry operation of the molecule.
3. In full (3N × 3N) matrix notation, check the effect of each symmetry operation and note down the character value corresponding to the un-shifted (or stationary) atom/vector in each case to obtain $\Gamma_{3N} = \chi(R)$.

In order to explain the first principle method clearly, let us consider the case of a bent (angular) molecule such as water for example and determine its overall reducible representation. Following the above procedure, we can write

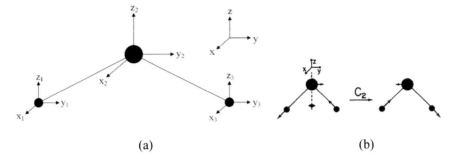

(a)   (b)

Fig. 4.8 a Displacement vectors on each atom of H_2O, **b** C_2 operation

The point group of H_2O (known): C_{2v}
Symmetry operations in the point group: E, C_2, $\sigma_{(xz)}$, and $\sigma_{(yz)}$
Number of classes in the point group: 4
Number of atoms (N) in H_2O: 3
Since N $= 3$, a (3N $\times$ 3N) $= 9 \times 9$ transformation matrix is required to represent the reducible representation of each symmetry operation.

Figure 4.8 shows the displacement vectors on each atom, representing the three degrees of freedom of that atom. Let the molecule lie on the yz plane and the x-vector comes out of the (plane) page.

Now, let us see the effect of the four symmetry operations of the point group C_{2v} on the nine individual vectors (three atoms). The identity operation, E, leaves all the nine vectors (3 atoms) unmoved, i.e.,

$$x_1 \rightarrow x_1, x_2 \rightarrow x_2, x_3 \rightarrow x_3$$
$$y_1 \rightarrow y_1, y_2 \rightarrow y_2, y_3 \rightarrow y_3$$
$$z_1 \rightarrow z_1, z_2 \rightarrow z_2, z_3 \rightarrow z_3$$

In a full matrix notation, this is written as

$$
E. \begin{bmatrix} x_1 \\ y_1 \\ z_1 \\ x_2 \\ y_2 \\ z_2 \\ x_3 \\ y_3 \\ z_3 \end{bmatrix} = \begin{bmatrix} 1\,0\,0\;0\,0\,0\;0\,0\,0 \\ 0\,1\,0\;0\,0\,0\;0\,0\,0 \\ 0\,0\,1\;0\,0\,0\;0\,0\,0 \\ 0\,0\,0\;1\,0\,0\;0\,0\,0 \\ 0\,0\,0\;0\,1\,0\;0\,0\,0 \\ 0\,0\,0\;0\,0\,1\;0\,0\,0 \\ 0\,0\,0\;0\,0\,0\;1\,0\,0 \\ 0\,0\,0\;0\,0\,0\;0\,1\,0 \\ 0\,0\,0\;0\,0\,0\;0\,0\,1 \end{bmatrix} \begin{bmatrix} x_1 \\ y_1 \\ z_1 \\ x_2 \\ y_2 \\ z_2 \\ x_3 \\ y_3 \\ z_3 \end{bmatrix} = \begin{bmatrix} x_1 \\ y_1 \\ z_1 \\ x_2 \\ y_2 \\ z_2 \\ x_3 \\ y_3 \\ z_3 \end{bmatrix}
$$

Here, all the three atoms of H_2O remain unmoved (stationary) after the identity operation, the corresponding character value of the full matrix is

$$\chi(E) = 9$$

The C_2 operation (Fig. 4.8b) exchanges the two hydrogen atoms and reverses the sense of all x and y vectors. This gives us

$$x_1 \to -x_3, x_2 \to -x_2, x_3 \to -x_1$$
$$y_1 \to -y_3, y_2 \to -y_2, y_3 \to -y_1$$
$$z_1 \to z_3, z_2 \to z_2, z_3 \to z_1$$

In a full matrix notation, this is written as

$$C_2 \cdot \begin{bmatrix} x_1 \\ y_1 \\ z_1 \\ x_2 \\ y_2 \\ z_2 \\ x_3 \\ y_3 \\ z_3 \end{bmatrix} = \begin{bmatrix} 0 & 0 & 0 & 0 & 0 & 0 & -1 & 0 & 0 \\ 0 & 0 & 0 & 0 & 0 & 0 & 0 & -1 & 0 \\ 0 & 0 & 0 & 0 & 0 & 0 & 0 & 0 & 1 \\ 0 & 0 & 0 & -1 & 0 & 0 & 0 & 0 & 0 \\ 0 & 0 & 0 & 0 & -1 & 0 & 0 & 0 & 0 \\ 0 & 0 & 0 & 0 & 0 & 1 & 0 & 0 & 0 \\ -1 & 0 & 0 & 0 & 0 & 0 & 0 & 0 & 0 \\ 0 & -1 & 0 & 0 & 0 & 0 & 0 & 0 & 0 \\ 0 & 0 & 1 & 0 & 0 & 0 & 0 & 0 & 0 \end{bmatrix} \begin{bmatrix} x_1 \\ y_1 \\ z_1 \\ x_2 \\ y_2 \\ z_2 \\ x_3 \\ y_3 \\ z_3 \end{bmatrix} = \begin{bmatrix} -x_3 \\ -y_3 \\ z_3 \\ -x_2 \\ -y_2 \\ z_2 \\ -x_1 \\ -y_1 \\ z_1 \end{bmatrix}$$

Here, the only central oxygen atom which remains stationary after the C_2 operation contributes to the overall character of the full matrix. Thus, the corresponding character value of the matrix is

$$\chi(C_2) = -1$$

The $\sigma_{(xz)}$ operation also exchanges the two hydrogen atoms like the C_2 operation shown in Fig. 4.8b. However, it reverses the sense of all y vectors only. This gives us

$$x_1 \to x_3, x_2 \to x_2, x_3 \to x_1$$
$$y_1 \to -y_3, y_2 \to -y_2, y_3 \to -y_1$$
$$z_1 \to z_3, z_2 \to z_2, z_3 \to z_1$$

In a full matrix notation, this is written as

$$\sigma_{(xz)} \cdot \begin{bmatrix} x_1 \\ y_1 \\ z_1 \\ x_2 \\ y_2 \\ z_2 \\ x_3 \\ y_3 \\ z_3 \end{bmatrix} = \begin{bmatrix} 0 & 0 & 0 & 0 & 0 & 0 & 1 & 0 & 0 \\ 0 & 0 & 0 & 0 & 0 & 0 & 0 & -1 & 0 \\ 0 & 0 & 0 & 0 & 0 & 0 & 0 & 0 & 1 \\ 0 & 0 & 0 & 1 & 0 & 0 & 0 & 0 & 0 \\ 0 & 0 & 0 & 0 & -1 & 0 & 0 & 0 & 0 \\ 0 & 0 & 0 & 0 & 0 & 1 & 0 & 0 & 0 \\ 1 & 0 & 0 & 0 & 0 & 0 & 0 & 0 & 0 \\ 0 & -1 & 0 & 0 & 0 & 0 & 0 & 0 & 0 \\ 0 & 0 & 1 & 0 & 0 & 0 & 0 & 0 & 0 \end{bmatrix} \begin{bmatrix} x_1 \\ y_1 \\ z_1 \\ x_2 \\ y_2 \\ z_2 \\ x_3 \\ y_3 \\ z_3 \end{bmatrix} = \begin{bmatrix} x_3 \\ -y_3 \\ z_3 \\ x_2 \\ -y_2 \\ z_2 \\ x_1 \\ -y_1 \\ z_1 \end{bmatrix}$$

In this case too, the only central oxygen atom which remains stationary after the $\sigma_{(xz)}$ operation contributes to the overall character of the matrix. Thus, the corresponding character value of the matrix is

$$\chi\left(\sigma_{(xz)}\right) = 1$$

Finally, the $\sigma_{(yz)}$ operation leaves all the atoms unmoved but reverses the sense of all x vectors. This gives us

$$x_1 \rightarrow -x_1, x_2 \rightarrow -x_2, x_3 \rightarrow -x_3$$
$$y_1 \rightarrow y_1, y_2 \rightarrow y_2, y_3 \rightarrow y_3$$
$$z_1 \rightarrow z_1, z_2 \rightarrow z_2, z_3 \rightarrow z_3$$

In a full matrix notation, this is written as

$$
\sigma_{(yz)} \cdot
\begin{bmatrix} x_1 \\ y_1 \\ z_1 \\ x_2 \\ y_2 \\ z_2 \\ x_3 \\ y_3 \\ z_3 \end{bmatrix}
=
\begin{bmatrix}
-1 & 0 & 0 & 0 & 0 & 0 & 0 & 0 & 0 \\
0 & 1 & 0 & 0 & 0 & 0 & 0 & 0 & 0 \\
0 & 0 & 1 & 0 & 0 & 0 & 0 & 0 & 0 \\
0 & 0 & 0 & -1 & 0 & 0 & 0 & 0 & 0 \\
0 & 0 & 0 & 0 & 1 & 0 & 0 & 0 & 0 \\
0 & 0 & 0 & 0 & 0 & 1 & 0 & 0 & 0 \\
0 & 0 & 0 & 0 & 0 & 0 & -1 & 0 & 0 \\
0 & 0 & 0 & 0 & 0 & 0 & 0 & 1 & 0 \\
0 & 0 & 0 & 0 & 0 & 0 & 0 & 0 & 1
\end{bmatrix}
\begin{bmatrix} x_1 \\ y_1 \\ z_1 \\ x_2 \\ y_2 \\ z_2 \\ x_3 \\ y_3 \\ z_3 \end{bmatrix}
=
\begin{bmatrix} -x_1 \\ y_1 \\ z_1 \\ -x_2 \\ y_2 \\ z_2 \\ -x_3 \\ y_3 \\ z_3 \end{bmatrix}
$$

Here, all the three atoms remain stationary during the $\sigma_{(yz)}$ operation, hence all the three atoms contribute to the overall character of the full matrix. Thus, the corresponding character value of the matrix is

$$\chi\left(\sigma_{(yz)}\right) = 3$$

Therefore, the overall character of the point group C_{2v} for the reducible representation $\Gamma_{3N} = \chi(R)$ is

$$\chi(E) = 9, \quad \chi(C_2) = -1, \quad \chi\left(\sigma_{(xz)}\right) = 1, \quad \chi\left(\sigma_{(yz)}\right) = 3$$

Un-shifted (or Stationary) Atom Method

From Sect. 4.4, we know the values of $\chi_s(R)$ per stationary atom for different symmetry operations. Therefore, we are left to determine the following:

1. Note down the character values $\chi_s(R)$ for all symmetry operations of the point group of the given molecule.

2. Count the number of stationary (un-shifted) atoms $N_s(R)$ for each symmetry operation (using suitable diagram) of the given molecule (point group).
3. Determine the overall character of the molecule (point group) $\chi(R)$ by using the formula

$$\Gamma_{3N} = \chi(R) = \chi_s(R) \times N_s(R) \tag{4.26}$$

where Γ_{3N} includes all the three degrees of freedom of the molecule, i.e.,

$$\Gamma_{3N} = \Gamma_{tr} + \Gamma_{rot} + \Gamma_{vib}$$

Let us again take the case of water molecule and determine its overall reducible representation. Let us write the character values $\chi_s(R)$ per stationary (un-shifted) atom for the symmetry operations, E, C_2, $\sigma_{(xz)}$, and $\sigma_{(yz)}$ and count the corresponding number of stationary atoms $N_s(R)$ in each case. Then using Eq. 4.26, we can obtain the overall character value of water molecule as

$$\chi(E) = 3 \times 3 = 9, \ \chi(C_2) = -1 \times 1 = -1,$$
$$\chi\left(\sigma_{(xz)}\right) = 1 \times 1 = 1 \text{ and } \chi\left(\sigma_{(yz)}\right) = 1 \times 3 = 3$$

The complete information can also be written in a tabulated form as

C_{2v}	E	C_2	σ_{xz}	σ_{yz}
$\chi_s(R)$	3	-1	1	1
$N_s(R)$	3	1	1	3
Γ_{3N}	9	-1	1	3

It is clear that we obtain the same overall character values from the two methods, i.e., the first principle method and un-shifted (or stationary) atom method.

4.6 Representations of Vibrational Modes of Nonlinear Molecules

There are two methods in use to determine the representations of vibrational modes of nonlinear molecules (they are similar in principle but different in approach). They are (i) by using standard reduction formula, and (ii) by preparing a tabular worksheet.

Using Standard Reduction Formula

Representations of normal modes of vibration of a given nonlinear molecule can be obtained in general when its reducible representation is decomposed (split) into a number of irreducible representations (equal to the order of the point group) by using the standard reduction formula

$$n_i = \frac{1}{h} \sum_{R_p} g_p \chi(R_p) \chi_i(R_p) \tag{4.27}$$

where

n_i	the number of times the irreducible representation, appearing in the reducible representation.
h	order of the point group.
g_p	number of operations in the pth class.
$\chi(R_p)$	character of the matrix corresponding to an operation R_p in reducible representation.
$\chi_i(R_p)$	character of the matrix corresponding to an operation R_p in irreducible representation.
R_p	symmetry operation in the pth class.

In order to decompose a reducible representation (obtained for a given molecule) into a number of irreducible representations, we need the following information:

1. Order of the point group (h) and the number of classes in the group.
2. Number of symmetry operation(s) in the pth class, g_p.
3. Overall character value of the reducible representation, $\Gamma_{3N} = \chi(R)$.
4. The character table of point group for the given molecule to get the irreducible representation character values, $\chi_i(R_p)$.

Let us take the case of water molecule (H_2O) again to obtain its normal modes of vibration; its reducible representation is already determined in the preceding section. Therefore, keeping all the information mentioned above ready, let us use the standard reduction formula (Eq. 4.27) for the required representations, i.e.,

$$n_{A1} = \frac{1}{4}[(1 \times 9 \times 1) + (1 \times -1 \times 1) + (1 \times 1 \times 1) + (1 \times 3 \times 1)] = \frac{1}{4}(9 + 3) = 3$$

$$n_{A2} = \frac{1}{4}[(1 \times 9 \times 1) + (1 \times -1 \times 1) + (1 \times 1 \times -1) + (1 \times 3 \times -1)]$$
$$= \frac{1}{4}(9 - 5) = 1$$

$$n_{B1} = \frac{1}{4}[(1 \times 9 \times 1) + (1 \times -1 \times -1) + (1 \times 1 \times 1) + (1 \times 3 \times -1)]$$
$$= \frac{1}{4}(11 - 3) = 2$$

$$n_{B2} = \frac{1}{4}[(1 \times 9 \times 1) + (1 \times -1 \times -1) + (1 \times 1 \times -1) + (1 \times 3 \times 1)]$$
$$= \frac{1}{4}(13 - 1) = 3$$

This gives us

$$\Gamma_{3N} = 3A_1 + A_2 + 2B_1 + 3B_2 = \Gamma_{tr} + \Gamma_{rot} + \Gamma_{vib}$$

Now, looking at the transformation properties in the character table of C_{2v}, we obtain

$$\Gamma_{tr} = A_1 + B_1 + B_2 \text{ and } \Gamma_{rot} = A_2 + B_1 + B_2$$
$$\Gamma_{vib} = \Gamma_{3N} - (\Gamma_{tr} + \Gamma_{rot})$$
$$= 3A_1 + A_2 + 2B_1 + 3B_2$$
$$- (A_1 + B_1 + B_2 + A_2 + B_1 + B_2)$$
$$= 2A_1 + B_2$$

Tabular Worksheet Method

The process of obtaining the reducible representation for the given molecule (say water) and splitting it into a number of irreducible representations together can be put in a tabular worksheet form. The construction of such a table is a two-step process. The first step is to determine the overall reducible representation of the molecule and then determine the irreducible representations for the vibrational modes. In this case too, all the information (all four points) mentioned above is required. All necessary steps used to obtain the reducible representation for the given molecule in tabular worksheet form are shown below.

C_{2v}	E	C_2	$\sigma_{(xz)}$	$\sigma_{(yz)}$		
$\chi_s(R)$	3	-1	1	1		
$N_s(R)$	3	1	1	3		
Γ_{3N}	9	-1	1	3	Σ	$n_i = \dfrac{\Sigma}{4}$
A_1	9	-1	1	3	12	3
A_2	9	-1	-1	-3	04	1
B_1	9	1	1	-3	08	2

Looking at the last column of the tabular worksheet, we observe that it gives the same values of irreducible representations as obtained above from the first (standard reduction) method. The rest of the calculations remain the same.

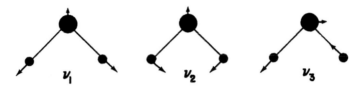

Fig. 4.9 Normal modes of vibration of water (H_2O) molecule

The above calculation shows that the representations obtained for water molecule will have three normal modes of vibration with which the molecule can vibrate. They are shown in Fig. 4.9. Clearly, there are two totally symmetric (A_1) modes and one asymmetric (B_2) mode. The two symmetric modes will vibrate with frequencies ν_1 (a symmetric stretching mode) and ν_2 (a symmetric bending mode), respectively. On the other hand, the third B_2 mode will vibrate with frequency ν_3 (also called the asymmetric stretching mode). Although the B_2 mode is less symmetric as compared to A_1 modes, the frequency ν_3 is higher than both ν_1 and ν_2. The relative ordering of observed frequencies is found to be similar in other molecules with the same point group, such as the SO_2 molecule.

In our further discussion, we shall use both methods to determine normal modes of vibration.

4.7 Vibrational Modes in Some Nonlinear Molecules

In the following, we shall determine the normal modes of vibration of some common nonlinear molecules as illustrative examples by using the two methods discussed above. This exercise will make the subject more understandable to the readers.

Example 4.3 Determine the representations of vibrational modes of hydrochlorus acid (HOCl) an angular molecule by using the standard reduction formula. The point group of the molecule is C_s.

Solution: *Given*: HOCl molecule; it has 3 atoms (i.e., $N = 3$), point group is C_s, and modes of vibration = ?

The geometry of the HOCl molecule is provided in Table 1.10. The symmetry operations of the point group C_s are E and σ_h, order of the point group, h = 2, and number of classes in the point group = 2 (because each symmetry operation has its own class).

The character values of the symmetry operations obtained per stationary atom are

$$\chi_s(E) = 3, \quad \chi_s(\sigma_h) = 1$$

The number of stationary atoms after carrying out each of the above symmetry operations is

$$N_s(E) = 3, \quad N_s(\sigma_h) = 3$$

Therefore, the overall character values $\chi(R) = \chi_s(R) \times N_s(R)$ from Eq. 4.26 for the HOCl molecule are

$$\chi(E) = 3 \times 3 = 9, \quad \chi(\sigma_h) = 1 \times 3 = 3$$

This can be written in a tabular form as

C_s	E	σ_h
$\chi_s(R)$	3	1
$N_s(E)$	3	3
Γ_{3N}	9	3

Now, using the standard reduction formula (Eq. 4.27) in combination with all required information mentioned above including the character table of the point group C_s, we obtain

$$n_{A'} = \frac{1}{2}[(1 \times 9 \times 1) + (1 \times 3 \times 1)] = \frac{1}{2}(9 + 3) = 6$$

$$n_{A''} = \frac{1}{2}[(1 \times 9 \times 1) + (1 \times 3 \times -1)] = \frac{1}{2}(9 - 3) = 3$$

This gives us

$$\Gamma_{3N} = 6A' + 3A''$$

Now, looking at the transformation properties in the character table of C_s, we obtain

$$\Gamma_{tr} = 2A' + A'' \text{ and } \Gamma_{rot} = 2A'' + A'$$

Therefore,

$$\begin{aligned} \Gamma_{vib} &= \Gamma_{3N} - (\Gamma_{tr} + \Gamma_{rot}) \\ &= 6A' + 3A'' - (2A' + A'' + 2A'' + A') \\ &= 3A' \end{aligned}$$

The representations obtained for HOCl molecule suggest that there are three normal modes of vibration. They all vibrate in symmetric (A') modes w.r.t. horizontal mirror plane.

Example 4.4 Determine the representations of vibrational modes of NH_3 (a pyramidal) molecule by using the standard reduction formula. The point group of the molecule is C_{3v}.

Solution: *Given:* NH_3 molecule; it has 4 atoms (i.e., $N = 4$), point group is C_{3v}, and modes of vibration = ?

The geometry of the NH_3 molecule is provided in Table 1.10. The symmetry operations of the point group C_{3v} are E, $2C_3$, and $3\sigma_v$, order of the point group, $h = 6$, and number of classes in the point group = 3.

The character values of the symmetry operations obtained per stationary atom are

$$\chi_s(E) = 3, \quad \chi_s(C_3) = 0 \text{ and } \chi_s(\sigma_v) = 1$$

The number of stationary atoms after carrying out each of the above symmetry operations is

$$N_s(E) = 4, \quad N_s(C_3) = 1 \text{ and } N_s(\sigma_v) = 2$$

Therefore, the overall character values $\chi(R) = \chi_s(R) \times N_s(R)$ from Eq. 4.26 for NH_3 molecule are

$$\chi(E) = 3 \times 4 = 12, \quad \chi(C_3) = 0 \times 1 = 0 \text{ and } \chi(\sigma_v) = 1 \times 2 = 2$$

Now, using the standard reduction formula (Eq. 4.27) in combination with all required information mentioned above including the character table of the point group C_{3v}, we obtain

$$n_{A1} = \frac{1}{6}[(1 \times 12 \times 1) + (2 \times 0 \times 1) + (3 \times 2 \times 1)] = \frac{1}{6}(12 + 0 + 6) = 3$$

$$n_{A2} = \frac{1}{6}[(1 \times 12 \times 1) + (2 \times 0 \times 1) + (3 \times 2 \times -1)] = \frac{1}{6}(12 + 0 - 6) = 1$$

$$n_E = \frac{1}{6}[(1 \times 12 \times 2) + (2 \times 0 \times -1) + (3 \times 2 \times 0)] = \frac{1}{6}(24 - 0 + 0) = 4$$

This gives us

$$\Gamma_{3N} = 3A_1 + A_2 + 4E = \Gamma_{tr} + \Gamma_{rot} + \Gamma_{vib}$$

Now, looking at the transformation properties in the character table of the point group C_{3v}, we obtain

$$\Gamma_{tr} = A_1 + E \text{ and } \Gamma_{rot} = A_2 + E$$

Therefore,

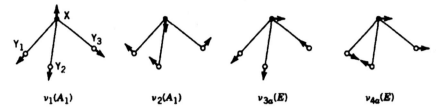

$$v_1(A_1) \qquad v_2(A_1) \qquad v_{3a}(E) \qquad v_{4a}(E)$$

Fig. 4.10 Normal modes of vibration of NH$_3$ molecule

$$\Gamma_{vib} = \Gamma_{3N} - (\Gamma_{tr} + \Gamma_{rot})$$
$$= 3A_1 + A_2 + 4E - (A_1 + E + A_2 + E)$$
$$= 2A_1 + 2E$$

The representations obtained for the NH$_3$ molecule suggest that there are in all six normal modes with which the molecule will vibrate but are limited to four discrete energies and hence frequencies. There are two symmetric (A$_1$) modes with frequencies v_1 as symmetric stretching and v_2 as symmetric bending. Further, one pair of modes, v_3 (E), is doubly degenerate and gives rise to a single frequency. Similarly, another pair of modes, v_4 (E), is also doubly degenerate and gives rise to another single frequency. They are shown in Fig. 4.10. The relative ordering of the observed frequencies is similar in other molecules with the same point groups, such as $(CO_3)^{2-}$.

Example 4.5 Determine the representations of vibrational modes of trans-dinitrogendifluoride (N$_2$F$_2$) molecule by using the tabular worksheet method. The point group of the molecule is C$_{2h}$.

Solution: *Given:* N$_2$F$_2$ molecule; it has 4 atoms (i.e., N = 4), point group is C$_{2h}$, and modes of vibration = ?

The geometry of the N$_2$F$_2$ molecule is provided in Table 1.10. The symmetry operations of the point group C$_{2h}$ are E, C$_2$, i, and σ_h, order of the point group, h = 4, and number of classes in the point group = 4 (because each symmetry operation has its own class).

The character values of the symmetry operations obtained per stationary atom are

$$\chi_s(E) = 3, \quad \chi_s(C_2) = -1, \quad \chi_s(i) = -3 \text{ and } \chi_s(\sigma_h) = 1$$

The number of stationary atoms after carrying out each of the above symmetry operations is

$$N_s(E) = 4, \quad N_s(C_2) = 0, \quad N_s(i) = 0 \text{ and } N_s(\sigma_h) = 4$$

Further, making use of the character table and the information mentioned above, the tabular form of the worksheet can be prepared as given below.

C_{2h}	E	C_2	i	σ_h		
$\chi_s(R)$	3	-1	-3	1		
$N_s(R)$	4	0	0	4		
Γ_{3N}	12	0	0	4	Σ	$n_i = \dfrac{\Sigma}{4}$
A_g	12	0	0	4	16	4
B_g	12	0	0	-4	08	2
A_u	12	0	0	-4	08	2
B_u	12	0	0	4	16	4

This gives us

$$\Gamma_{3N} = 4A_g + 2B_g + 2A_u + 4B_u$$

Now, looking at the transformation properties in the character table of the point group C_{2h}, we obtain

$$\Gamma_{tr} = A_u + 2B_u \text{ and } \Gamma_{rot} = A_g + 2B_g$$

Therefore,

$$\begin{aligned}
\Gamma_{vib} &= \Gamma_{3N} - (\Gamma_{tr} + \Gamma_{rot}) \\
&= 4A_g + 2B_g + 2A_u + 4B_u - (A_u + 2B_u + A_g + 2B_g) \\
&= 3A_g + A_u + 2B_u
\end{aligned}$$

The representations obtained for the N_2F_2 molecule suggest that there are in all six normal modes with which the molecule will vibrate. Three of them are symmetric (A_g) modes w.r.t. inversion, one (A_u) mode is symmetric w.r.t. C_2 (principal) axis but asymmetric with inversion, while two (B_u) modes are asymmetric to both C_2 and i.

Example 4.6 Determine the representations of vibrational modes of formaldehyde (H_2CO) molecule by using the tabular worksheet method. The point group of the molecule is C_{2v}.

Solution: *Given:* H_2CO molecule; it has 4 atoms (i.e., $N = 4$), point group is C_{2v}, and modes of vibration $= ?$

The geometry of the H_2CO molecule is provided in Table 1.10. The symmetry operations of the point group C_{2v} are E, C_2, $\sigma_{(xz)}$, and $\sigma_{(yz)}$, order of the point group, h $= 4$, and number of classes in the point group $= 4$ (because each symmetry operation is a class of its own).

The character values of the symmetry operations obtained per stationary atom are

$$\chi_s(E) = 3, \quad \chi_s(C_2) = -1, \quad \chi_s(\sigma_{(xz)}) = 1 \text{ and } \chi_s(\sigma_{(yz)}) = 1$$

The number of stationary atoms after carrying out each of the above symmetry operations is

$$N_s(E) = 4, \quad N_s(C_2) = 2, \quad N_s(\sigma_{(xz)}) = 2 \text{ and } N_s(\sigma_{(yz)}) = 4$$

Further, making use of the character table and the information mentioned above, the tabular form of the worksheet can be prepared as given below.

C_{2v}	E	C_2	$\sigma_{(xz)}$	$\sigma_{(yz)}$		
$\chi_s(R)$	3	-1	1	1		
$N_s(R)$	4	2	2	4		
Γ_{3N}	12	-2	2	4	Σ	$n_i = \dfrac{\Sigma}{4}$
A_1	12	-2	2	4	16	4
A_2	12	-2	-2	-4	04	1
B_1	12	2	2	-4	12	3
B_2	12	2	-2	4	16	4

This gives us

$$\Gamma_{3N} = 4A_1 + 3B_1 + A_2 + 4B_2$$

Now, looking at the transformation properties in the character table of the point group C_{2v}, we obtain

$$\Gamma_{tr} = A_1 + B_1 + B_2 \text{ and } \Gamma_{rot} = A_2 + B_1 + B_2$$

Therefore,

$$\begin{aligned}
\Gamma_{vib} &= \Gamma_{3N} - (\Gamma_{tr} + \Gamma_{rot}) \\
&= 4A_1 + 3B_1 + A_2 + 4B_2 - (A_1 + B_1 + B_2 + A_2 + B_1 + B_2) \\
&= 3A_1 + B_1 + 2B_2
\end{aligned}$$

The representations obtained for the formaldehyde (H_2CO) molecule suggest that there are in all six normal modes with which the molecule will vibrate. Three of them are symmetric (A_1) modes w.r.t. vertical mirror, one (B_1) mode is symmetric w.r.t. C_2 (principal) axis but asymmetric with inversion, while two (B_2) modes are asymmetric to both C_2 and vertical mirror.

Example 4.7 Determine the representations of vibrational modes of Cis- and Trans-forms 1, 2-dichloroethylene $(CHCl)_2$ molecule by using the standard reduction formula. The point groups of Cis- and Trans-forms of the molecule are C_{2v} and C_{2h}, respectively.

Fig. 4.11 Cis-form of dichloroethylene

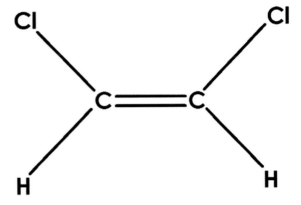

Solution: *Given*: Cis- and Trans-forms of 1, 2-dichloroethylene $(CHCl)_2$ molecule; it has 6 atoms (i.e., N = 6), point groups are respectively C_{2v} and C_{2h}, and modes of vibration = ?

Since the point groups of Cis- and Trans-forms are different, let us take the two cases separately.

(a) Cis-form of the molecule: Its chemical formula is shown in Fig. 4.11. The symmetry operations of the point group C_{2v} are E, C_2, $\sigma_{(xz)}$, and $\sigma_{(yz)}$, order of the point group, h = 4, and number of classes in the point group = 4 (because each symmetry operation is a class of its own).

The character values of the symmetry operations obtained per stationary atom are

$$\chi_s(E) = 3, \quad \chi_s(C_2) = -1, \quad \chi_s\left(\sigma_{(xz)}\right) = 1 \text{ and } \chi_s\left(\sigma_{(yz)}\right) = 1$$

The number of stationary atoms after carrying out each of the above symmetry operations is

$$N_s(E) = 6, \quad N_s(C_2) = 0, \quad N_s\left(\sigma_{(xz)}\right) = 0 \text{ and } N_s\left(\sigma_{(yz)}\right) = 6$$

Further, making use of the character table and the information mentioned above, and using a standard formula, we obtain

$$n_{A1} = \frac{1}{4}[(1 \times 18 \times 1) + (1 \times 0 \times 1) + (1 \times 0 \times 1) + (1 \times 6 \times 1)]$$
$$= \frac{1}{4}(18 + 6) = 6$$

$$n_{A2} = \frac{1}{4}[(1 \times 18 \times 1) + 0 + 0 + (1 \times 6 \times -1)] = \frac{1}{4}(18 - 6) = 3$$

$$n_{B1} = \frac{1}{4}[(1 \times 18 \times 1) + 0 + 0 + (1 \times 6 \times -1)] = \frac{1}{4}(18 - 6) = 3$$

$$n_{B2} = \frac{1}{4}[(1 \times 18 \times 1) + 0 + 0 + (1 \times 6 \times 1)] = \frac{1}{4}(18 + 6) = 6$$

This gives us

$$\Gamma_{3N} = 6A_1 + 3B_1 + 3A_2 + 6B_2$$

Now, looking at the transformation properties in the character table of the point group C_{2v}, we obtain

$$\Gamma_{tr} = A_1 + B_1 + B_2 \text{ and } \Gamma_{rot} = A_2 + B_1 + B_2$$

Therefore,

$$\begin{aligned}
\Gamma_{vib} &= \Gamma_{3N} - (\Gamma_{tr} + \Gamma_{rot}) \\
&= 6A_1 + 3B_1 + 3A_2 + 6B_2 - (A_1 + B_1 + B_2 + A_2 + B_1 + B_2) \\
&= 5A_1 + 2A_2 + B_1 + 4B_2
\end{aligned}$$

In all, there are twelve symmetric and asymmetric modes of vibration.

(b) Trans-form of the molecule: Its chemical formula is shown in Fig. 4.12; it has 6 atoms (i.e., $N = 6$). The symmetry operations of the point group C_{2h} are E, C_2, i, and σ_h, order of the point group, $h = 4$, and number of classes in the point group $= 4$ (because each symmetry operation is a class of its own), hence there are four irreducible representations.

The character values of the symmetry operations obtained per stationary atom are

$$\chi_s(E) = 3, \quad \chi_s(C_2) = -1, \quad \chi_s(i) = -3 \text{ and } \chi_s(\sigma_h) = 1$$

Fig. 4.12 Trans-form of dichloroethylene

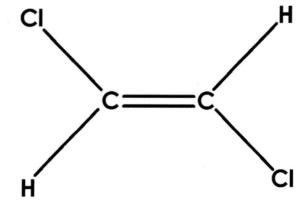

The number of stationary atoms after carrying out each of the above symmetry operations is

$$N_s(E) = 6, \quad N_s(C_2) = 0, \quad N_s(i) = 0 \text{ and } N_s(\sigma_h) = 6$$

Further, making use of the character table and the information mentioned above, and using a standard formula, we obtain

$$n_{A_g} = \frac{1}{4}[(1 \times 18 \times 1) + (1 \times 0 \times 1) + (1 \times 0 \times 1) + (1 \times 6 \times 1)] = \frac{1}{4}(18 + 6) = 6$$

$$n_{A_u} = \frac{1}{4}[(1 \times 18 \times 1) + 0 + 0 + (1 \times 6 \times -1)] = \frac{1}{4}(18 - 6) = 3$$

$$n_{B_g} = \frac{1}{4}[(1 \times 18 \times 1) + 0 + 0 + (1 \times 6 \times -1)] = \frac{1}{4}(18 - 6) = 3$$

$$n_{B_u} = \frac{1}{4}[(1 \times 18 \times 1) + 0 + 0 + (1 \times 6 \times 1)] = \frac{1}{4}(18 + 6) = 6$$

This gives us

$$\Gamma_{3N} = 6A_g + 3B_g + 3A_u + 6B_u$$

Now, looking at the transformation properties in the character table of the point group C_{2h}, we obtain

$$\Gamma_{tr} = A_u + 2B_u \text{ and } \Gamma_{rot} = A_g + 2B_g$$

Therefore,

$$\begin{aligned}
\Gamma_{vib} &= \Gamma_{3N} - (\Gamma_{tr} + \Gamma_{rot}) \\
&= 6A_g + 3B_g + 3A_u + 6B_u - \left(A_u + 2B_u + A_g + 2B_g\right) \\
&= 5A_g + 2A_u + B_g + 4B_u
\end{aligned}$$

In this case too, there are a total of twelve symmetric and asymmetric modes of vibration. However, the modes of vibration in the two cases are slightly different. In the previous case, the symmetric or asymmetric mode was w.r.t. mirror plane, but in this case the same is w.r.t. inversion.

Example 4.8 Determine the representations of vibrational modes of the Allene (C_3H_4) molecule by using the tabular worksheet method. The point group of the molecule is D_{2d}.

Solution: *Given:* C_3H_4 molecule; it has 7 atoms (i.e., N = 7), point group is D_{2d}, and modes of vibration = ?

The symmetry elements of the C_3H_4 molecule are shown in Fig. 4.13. The

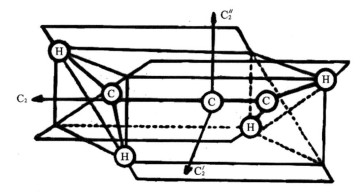

Fig. 4.13 symmetry elements of C_3H_4 molecule

symmetry operations of the point group D_{2d} are E, $2S_4$, C_2, 2 C'_2, and $2\sigma_d$, order of the point group, h = 8, and number of classes in the point group = 5.

The character values of the symmetry operations obtained per stationary atom are

$$\chi_s(E) = 3, \quad \chi_s(S_4) = -1, \quad \chi_s(C_2) = -1, \quad \chi_s(C'_2) = -1 \text{ and } \chi_s(\sigma_d) = 1$$

The number of stationary atoms after carrying out each of the above symmetry operations is

$$N_s(E) = 7, \quad N_s(C_2) = 1, \quad N_s(S_4) = 3, \quad N_s(C'_2) = 1 \text{ and } N_s(\sigma_d) = 5$$

Further, making use of the character table and the information mentioned above, the tabular form of the worksheet can be prepared as given below.

D_{2d}	E	$2S_4$	C_2	$2C'_2$	$2\sigma_d$		
$\chi_s(R)$	3	-1	-1	-1	1		
$N_s(R)$	7	1	3	1	5		
Γ_{3N}	21	-1	-3	-1	5	Σ	$n_i = \dfrac{\Sigma}{8}$
A_1	21	-2	-3	-2	10	24	3
A_2	21	-2	-3	2	-10	08	1
B_1	21	2	-3	-2	-10	08	1
B_2	21	2	-3	2	10	32	4
E	42	0	6	0	0	48	6

This gives us

$$\Gamma_{3N} = 3A_1 + B_1 + A_2 + 4B_2 + 6E$$

Now, looking at the transformation properties in the character table of the point group D_{2d}, we obtain

$$\Gamma_{tr} = B_2 + E \text{ and } \Gamma_{rot} = A_2 + E$$

Therefore,

$$
\begin{aligned}
\Gamma_{vib} &= \Gamma_{3N} - (\Gamma_{tr} + \Gamma_{rot}) \\
&= 3A_1 + B_1 + A_2 + 4B_2 + 6E - (B_2 + E + A_2 + E) \\
&= 3A_1 + B_1 + 3B_2 + 4E
\end{aligned}
$$

The representations obtained for allene (C_3H_4) molecule suggest that there are in all fifteen normal modes with which the molecule will vibrate, where four (E) modes are doubly degenerate. There are three symmetric (A_1) modes w.r.t. C_2' or σ_d (dihedral mirror), one (B_1) asymmetric mode w.r.t. C_2 (principal) axis, and three (B_2) asymmetric modes w.r.t. both C_2 and σ_d, respectively.

Example 4.9 Determine the representations of vibrational modes of square planar XY_4 type molecule such as XeF_4 or $(PtCl_4)^{-2}$ by using the tabular worksheet method. The point group of the molecule is D_{4h}.

Solution: *Given*: Square planar XY_4 type such as XeF_4 or $(PtCl_4)^{-2}$ molecule; it has 5 atoms (i.e., $N = 5$), point group is D_{4h}, and modes of vibration $= ?$

The geometry of the molecule is shown in Fig. 4.14. The symmetry operations of the point group D_{4h} are E, $2C_4$, C_2, $2\,C_2'$, $2\,C_2''$, i, $2S_4$, σ_h, $2\sigma_v$, and $2\sigma_d$, order of the point group, $h = 16$, and number of classes in the point group $= 10$.

The character values of the symmetry operations obtained per stationary atom are

$$\chi_s(E) = 3, \chi_s(C_4) = 1, \chi_s(C_2) = -1, \chi_s(C_2') = -1, \chi_s(C_2'') = -1, \chi_s(i) = -3,$$
$$\chi_s(S_4) = -1, \chi_s(\sigma_h) = 1, \chi_s(\sigma_v) = 1 \text{ and } \chi_s(\sigma_d) = 1$$

The number of stationary atoms after carrying out each of the above symmetry operations is

$$N_s(E) = 5, N_s(C_4) = 1, N_s(C_2) = 1, N_s(C_2') = 3, N_s(C_2'') = 1,$$

$$N_s(i) = 1, N_s(S_4) = 1, N_s(\sigma_h) = 5, N_s(\sigma_v) = 3 \text{ and } N_s(\sigma_d) = 1$$

Further, making use of the character table and the information mentioned above, the tabular form of the worksheet can be prepared as given below.

D_{4h}	E	$2C_4$	C_2	$2C_2'$	$2C_2''$	i	$2S_4$	σ_h	$2\sigma_v$	$2\sigma_d$		
$\chi_s(R)$	3	1	-1	-1	-1	-3	-1	1	1	1		
$N_s(R)$	5	1	1	3	1	1	1	5	3	1		
Γ_{3N}	15	1	-1	-3	-1	-3	-1	5	3	1	Σ	$n_i = \dfrac{\Sigma}{16}$
A_{1g}	15	2	-1	-6	-2	-3	-2	5	6	2	16	1
A_{2g}	15	2	-1	6	2	-3	-2	5	-6	-2	16	1
B_{1g}	15	-2	-1	-6	2	-3	2	5	6	-2	16	1
B_{2g}	15	-2	-1	6	-2	-3	2	5	-6	2	16	1
E_g	30	0	2	0	0	-6	0	-10	0	0	16	1
A_{1u}	15	2	1	-6	-2	3	2	-5	-6	-2	00	0
A_{2u}	15	2	1	6	2	3	2	-5	6	2	32	2
B_{1u}	15	-2	1	-6	2	3	-2	-5	-6	2	00	0
B_{2u}	15	-2	1	6	-2	3	-2	-5	6	-2	16	1
E_u	30	0	2	0	0	6	0	10	0	0	48	3

This gives us

$$\Gamma_{3N} = A_{1g} + A_{2g} + B_{1g} + B_{2g} + E_g + 2A_{2u} + B_{2u} + 3E_u$$

Now, looking at the transformation properties in the character table of the point group D_{4h}, we obtain

$$\Gamma_{tr} = A_{2u} + E_u \text{ and } \Gamma_{rot} = A_{2g} + E_g$$

Therefore,

$$\begin{aligned}
\Gamma_{vib} &= \Gamma_{3N} - (\Gamma_{tr} + \Gamma_{rot}) \\
&= A_{1g} + A_{2g} + B_{1g} + B_{2g} + E_g + 2A_{2u} + B_{2u} + 3E_u \\
&\quad - (A_{2u} + E_u + A_{2g} + E_g) \\
&= A_{1g} + B_{1g} + B_{2g} + E_g + A_{2u} + B_{2u} + 2E_u
\end{aligned}$$

The representations obtained for XeF$_4$ or $(PtCl_4)^{-2}$ molecule suggest that there are in all nine normal modes with which the molecule will vibrate, where two (E_u) modes are doubly degenerate.

Here, the symmetries with the subscript "g" are symmetric while that with the subscript "u" are asymmetric w.r.t. inversion. Similarly, the symmetries with the subscript "1" are symmetric while the symmetries with the subscript "2" are asymmetric w.r.t. twofold principal axis.

Example 4.10 Determine the representations of vibrational modes of triangular planar BF$_3$ molecule by using the tabular worksheet method. The point group of the molecule is D_{3h}.

Fig. 4.14 Geometry of
square planar $(PtCl_4)^{-2}$
molecule

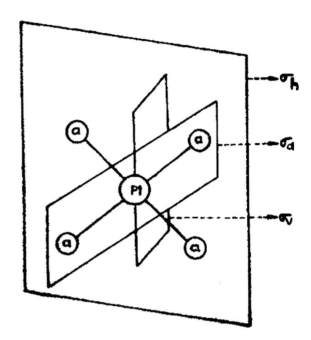

Solution: *Given*: Triangular planar BF_3 molecule; it has 4 atoms (i.e., $N = 4$), point
group is D_{3h}, and modes of vibration = ?

The geometry of the molecule is shown in Fig. 4.15. The symmetry operations of
the point group D_{3h} are E, $2C_3$, $3C_2$, $2S_3$, σ_h, and $3\sigma_v$, order of the point group, h =
12, and number of classes in the point group = 6.

The character values of the symmetry operations obtained per stationary atom are

Fig. 4.15 Triangular planar
BF_3 molecule

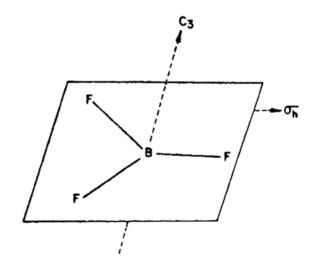

$\chi_s(E) = 3$, $\chi_s(C_3) = 0$, $\chi_s(C_2) = -1$, $\chi_s(\sigma_h) = 1$, $\chi_s(S_3) = -2$ and $\chi_s(\sigma_v) = 1$

The number of stationary atoms after carrying out each of the above symmetry operations is

$N_s(E) = 4$, $N_s(C_3) = 1$, $N_s(C_2) = 2$, $N_s(\sigma_h) = 4$, $N_s(S_3) = 1$ and $N_s(\sigma_v) = 2$

Let us write the character values per stationary atom for all symmetry operations and their corresponding number of stationary atoms in each case to obtain the overall character of the molecule.

D_{3h}	E	$2C_3$	$3C_2$	σ_h	$2 S_3$	$3\sigma_v$
$\chi_s(R)$	3	0	−1	1	−2	1
$N_s(R)$	4	1	2	4	1	2
Γ_{3N}	12	0	−2	4	−2	2

Further, making use of the character table and the information mentioned above, the tabular form of the worksheet can be prepared as given below.

D_{3h}	E	$2C_3$	$3C_2$	σ_h	$2S_3$	$3\sigma_v$		
$\chi_s(R)$	3	0	-1	1	-2	1		
$N_s(R)$	4	1	2	4	1	2		
Γ_{3N}	12	0	-2	4	-2	2	Σ	$n_i = \dfrac{\Sigma}{12}$
A_1'	12	0	-6	4	-4	6	12	1
A_2'	12	0	6	4	-4	-6	12	1
E'	24	0	0	8	4	0	36	3
A_1''	12	0	-6	-4	4	-6	00	0
A_2''	12	0	6	-4	4	6	24	2
E''	24	0	0	-8	-4	0	12	1

This gives us

$$\Gamma_{3N} = A_1' + A_2' + 3E' + 2A_2'' + E''$$

Now, from the character table of D_{3h}, we obtain

$$\Gamma_{tr} = A_2'' + E' \text{ and } \Gamma_{rot} = A_2' + E''$$

Therefore,

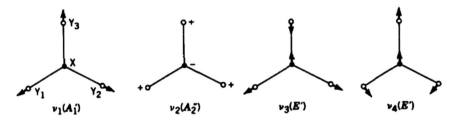

Fig. 4.16 Normal modes of vibration of BF_3 molecule

$$\Gamma_{vib} = \Gamma_{3N} - (\Gamma_{tr} + \Gamma_{rot})$$
$$= A_1' + A_2' + 3E' + 2A_2'' + E'' - (A_2'' + E' + A_2' + E'')$$
$$= A_1' + 2E' + A_2''$$

The representations obtained for BF_3 molecule suggest that there are in all six normal modes of vibration but are limited to four discrete energies and hence four fundamental frequencies. In this case, completely symmetric (A_1') mode is assigned with the frequency v_1 and the asymmetric (A_2'') mode (w.r.t. σ_h) is assigned with the frequency v_2, respectively. Further, one pair of modes, symmetric with σ_h, is doubly degenerate and gives rise to a single frequency v_3 (E'). Similarly, another pair of modes, also symmetric with σ_h, is doubly degenerate and gives rise to another single frequency v_4 (E'). They are shown in Fig. 4.16. The relative ordering of observed frequencies is similar in other molecules with the same point group, such as the BCl_3 molecule.

Example 4.11 Determine the representations of vibrational modes of diborane (B_2H_6) molecule by using the tabular worksheet method. The point group of the molecule is D_{2h}.

Solution: *Given*: B_2H_6 (diborane) molecule; it has 8 atoms (i.e., N = 8), point group is D_{2h}, and modes of vibration = ?

The geometry of the molecule is shown in Fig. 4.17. The symmetry operations of the point group D_{2h} are E, C_{2z}, C_{2y}, C_{2x}, i, $\sigma_{(xy)}$, $\sigma_{(xz)}$, and $\sigma_{(yz)}$, order of the point group, h = 8, and number of classes in the point group = 8 (because each symmetry operation is a class of its own).

The character values of the symmetry operations obtained per stationary atom are

$$\chi_s(E) = 3, \ \chi_s(C_{2z}) = -1, \ \chi_s(C_{2y}) = -1, \ \chi_s(C_{2x}) = -1, \ \chi_s(i) = -3,$$

Fig. 4.17 Geometry of B_2H_6 (diborane) molecule

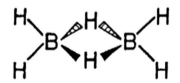

$$\chi_s(\sigma_{(xy)}) = 1, \ \chi_s(\sigma_{(xz)}) = 1 \text{ and } \chi_s(\sigma_{(yz)}) = 1$$

The number of stationary atoms after carrying out each of the above symmetry operations is

$$N_s(E) = 8, \ N_s(C_{2z}) = 0, \ N_s(C_{2y}) = 2, \ N_s(C_{2x}) = 2, \ N_s(i) = 0, \ N_s(\sigma_{(xy)}) = 4,$$

$$N_s(\sigma_{(xz)}) = 6 \text{ and } N_s(\sigma_{(yz)}) = 2$$

Further, making use of the character table and the information mentioned above, the tabular form of the worksheet can be prepared as given below.

D_{2h}	E	C_{2z}	C_{2y}	C_{2x}	i	$\sigma_{(xy)}$	$\sigma_{(xz)}$	$\sigma_{(yz)}$		
$\chi_s(R)$	3	-1	-1	-1	-3	1	1	1		
$N_s(R)$	8	0	2	2	0	4	6	2		
Γ_{3N}	24	0	-2	-2	0	4	6	2	Σ	$n_i = \dfrac{\Sigma}{8}$
A_g	24	0	-2	-2	0	4	6	2	32	4
B_{1g}	24	0	2	2	0	4	-6	-2	24	3
B_{2g}	24	0	-2	2	0	-4	6	-2	24	3
B_{3g}	24	0	2	-2	0	-4	-6	2	16	2
A_u	24	0	-2	-2	0	-4	-6	-2	08	1
B_{1u}	24	0	2	2	0	-4	6	2	32	4
B_{2u}	24	0	-2	2	0	4	-6	2	24	3
B_{3u}	24	0	2	-2	0	4	6	-2	32	4

This gives us

$$\Gamma_{3N} = 4A_g + 3B_{1g} + 3B_{2g} + 2B_{3g} + A_u + 4B_{1u} + 3B_{2u} + 4B_{3u}$$

Now, looking at the transformation properties in the character table of the point group D_{2h}, we obtain

$$\Gamma_{tr} = B_{1u} + B_{2u} + B_{3u} \text{ and } \Gamma_{rot} = B_{1g} + B_{2g} + B_{3g}$$

Therefore,

$$\begin{aligned}
\Gamma_{vib} &= \Gamma_{3N} - (\Gamma_{tr} + \Gamma_{rot}) \\
&= 4A_g + 3B_{1g} + 3B_{2g} + 2B_{3g} + A_u + 4B_{1u} + 3B_{2u} + 4B_{3u} \\
&\quad - (B_{1u} + B_{2u} + B_{3u} + B_{1g} + B_{2g} + B_{3g})
\end{aligned}$$

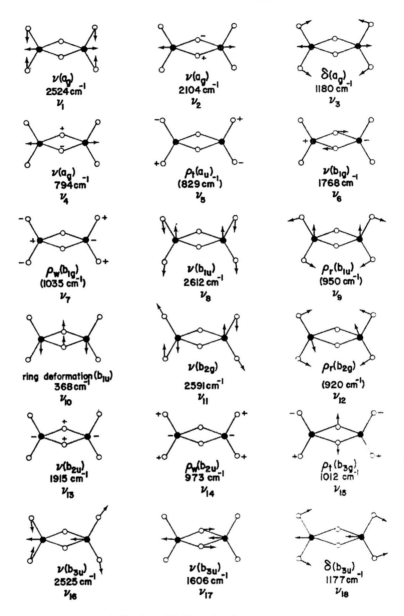

Fig. 4.18 Normal modes of vibration of B_2H_6 molecule

$$= 4A_g + 2B_{1g} + 2B_{2g} + B_{3g} + A_u + 3B_{1u} + 2B_{2u} + 3B_{3u}$$

The representations obtained for the B_2H_6 molecule suggest that there are in all eighteen normal modes of vibration with which the molecule will vibrate. They are shown in Fig. 4.18. This molecule is used to illustrate numbering and descriptive naming (such as rocking, wagging, and twisting) of different modes of vibration.

The convention for numbering the normal modes of vibration is to put the totally symmetric modes on the top of the list and then put others in the descending order of symmetry.

Example 4.12 Determine the representations of vibrational modes of a tetrahedral XY_4 (e.g., CH_4, CCl_4, SO_4^{2-}, SF_4, and PO_4) molecule by using the tabular worksheet method. The point group of the molecule is T_d.

Solution: *Given:* XY_4 molecule; it has 5 atoms (i.e., $N = 5$), point group is T_d, and modes of vibration $= ?$

The geometry of the tetrahedral XY_4 molecule is provided in Table 1.10. The symmetry operations of the point group T_d are E, $8C_3$, $3C_2$, $6S_4$, and $6\sigma_d$, order of the point group, $h = 24$, and number of classes in the point group $= 5$.

The character values of the symmetry operations obtained per stationary atom are

$$\chi_s(E) = 3, \ \chi_s(C_3) = 0, \ \chi_s(C_2) = -1, \ \chi_s(S_4) = -1 \text{ and } \chi_s(\sigma_d) = 1$$

The number of stationary atoms after carrying out each of the above symmetry operations is

$$N_s(E) = 5, N_s(C_3) = 2, N_s(C_2) = 1, N_s(S_4) = 1 \text{ and } N_s(\sigma_d) = 3$$

Further, making use of the character table and the information mentioned above, the tabular form of the worksheet can be prepared as given below.

T_d	E	$8C_3$	$3C_2$	$6S_4$	$6\sigma_d$		
$\chi_s(R)$	3	0	-1	-1	1		
$N_s(R)$	5	2	1	1	3		
Γ_{3N}	15	0	-1	-1	3	Σ	$n_i = \dfrac{\Sigma}{24}$
A_1	15	0	-3	-6	18	24	1
A_2	15	0	-3	6	-18	00	0
E	30	0	-6	0	0	24	1
T_1	45	0	3	-6	-18	24	1
T_2	45	0	3	6	18	72	3

This gives us

$$\Gamma_{3N} = A_1 + E + T_1 + 3T_2$$

Now, looking at the transformation properties in the character table of the point group T_d, we obtain

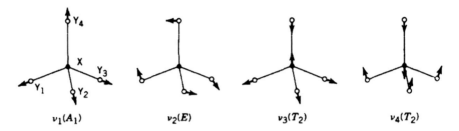

$$v_1(A_1) \qquad\qquad v_2(E) \qquad\qquad v_3(T_2) \qquad\qquad v_4(T_2)$$

Fig. 4.19 Normal modes of vibration of (XY_4) molecule

$$\Gamma_{tr} = T_2 \text{ and } \Gamma_{rot} = T_1$$

Therefore,

$$\begin{aligned}
\Gamma_{vib} &= \Gamma_{3N} - (\Gamma_{tr} + \Gamma_{rot}) \\
&= A_1 + E + T_1 + 3T_2 - (T_2 + T_1) \\
&= A_1 + E + 2T_2
\end{aligned}$$

The representations obtained for tetrahedral (XY_4) molecule suggest that there are in all nine normal modes of vibration but are limited to four energies and hence four fundamental frequencies. They are shown in Fig. 4.19. The frequency v_1 is a symmetric stretching mode. The frequency v_2 is doubly degenerate and arises from two bending deformations. The frequency v_3 is triply degenerate and arises from three asymmetric stretching modes. Finally, the frequency v_4 is also triply degenerate and arises from three bending motions.

Example 4.13 Determine the representations of vibrational modes of fullerene (C_{60}) molecule by using the tabular worksheet method. The point group of the molecule is I_h.

Solution: *Given*: Fullerene (C_{60}) molecule; it has 60 atoms (i.e., $N = 60$), point group is I_h, and modes of vibration = ?

The geometry and symmetry of the molecule are shown in Fig. 4.20. The symmetry operations of the point group I_h are

$$E, 12C_5, 12C_5^2, 20C_3, 15C_2, i, 12S_{10}, 12S_{10}^3, 20S_6, 15\sigma_2$$

Order of the point group, $h = 120$, and number of classes in the point group = 10. Hence, the number of irreducible representations = 10.

The fullerene C_{60} is a rare example of the most symmetric molecule belonging to the icosahedral point group I_h. The $3N - 6$ rule suggests the possibility of 174 normal modes of vibration for fullerene molecule and on the face of it, the problem of deriving the same appears to be a formidable task. The character table also looks not very friendly. The group of order $h = 120$ with 10 irreducible representations including fourfold (G) and fivefold (H) degenerate modes has not been encountered in other molecular point groups so far. However, looking at the geometry and symmetry

Fig. 4.20 The Fullerene (C_{60}) molecule

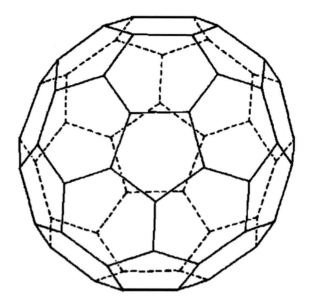

of the molecule, the task of finding the normal modes of vibration in fullerene turned out to be simple because none of the rotational symmetry axes are passing through any atomic position, hence their corresponding character values are zero. Similarly, the character under inversion is also zero. Accordingly, each member within the white rectangular part of the tabular worksheet surrounded by a shaded region should be treated as zero.

Taking these aspects into account, the character values of the symmetry operations obtained per stationary atom are

$$\chi_s(E) = 3,\ \chi_s(C_5) = 0,\ \chi_s(C_5^2) = 0,\ \chi_s(C_3) = 0,\ \chi_s(C_2) = 0,\ \chi_s(i) = 0$$
$$\chi_s(S_{10}) = 0,\ \chi_s(S_{10}^3) = 0,\ \chi_s(S_6) = 0 \text{ and } \chi_s(\sigma_d) = 1$$

Similarly, the number of stationary atoms after carrying out each of the above symmetry operations is

$$N_s(E) = 60,\ N_s(C_5) = 0,\ N_s(C_5^2) = 0,\ N_s(C_3) = 0,\ N_s(C_2) = 0,\ N_s(i) = 0$$
$$N_s(S_{10}) = 0,\ N_s(S_{10}^3) = 0,\ N_s(S_6) = 0 \text{ and } N_s(\sigma_d) = 4$$

Further, making use of the character table and the information mentioned above, the tabular form of the worksheet can be prepared as given below.

I_h	E	$12C_5$	$12C_5^2$	$20C_3$	$15C_2$	i	$12S_{10}$	$12S_{10}^3$	$20S_6$	$15\sigma_d$	Σ	$n_i=\dfrac{\Sigma}{120}$	Transformation Properties
$\chi_s(R)$	3									1			
$N_s(R)$	60									4			
Γ_{3N}	180	0	0	0	0	0	0	0	0	4			
A_g	1	1	1	1	1	1	1	1	1	1	240	2	$x^2+y^2+z^2$
T_{1g}	3	$2c\frac{\pi}{5}$	$2c\frac{3\pi}{5}$	0	-1	3	$2c\frac{3\pi}{5}$	$2c\frac{\pi}{5}$	0	-1	480	4	(R_x, R_y, R_z)
T_{2g}	3	$2c\frac{3\pi}{5}$	$2c\frac{\pi}{5}$	0	-1	3	$2c\frac{\pi}{5}$	$2c\frac{3\pi}{5}$	0	-1	480	4	
G_g	4	-1	-1	1	0	4	-1	-1	1	0	720	6	
H_g	5	0	0	-1	1	5	0	0	-1	1	960	8	$(2z^2-x^2-y^2,\ x^2-y^2)$
A_u	1	1	1	1	1	-1	-1	-1	-1	-1	120	1	
T_{1u}	3	$2c\frac{\pi}{5}$	$2c\frac{3\pi}{5}$	0	-1	-3	$-2c\frac{3\pi}{5}$	$-2c\frac{\pi}{5}$	0	1	600	5	(x, y, z)
T_{2u}	3	$2c\frac{3\pi}{5}$	$2c\frac{\pi}{5}$	0	-1	-3	$-2c\frac{\pi}{5}$	$-2c\frac{3\pi}{5}$	0	1	600	5	
G_u	4	-1	-1	1	0	-4	1	1	-1	0	720	6	
H_u	5	0	0	-1	1	-5	0	0	1	-1	840	7	
Γ_{3N}	3	0	0	0	0	0	0	0	0	1			

This gives us

$$\Gamma_{3N} = 2A_g + 4T_{1g} + 4T_{2g} + 6G_g + 8H_g + 1A_u + 5T_{1u} + 5T_{2u} + 6G_u + 7H_u$$

From character table, we find

$$\Gamma_{tr} = T_{1u} \text{ and } \Gamma_{rot} = T_{1g}$$

Therefore,

$$
\begin{aligned}
\Gamma_{vib} &= \Gamma_{3N} - (\Gamma_{tr} + \Gamma_{rot}) \\
&= 2A_g + 4T_{1g} + 4T_{2g} + 6G_g + 8H_g + 1A_u + 5T_{1u} + 5T_{2u} \\
&\quad + 6G_u + 7H_u - (T_{1u} + T_{1g}) \\
&= 2A_g + 3T_{1g} + 4T_{2g} + 6G_g + 8H_g + 1A_u + 4T_{1u} + 5T_{2u} + 6G_u + 7H_u
\end{aligned}
$$

Group theoretical calculation shows that there are 9 vibrational modes, out of which many are degenerate and that there are 46 distinct vibrational frequencies with which the fullerene molecule can vibrate. Among the vibrational modes, four triply degenerate T_{1u} modes are IR active and ten modes ($2 A_g + 8 H_g$) are Raman active.

4.8   Vibrational Modes in Some Linear Molecules

The reduction process as applied to nonlinear molecules cannot be applied to linear molecules because division by infinite order of the group makes everything redundant. Therefore, to solve the problem of linear molecules, different methods were put forward to split the reducible representations into various irreducible representations. For example, based on the group-subgroup relationships (such as $C_{\infty v}$ and $D_{\infty h}$ are considered special cases of the family of groups C_{nv} and D_{nh}, respectively), a practical method for reducing the representations of infinite order groups was proposed by Strommen and Lippincot in 1972.

Here, we will use the simplest inspection technique to accomplish the reduction process. For this purpose, while solving problems, we proceed according to the following steps and obtain the vibrational modes using the tabular worksheet method.

1. Determine the overall character of the molecule (point group) by using the formula

$$\Gamma_{3N} = \chi(R) = \chi_s(R) \times N_s(R)$$

as before. Here, $\chi_s(R)$ is obtained from Table 4.2 and $N_s(R)$ from the given symmetry operations of the molecule.

Fig. 4.21 Geometry and
symmetry of CO molecule

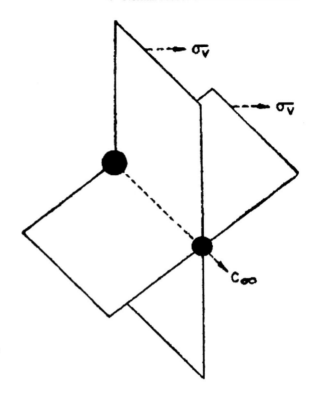

2.   Subtract Γ_{tr} and Γ_{rot} from Γ_{3N} to obtain Γ_{vib}, i.e.,
      $$\Gamma_{vib} = \Gamma_{3N} - (\Gamma_{tr} + \Gamma_{rot}) \text{ as before}$$
3.   Reduce Γ_{vib} into maximum possible symmetry species through inspection.

Example 4.14 Determine the representations of vibrational modes of carbon
monoxide (CO) molecule using tabular worksheet method and inspection technique.
The point group of the molecule is $C_{\infty v}$.

Solution: *Given*: Carbon monoxide (CO) molecule; it has 2 atoms (i.e., $N = 2$), point
group is $C_{\infty v}$, and modes of vibration = ?
      The geometry along with the associated symmetry elements of the molecule is
shown in Fig. 4.21. The fundamental symmetry operations of the point group $C_{\infty v}$
are E, $2C_{\infty}^{\varphi}$, $\infty \sigma_v$.

      The character values of the fundamental symmetry operations per stationary atom
$\chi_s(R)$ are obtained from Table 4.2 for $N = 2$. Similarly, the number of stationary
atoms $N_s(R)$ for each symmetry operation is obtained by checking the molecular
orientation w.r.t. each symmetry operation. Keeping in view the above-mentioned
steps including transformation properties in the character table, we can prepare a
tabular worksheet as given below.

$C_{\infty v}$	E	$2C_\infty^\varphi$	$\infty \sigma_v$
$\chi_s(R)$	3	$1 + 2c\varphi$	1
$N_s(R)$	2	2	2
Γ_{3N}	6	$2 + 4c\varphi$	2
$\Gamma_{tr} = \Sigma^+ + \pi$	3	$1 + 2c\varphi$	1
$\Gamma_{rot} = \pi$	2	$2c\varphi$	0
$\Gamma_{tr} + \Gamma_{rot}$	5	$1 + 4c\varphi$	1
Γ_{vib}	1	1	1

This gives us

$$\Gamma_{vib} = \Sigma^+$$

which represents a completely symmetric mode. Therefore, the representation obtained for CO molecule suggests that there is only one normal mode of vibration with which the molecule will vibrate.

Example 4.15 Determine the representations of vibrational modes of nitrogen (N_2) molecule using tabular worksheet method and inspection technique. The point group of the molecule is $D_{\infty h}$.

Solution: *Given*: N_2 ($N\equiv N$) molecule; it has 2 atoms (i.e., $N = 2$), point group is $D_{\infty h}$, and modes of vibration = ?

The geometry along with the associated symmetry elements of the molecule is shown in Fig. 4.22. The fundamental symmetry operations of the point group $D_{\infty h}$ are E, $2C_\infty^\varphi$, $\infty \sigma_v$, i, $2S_\infty^\varphi$, ∞C_2.

The character values of the fundamental symmetry operations per stationary atom $\chi_s(R)$ are obtained from Table 4.2 for $N = 2$. Similarly, the number of stationary

Fig. 4.22 Geometry and symmetry of N_2 molecule

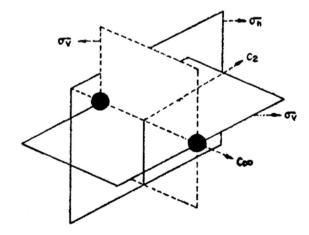

atoms $N_s(R)$ for each symmetry operation is obtained by checking the molecular orientation w.r.t. each symmetry operation. Keeping in view the above-mentioned steps including transformation properties in the character table, we can prepare a tabular worksheet as given below.

$D_{\infty h}$	E	$2C_\infty^\varphi$	$\infty\sigma_v$	i	$2S_\infty^\varphi$	∞C_2
$\chi_s(R)$	3	$1 + 2c\varphi$	1	-3	$-1 + 2c\varphi$	-1
$N_s(R)$	2	2	2	0	0	0
Γ_{3N}	6	$2 + 4c\varphi$	2	0	0	0
$\Gamma_{tr} = \Sigma_u^+ + \pi_u$	3	$1 + 2c\varphi$	1	-3	$-1 + 2c\varphi$	-1
$\Gamma_{rot} = \pi_g$	2	$2c\varphi$	0	2	$-2c\varphi$	0
$\Gamma_{tr} + \Gamma_{rot}$	5	$1 + 4c\varphi$	1	-1	-1	-1
Γ_{vib}	1	1	1	1	1	1

This gives us

$$\Gamma_{vib} = \Sigma^+$$

which represents a completely symmetric mode. Therefore, the representation obtained for the N_2 molecule suggests that there is only one normal mode of vibration with which the molecule will vibrate.

Example 4.16 Determine the representations of vibrational modes of HCN molecule using tabular worksheet method and inspection technique. The point group of the molecule is $C_{\infty v}$.

Solution: *Given*: HCN molecule; it has 3 atoms (i.e., N = 3), point group is $C_{\infty v}$, and modes of vibration = ?

The geometry along with the associated symmetry elements of the molecule is shown in Fig. 4.23. The fundamental symmetry operations of the point group $C_{\infty v}$ are E, C_∞^φ, $\infty\sigma_v$.

The character values of the fundamental symmetry operations per stationary atom $\chi_s(R)$ are obtained from Table 4.2 for N = 3. Similarly, the number of stationary atoms $N_s(R)$ for each symmetry operation is obtained by checking the molecular orientation w.r.t. each symmetry operation. Keeping in view the above-mentioned steps including transformation properties in the character table, we can prepare a tabular worksheet as given below.

$C_{\infty v}$	E	$2C_\infty^\varphi$	$\infty\sigma_v$
$\chi_s(R)$	3	$1 + 2c\varphi$	1
$N_s(R)$	3	3	3
Γ_{3N}	9	$3 + 6c\varphi$	3

(continued)

(continued)

$C_{\infty v}$	E	$2C_\infty^\varphi$	$\infty\sigma_v$
$\Gamma_{tr} = \Sigma^+ + \pi$	3	$1 + 2c\varphi$	1
$\Gamma_{rot} = \pi$	2	$2c\varphi$	0
$\Gamma_{tr} + \Gamma_{rot}$	5	$1 + 4c\varphi$	1
Γ_{vib}	4	$2 + 2c\varphi$	2

The resulting Γ_{vib} contains four normal modes of vibration and therefore needs to be decomposed to get the proper symmetry species. This is done through inspection of the character table as given below.

$C_{\infty v}$	E	$2C_\infty^\varphi$	$\infty\sigma_v$
Γ_{vib}	4	$2 + 2c\varphi$	2
$2\Sigma^+$	2	2	2
π	2	$2c\varphi$	0

$$\Rightarrow \Gamma_{vib} = 2\Sigma^+ + \pi$$

where $\Sigma^+ (\equiv A_1)$ is completely symmetric and $\pi (\equiv E_1)$ represents doubly degenerate mode. Therefore, the representation obtained for the HCN molecule suggests that there are three normal modes of vibration with which the molecule will vibrate.

Example 4.17 Determine the representations of vibrational modes of carbon dioxide (CO_2) molecule using tabular worksheet method and inspection technique. The point group of the molecule is $D_{\infty h}$.

Solution: *Given:* CO_2 molecule; it has 3 atoms (i.e., $N = 3$), point group is $D_{\infty h}$, and modes of vibration = ?

The geometry of the molecule is provided in Table 1.10. The fundamental symmetry operations of the point group $D_{\infty h}$ are E, $2C_\infty^\varphi$, $\infty\sigma_v$, i, $2S_\infty^\varphi$, ∞C_2.

The character values of the fundamental symmetry operations per stationary atom $\chi_s(R)$ are obtained from Table 4.2 for $N = 3$. Similarly, the number of stationary atoms $N_s(R)$ for each symmetry operation is obtained by checking the molecular orientation w.r.t. each symmetry operation. Keeping in view the above-mentioned steps including transformation properties in the character table, we can prepare a tabular worksheet as given below.

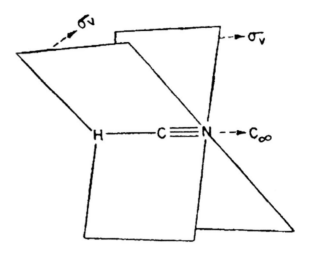

Fig. 4.23 Geometry and symmetry of HCN molecule

$D_{\infty h}$	E	$2C_\infty^\varphi$	$\infty\sigma_v$	i	$2S_\infty^\varphi$	∞C_2
$\chi_s(R)$	3	$1 + 2c\varphi$	1	-3	$-1 + 2c\varphi$	-1
$N_s(R)$	3	3	3	1	1	1
Γ_{3N}	9	$3 + 6c\varphi$	3	-3	$-1 + 2c\varphi$	-1
$\Gamma_{tr} = \Sigma_u^+ + \pi_u$	3	$1 + 2c\varphi$	1	-3	$-1 + 2c\varphi$	-1
$\Gamma_{rot} = \pi_g$	2	$2c\varphi$	0	2	$-2c\varphi$	0
$\Gamma_{tr} + \Gamma_{rot}$	5	$1 + 4c\varphi$	1	-1	-1	-1
Γ_{vib}	4	$2 + 2c\varphi$	2	-2	$2c\varphi$	0

The resulting Γ_{vib} contains four normal modes of vibration and therefore needs to be decomposed to get the proper symmetry species. This is done through inspection of the character table as given below.

$D_{\infty h}$	E	$2C_\infty^\varphi$	$\infty\sigma_v$	i	$2S_\infty^\varphi$	∞C_2
Γ_{vib}	4	$2 + 2c\varphi$	2	-2	$2c\,\varphi$	0
π_u	2	$2c\varphi$	0	-2	$2c\varphi$	0
$\Gamma_{vib} - \pi_u$	2	2	2	0	0	0
Σ_g^+	1	1	1	1	1	1
Σ_u^+	1	1	1	-1	-1	-1

This gives us

$$\Gamma_{vib} = \Sigma_g^+ + \Sigma_u^+ + \pi_u$$

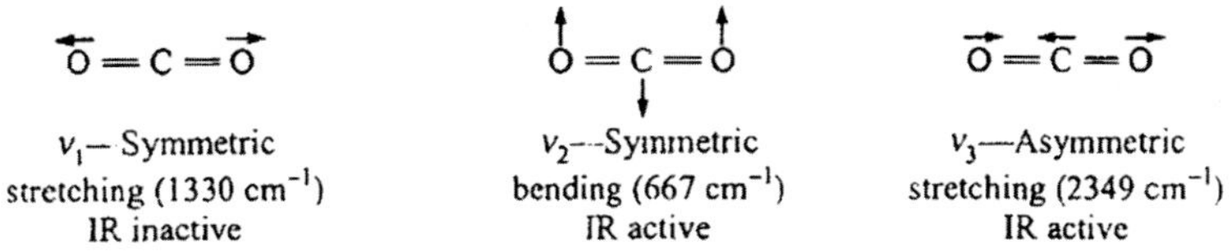

$\overleftrightarrow{O} = C = \overrightarrow{O}$ $\overset{\uparrow}{O} = C = \overset{\uparrow}{O}$ $\overrightarrow{O} = \overleftarrow{C} = \overrightarrow{O}$

v_1– Symmetric
stretching (1330 cm^{-1})
IR inactive

v_2–Symmetric
bending (667 cm^{-1})
IR active

v_3—Asymmetric
stretching (2349 cm^{-1})
IR active

Fig. 4.24 Modes of vibration of CO_2 molecule

H————C≡≡≡C————Na

Fig. 4.25 Geometry of C_2HNa molecule

The representations obtained for the CO_2 molecule suggest that there are four normal modes of vibration with which the molecule will vibrate. Here, Σ_g^+ and Σ_u^+ represent the symmetric stretching and the asymmetric stretching, respectively. On the other hand, π_u represents a doubly degenerate bending. They are shown in Fig. 4.24.

Example 4.18 Determine the representations of vibrational modes of sodium acetylide (C_2HNa) molecule using tabular worksheet method and inspection technique. The point group of the molecule is $C_{\infty v}$.

Solution: *Given:* C_2HNa molecule; it has 4 atoms (i.e., $N = 4$), point group is $C_{\infty v}$, and modes of vibration = ?

The geometry of the molecule is shown in Fig. 4.25. The fundamental symmetry operations of the point group $C_{\infty v}$ are E, $2C_\infty^\varphi$, $\infty\sigma_v$.

The character values of the fundamental symmetry operations per stationary atom $\chi_s(R)$ are obtained from Table 4.2 for $N = 4$. Similarly, the number of stationary atoms $N_s(R)$ for each symmetry operation is obtained by checking the molecular orientation w.r.t. each symmetry operation. Keeping in view the above-mentioned steps including transformation properties in the character table, we can prepare a tabular worksheet as given below.

$C_{\infty v}$	E	$2C_\infty^\varphi$	$\infty\sigma_v$
$\chi_s(R)$	3	$1 + 2c\varphi$	1
$N_s(R)$	4	4	4
Γ_{3N}	12	$4 + 8c\varphi$	4
$\Gamma_{tr} = \Sigma^+ + \pi$	3	$1 + 2c\varphi$	1
$\Gamma_{rot} = \pi$	2	$2c\varphi$	0
$\Gamma_{tr} + \Gamma_{rot}$	5	$1 + 4c\varphi$	1
Γ_{vib}	7	$3 + 4c\varphi$	3

The resulting Γ_{vib} contains seven normal modes of vibration and therefore needs to be decomposed to get the proper symmetry species. This is done through inspection of the character table as given below.

$C_{\infty v}$	E	$2C_\infty^\varphi$	$\infty\sigma_v$
Γ_{vib}	7	$3 + 4c\varphi$	3
$3\Sigma^+$	3	3	3
2π	4	$4c\varphi$	0

$$\Rightarrow \Gamma_{vib} = 3\Sigma^+ + 2\pi$$

where $\Sigma^+ (\equiv A_1)$ is completely symmetric and $\pi (\equiv E_1)$ represents doubly degenerate mode. Therefore, the representations obtained for the C_2HNa molecule suggest that there are seven normal modes of vibration with which the molecule will vibrate.

Example 4.19 Determine the representations of vibrational modes of acetylene (C_2H_2) molecule using tabular worksheet method and inspection technique. The point group of the molecule is $D_{\infty h}$.

Solution: *Given:* C_2H_2 molecule; it has 4 atoms (i.e., $N = 4$), point group is $D_{\infty h}$, and modes of vibration $= ?$

The geometry/symmetry of the molecule is shown in Fig. 4.26. The fundamental symmetry operations of the point group $D_{\infty h}$ are E, $2C_\infty^\varphi$, $\infty\sigma_v$, i, $2S_\infty^\varphi$, ∞C_2.

The character values of the fundamental symmetry operations per stationary atom $\chi_s(R)$ are obtained from Table 4.2 for $N = 4$. Similarly, the number of stationary atoms $N_s(R)$ for each symmetry operation is obtained by checking the molecular orientation w.r.t. each symmetry operation. Keeping in view the above-mentioned steps including transformation properties in the character table, we can prepare a tabular worksheet as given below.

$D_{\infty h}$	E	$2C_\infty^\varphi$	$\infty\sigma_v$	i	$2S_\infty^\varphi$	∞C_2
$\chi_s(R)$	3	$1 + 2c\varphi$	1	-3	$-1 + 2c\varphi$	-1
$N_s(R)$	4	4	4	0	0	0
Γ_{3N}	12	$4 + 8c\varphi$	4	0	0	0
$\Gamma_{tr} = \Sigma_u^+ + \pi_u$	3	$1 + 2c\varphi$	1	-3	$-1 + 2c\varphi$	-1
$\Gamma_{rot} = \pi_g$	2	$2c\varphi$	0	2	$-2c\varphi$	0
$\Gamma_{tr} + \Gamma_{rot}$	5	$1 + 4c\varphi$	1	-1	-1	-1
Γ_{vib}	7	$3 + 4c\varphi$	3	1	1	1

The resulting Γ_{vib} contains seven normal modes of vibration and therefore needs to be decomposed to get the proper symmetry species. This is done through inspection of the character table as given below.

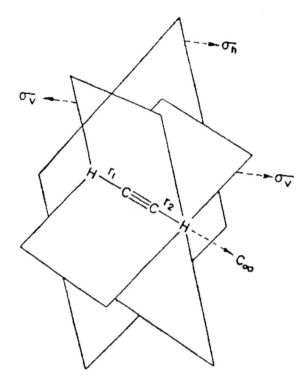

Fig. 4.26 Geometry and symmetry of C_2H_2 molecule

$D_{\infty h}$	E	$2C_\infty^\varphi$	$\infty\sigma_v$	i	$2S_\infty^\varphi$	∞C_2
Γ_{vib}	7	$3 + 4c\varphi$	3	1	1	1
Γ_P	3	3	3	1	1	1
Γ_Q	4	$4c\varphi$	0	0	0	0

where Γ_P and Γ_Q are still in composite form and therefore need to be decomposed further. Let us decompose each of them separately. The decomposition of Γ_P is given below.

$D_{\infty h}$	E	$2C_\infty^\varphi$	$\infty\sigma_v$	i	$2S_\infty^\varphi$	∞C_2
Γ_P	3	3	3	1	1	1
$2\Sigma_g^+$	2	2	2	2	2	2
Σ_u^+	1	1	1	-1	-1	-1

From the decomposition of Γ_P, we obtain

$$\Gamma_P = 2\Sigma_g^+ + \Sigma_u^+$$

Similarly, let us decompose Γ_Q as given below.

$D_{\infty h}$	E	$2C_{\infty}^{\varphi}$	$\infty\sigma_v$	i	$2S_{\infty}^{\varphi}$	∞C_2
Γ_Q	4	$4c\varphi$	0	0	0	0
π_g^+	2	$2c\varphi$	0	2	$-2c\varphi$	0
π_u	2	$2c\varphi$	0	-2	$2c\varphi$	0

From decomposition of Γ_Q, we obtain

$$\Gamma_Q = \pi_g^+ + \pi_u$$

Thus, finally we obtain

$$\Gamma_{vib} = \Gamma_P + \Gamma_Q = 2\Sigma_g^+ + \Sigma_u^+ + \pi_g^+ + \pi_u$$
$$= 2\Sigma_g^+ + \pi_g^+ + \Sigma_u^+ + \pi_u$$

Therefore, the representations obtained for the C_2H_2 molecule suggest that there are seven normal modes of vibration with which the molecule will vibrate.

Example 4.20 Determine the representations of vibrational modes of carbon suboxide (C_3O_2) molecule using tabular worksheet method and inspection technique. The point group of the molecule is $D_{\infty h}$.

Solution: *Given:* C_3O_2 molecule; it has 5 atoms (i.e., N = 5), point group is $D_{\infty h}$, and modes of vibration = ?

The geometry of the molecule is shown in Fig. 4.27. The fundamental symmetry operations of the point group $D_{\infty h}$ are E, $2C_{\infty}^{\varphi}$, $\infty\sigma_v$, i, $2S_{\infty}^{\varphi}$, ∞C_2.

The character values of the fundamental symmetry operations per stationary atom $\chi_s(R)$ are obtained from Table 4.2 for N = 5. Similarly, the number of stationary atoms $N_s(R)$ for each symmetry operation is obtained by checking the molecular orientation w.r.t. each symmetry operation. Keeping in view the above-mentioned steps including transformation properties in the character table, we can prepare a tabular worksheet as given below.

$D_{\infty h}$	E	$2C_{\infty}^{\varphi}$	$\infty\sigma_v$	i	$2S_{\infty}^{\varphi}$	∞C_2
$\chi_s(R)$	3	$1 + 2c\varphi$	1	-3	$-1 + 2c\varphi$	-1
$N_s(R)$	5	5	5	1	1	1
Γ_{3N}	15	$5 + 10c\varphi$	5	-3	$-1 + 2c\varphi$	-1

(continued)

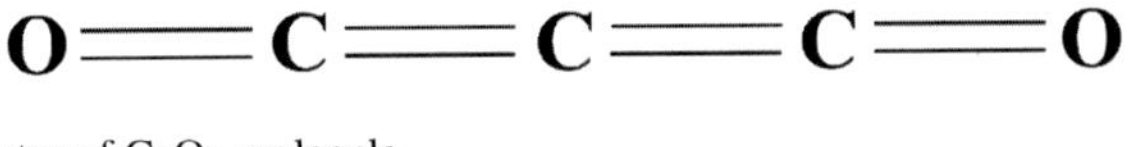

$$O=\!\!=\!C=\!\!=\!C=\!\!=\!C=\!\!=\!O$$

Fig. 4.27 Geometry of C_3O_2 molecule

(continued)

$D_{\infty h}$	E	$2C_\infty^\varphi$	$\infty\sigma_v$	i	$2S_\infty^\varphi$	∞C_2
$\Gamma_{tr} = \Sigma_u^+ + \pi_u$	3	$1 + 2c\varphi$	1	-3	$-1 + 2c\varphi$	-1
$\Gamma_{rot} = \pi_g$	2	$2c\varphi$	0	2	$-2c\varphi$	0
$\Gamma_{tr} + \Gamma_{rot}$	5	$1 + 4c\varphi$	1	-1	-1	-1
Γ_{vib}	10	$4 + 6c\varphi$	4	-2	$2c\varphi$	0

The resulting Γ_{vib} contains ten normal modes of vibration and therefore needs to be decomposed to get the proper symmetry species. This is done through inspection of the character table as given below.

$D_{\infty h}$	E	$2C_\infty^\varphi$	$\infty\sigma_v$	i	$2S_\infty^\varphi$	∞C_2
Γ_{vib}	10	$4 + 6c\varphi$	4	-2	$2c\varphi$	0
Γ_P	4	4	4	0	0	0
Γ_Q	6	$6c\varphi$	0	-2	$2c\varphi$	0

where Γ_P and Γ_Q are still in composite form and therefore need to be decomposed further. Let us decompose each of them separately. The decomposition of Γ_P is given below.

$D_{\infty h}$	E	$2C_\infty^\varphi$	$\infty\sigma_v$	i	$2S_\infty^\varphi$	∞C_2
Γ_P	4	4	4	0	0	0
$2\Sigma_g^+$	2	2	2	2	2	2
$2\Sigma_u^+$	2	2	2	-2	-2	-2

From the decomposition of Γ_P, we obtain

$$\Gamma_P = 2\Sigma_g^+ + 2\Sigma_u^+$$

Similarly, let us decompose Γ_Q as given below.

$D_{\infty h}$	E	$2C_\infty^\varphi$	$\infty\sigma_v$	i	$2S_\infty^\varphi$	∞C_2
Γ_Q	6	$6c\varphi$	0	-2	$2c\varphi$	0
π_g^+	2	$2c\varphi$	0	2	$-2c\varphi$	0
$2\pi_u$	4	$4c\varphi$	0	-4	$4c\varphi$	0

From decomposition of Γ_Q, we obtain

$$\Gamma_Q = \pi_g^+ + 2\pi_u$$

Thus, finally we obtain

$$\Gamma_{vib} = \Gamma_P + \Gamma_Q = 2\Sigma_g^+ + 2\Sigma_u^+ + \pi_g^+ + 2\pi_u$$
$$= 2\Sigma_g^+ + \pi_g^+ + 2\Sigma_u^+ + 2\pi_u$$

Therefore, the representations obtained for the C_3O_2 molecule suggest that there are ten normal modes of vibration with which the molecule will vibrate.

4.9 Summary

1. Molecular motions include translational motion, rotational motion, and vibrational motion.
2. A nonlinear (angular) molecule has six degrees of freedom, three each for translational and rotational and hence the number of vibrational degrees of freedom associated with this is given by

$$\nu_{vib} = 3N - 6$$

3. On the other hand, a linear molecule has five degrees of freedom, three translational and two rotational (because the rotation about the molecular axis does not bring any change in the atomic coordinates) and hence the number of vibrational degrees of freedom associated with this is given by

$$\nu_{vib} = 3N - 5$$

4. The character of a matrix in the reducible representation is related to the characters of the matrices in the irreducible representations as

$$\chi(R) = \sum_j n_j \chi_j(R)$$

5. The number of times the ith irreducible representation occurs in the corresponding reducible representation can be determined by the equation given by

$$n_i = \frac{1}{h} \sum_{R_p} g_p \chi(R_p) \chi_i(R_p)$$

where R_p = a symmetry operation in the pth class, h is the order (total number of independent symmetry elements) of the point group, and g_p refers to the number of operations in the pth class of the group.

6. The overall reducible representation of nonlinear molecules can be determined by using two different methods. They are the first principle method and un-shifted (or stationary) atom method.

7. In the first principle method, the following procedure is adopted.

    (i)   Determine the point group of the nonlinear molecule, if not known. Write its symmetries, number of classes, order of the point group, and the number of atoms (N) in the molecule.

    (ii) Construct a 3N × 3N matrix in the form of reducible representation for each symmetry operation of the molecule.

    (iii) In full (3N × 3N) matrix notation, check the effect of each symmetry operation on the atoms of the given molecule and note down the character value corresponding to the un-shifted (or stationary) atom/vector in each case to obtain $\Gamma_{3N} = \chi(R)$.

8. In the stationary (un-shifted) atoms method, the following procedure is adopted:

    (i)   Note down the character values $\chi_s(R)$ for all symmetry operations of the point group of the given molecule.

    (ii) Count the number of stationary (un-shifted) atoms $N_s(R)$ for each symmetry operation (using a suitable diagram) of the given molecule (point group).

    (iii) Determine the overall character of the molecule (point group) $\chi(R)$ by using the formula

$$\Gamma_{3N} = \chi(R) = \chi_s(R) \times N_s(R)$$

    where Γ_{3N} includes all the three degrees of freedom of the molecule, i.e.,

$$\Gamma_{3N} = \Gamma_{tr} + \Gamma_{rot} + \Gamma_{vib}$$

9. In order to decompose a reducible representation (obtained for a given molecule) into a number of irreducible representations, we need the following information:

    (i)   Order of the point group (h) and the number of classes in the group.

    (ii) Number of symmetry operation(s) in the pth class, g_p.

    (iii) Overall character value of the reducible representation, $\chi(R)$.

    (iv) The character table of the point group of the given molecule to get the irreducible representation character values, $\chi_i(R_p)$.

10. The following steps are used to obtain the vibrational modes in tabular worksheet and inspection method for linear molecules.

    (i)   Determine the overall character of the molecule (point group) by using the formula

$$\Gamma_{3N} = \chi(R) = \chi_s(R) \times N_s(R)$$

as before. Here, $\chi_s(R)$ is obtained from Table 4.2 and $N_s(R)$ from the given symmetry operations of the molecule.

(ii)   Subtract Γ_{tr} and Γ_{rot} from Γ_{3N} to obtain Γ_{vib}, i.e.,

(iii) $\Gamma_{vib} = \Gamma_{3N} - (\Gamma_{tr} + \Gamma_{rot})$ as before.

(iv)   Reduce Γ_{vib} into maximum possible symmetry species through inspection.

Chapter 5
Vibrational Spectroscopy of Molecules

5.1 Introduction

In earlier chapters, we studied in detail the fundamental aspects of symmetries and their matrix representations, the elements of group theory, construction of group multiplication tables, about reducible and irreducible representations of symmetry operations, construction of character tables for different point groups and their need to understand the subject of normal modes of vibration in different molecules, and then use the necessary knowledge to derive the modes of vibration theoretically for different molecules of different symmetries.

On the other hand, in the present chapter we will study the theoretical and experimental aspects of vibrational spectroscopy in the infrared (IR) region, important for understanding the molecular vibrations in different crystalline samples. Our main focus will be to discuss the fundamental theories and the related experimental methods such as IR, FTIR, Raman, and FT Raman techniques to obtain vibrational spectra for any given crystal/molecular sample. The addition of this chapter will not only make it possible to compare the experimental results with the theoretically calculated modes of vibration for the given samples but also facilitate in understanding the basic physics involved in the explanation of various spectra.

5.2 Some Useful Observations Concerning Molecular Vibrations

Some useful observations made during the study of molecular vibrations whose frequencies lie in the infrared (IR) region of the electromagnetic spectrum are presented as follows:

1. In a molecule, all motions of the atoms relative to each other are a superposition of the so-called normal modes of vibration in which all the atoms vibrate in the same phase and with the same frequency.

© The Author(s), under exclusive license to Springer Nature Singapore Pte Ltd. 2022   251
M. A. Wahab, *Symmetry Representations of Molecular Vibrations*, Springer Series
in Chemical Physics 126, https://doi.org/10.1007/978-981-19-2802-4_5

2.   All normal modes of vibration are either symmetric or asymmetric (anti-symmetric) w.r.t. the symmetry operations of the molecule. That is, each of them can be characterized by a symmetry species (irreducible representation).

3.   If in a vibrating system, either the force constant is reduced or the mass is increased, then all vibrational frequencies are either reduced or remain constant, but no frequency is increased (called Rayleigh's rule).

4.   Fundamentally, there are two types of molecular vibrations. The first type is the stretching vibrations, which are due to a change in bond length. The second type includes the bending and torsional vibrations, which are due to the change in bond angles. Bending vibrations deform mainly bond angles at one atom, while torsional vibrations change the dihedral angles between planes defined by ligands at both atoms of a bond.

5.   The number of vibrational degrees of freedom for a nonlinear molecule with N atoms is $\nu_{vib} = 3N - 6$ and for a linear molecule, this number is $\nu_{vib} = 3N - 5$. A nonlinear and non-cyclic molecule produces stretching, bending, and torsional vibrations, respectively, for different degrees of freedom. However, the observed normal modes of vibration are a linear combination of these degrees of freedom. In addition, combinations of normal modes may produce harmonics, sum, and difference "tones" usually of weak bands.

6.   When two oscillators of different frequencies are coupled, then the lower frequency oscillator will oscillate with a further lower frequency and the higher frequency oscillator will oscillate with a further higher frequency. This effect is more pronounced if the difference in the original frequencies is small.

7.   When a molecule is already in a vibrational exited state, it undergoes further excitation, and a hot band is observed where an absorption line is weak.

8.   An overtone (due to deformation) vibration which has about the same frequency as a fundamental (stretching) vibration of the same bonds may combine with the fundamental to produce two bands of nearly equal intensity. This effect, which may produce supernumerary bands, is called Fermi resonance. A necessary condition for Fermi resonance is that the overtone and the fundamental must belong to the same symmetry species.

9.   If a molecule contains a bond or a functional group (taken as an isolated unit) whose vibrational frequency differs considerably from those of neighboring groups, then there will always be some vibrations which are primarily localized on this specific bond or functional group. Such vibrations therefore exhibit frequencies as well as IR and Raman intensities which are characteristic of this particular bond functional group.

10. Molecular aggregates such as crystals or complexes behave like "super molecule" in which the vibrations of the individual component are coupled. In the first approximation, the normal vibrations are not coupled; they do not interact. However, the elasticity of bonds does not strictly follow Hooke's law, therefore overtones and combinations of normal vibrations appear.

5.3   General Survey of Vibrational Spectroscopy

Some general but important points related to vibrational spectroscopy are presented below to get acquainted with the subject that we are going to discuss.

1.   Vibrational spectroscopy involves many methods, the most important of which being infrared (IR) and Raman spectroscopy.
2.   The IR and Raman spectra of two molecules differing in terms of their constituents, isotopic distributions, configurations, conformations, and optical activities are found to be different.
3.   Different substances can be identified by their IR and Raman spectra; they are interpreted like fingerprints.
4.   The IR and Raman spectra of particular groups of atoms showing certain typical bonds defined by definite ranges of frequencies and intensities (i.e., exhibiting characteristic vibrations) may be employed to explain the molecular structure of the group of atoms.
5.   Vibrational spectra of different substances in any state (phase) can be recorded. The spectra of one substance in different phases are similar, but they differ from one another in terms of activity and intensity of vibrations, frequency, half-width, and fine structure of the bands.
6.   The spectrum generated by the molecular vibrations lies in the infrared (absorption) region of the electromagnetic spectrum. The same range is also involved in the scattering process of Raman spectra.
7.   Depending on the symmetry (point group) of the molecule, its modes of vibration and whether the same is active or forbidden in IR and Raman can be determined.
8.   The time needed to record a vibrational spectrum is just a few seconds or minutes.

5.4   Infrared (IR) Spectral Region

The infrared (IR) spectral region lies between the visible and the microwave regions of the electromagnetic spectrum. This region of the spectrum is particularly more important from chemists' point of view for their studies on molecular vibrations. Two very important techniques involving IR and Raman spectra are used by all those in the field of materials science. Some of the physical quantities and their infrared ranges are provided in Table 5.1.

Further, the IR region of the electromagnetic spectrum is conveniently divided into three regions in terms of wavenumbers (which is the commonly used unit of frequency/energy in vibrational spectroscopy). They are

1.   Near-infrared region (NIR): 12,800–4000 cm^{-1}.
2.   Mid-infrared region (MIR): 4000–400 cm^{-1}.
3.   Far-infrared region (FIR): 400–10 cm^{-1}.

Table 5.1 Some physical quantities in infrared range

Physical quantity	Infrared range	Unit
Wavelength (λ)	7.80×10^{-7} to 3×10^{-4}	m
Frequency (v)	3.84×10^{14} to 1×10^{12}	Hz
Energy (E)	1.5×10^5 to 400	Jmol^{-1}
Wavenumber ($\bar{v}$)	12,800 to 50	cm^{-1}

The absorptions observed in the NIR region (12,800–4000 cm^{-1}) are in general overtones or combinations of fundamental stretching bends which occur in the region (3000–1700 cm^{-1}). The resulting bends in the NIR region are usually weak in intensity and also decrease by a factor of 10 from one overtone to the next. Further, the bands in the NIR region are often overlapped, making them less useful than bands observed in the MIR region for qualitative analysis. However, this region can be exploited for quantitative analysis as compared to the MIR region in terms of the functional groups in the molecules.

It is the MIR region (4000–400 cm^{-1}) within which most of the molecular transitions take place and hence is the most commonly employed region in the field of research to study molecular vibrations in different crystal and molecular samples.

FIR region (400–10 cm^{-1}) is important for the molecules containing heavy atoms, molecular skeleton vibration molecular torsion, and crystal lattice vibrations.

5.5 Theory of IR Absorption

A brief description of classical and quantum concepts regarding the fundamental principles of IR absorption is presented below.

IR Absorption

A polychromatic IR light is allowed to pass through a molecular sample, and the infrared spectrum is usually recorded by measuring the transmittance of light quanta. The frequencies of the absorption bands v_s are found to be proportional to the energy difference ($\Delta E = hv_s$) between vibrational ground and excited states as shown in Fig. 5.1. The absorption bands due to vibrational transitions lie in the wavelength region $\lambda = 2.5$ to 1000×10^{-6} m (approximately equivalent to 4000–10 cm^{-1} in terms of wavenumbers, where $\bar{v} = \frac{1}{\lambda}$). Here, we observe that the sample merely absorbs certain frequencies (energies) of infrared light. Let us try to understand on a molecular level how this phenomenon occurs.

Let us consider a hetero-nuclear diatomic molecular system having equal and opposite charges, $+q$ and $-q$ separated by a distance r and forming a dipole. The moment of this dipole can be defined as

$$\vec{\mu} = q\vec{r} \tag{5.1}$$

Fig. 5.1 Energy difference between the ground and excited states

A typical molecular dipole can be taken as a combination of a proton and an electron separated by ~1 Å; the magnitude of the moment of such a dipole will be

$$\mu = 1.602 \times 10^{-19} \times 10^{-10} \text{ Coulomb-m}$$
$$= 1.602 \times 10^{-29} \text{ C-m}$$

In cgs unit, this will correspond to

$$\mu = 4.803 \times 10^{-18} \text{ esul-cm}$$
$$= 4.803 \text{ D}$$
$$\text{where } 1D = 10^{-18} \text{ esu-cm.}$$

In Eq. 5.1, μ is a vector and can be defined in terms of its components as

$$\vec{\mu} = \hat{\mu}_x + \hat{\mu}_y + \hat{\mu}_z$$

Now, suppose the same hetero-nuclear diatomic molecule vibrates with a particular frequency. The moment of the dipole also oscillates with its equilibrium value as the two atoms along with their net charges move back and forth during which the moment of the dipole changes continuously. As a matter of principle, we know that the oscillating dipole can absorb energy from an oscillating electric field only if the field also oscillates with the same frequency. Therefore, the absorption of energy from the light wave by an oscillating dipole in this manner is a molecular explanation of infrared spectroscopy.

This absorption problem can be understood in terms of quantum mechanics. Accordingly, the transition moment for the fundamental of a normal mode (i.e., the transition $v = 0 \rightarrow v = 1$) can be written as

$$M(0, 1) = \int_{-\infty}^{\infty} \Psi_0 \, \vec{\mu} \, \Psi_1 \, d\tau \tag{5.2}$$

where ψ_0 and ψ_1 are wave functions for the ground and excited states, respectively. $\vec{\mu}$ is the oscillating electric dipole moment vector as a function of normal coordinates, q_k of the vibrational mode and must change during the vibration if the transition in Eq. 5.2 is to be allowed. $d\tau$ is a generalized differential such as dx dy dz in Cartesian coordinates. The dipole moment μ can be expanded in a Taylor series as

$$\mu = \mu_0 + \left(\frac{\partial \mu}{\partial q_k}\right)_0 q_k + \frac{1}{2}\left(\frac{\partial^2 \mu}{\partial q_k^2}\right)_0 q_k^2 + \cdots \tag{5.3}$$

where μ_0 is a constant independent of vibration.

Neglecting higher order terms and substituting only the first two terms for μ in Eq. 5.2, we obtain

$$M(0, 1) = \mu_0 \int_{-\infty}^{\infty} \Psi_0 \Psi_1 \, d\tau + \left(\frac{\partial \mu}{\partial q_k}\right)_0 \int_{-\infty}^{\infty} \Psi_0 \, q_k \, \Psi_1 \, d\tau \tag{5.4}$$

The first term in Eq. 5.4 is zero because of the Orthogonality condition. However, the conditions for the second term to be nonzero are

(i) $\left(\frac{\partial \mu}{\partial q_k}\right)_0$ must be finite at least for one component of the dipole moment. That is, the said component of the vibration must bring a change in the dipole moment.

(ii) The integral $\int_{-\infty}^{\infty} \psi_0 \, q_k \, \psi_1 \, d\tau$ must be finite. This is possible only if the vibrational quantum number change $\Delta v = \pm 1$ under harmonic approximation. For an anharmonic oscillator, $\Delta v = \pm 1, \pm 2, \pm 3, \ldots$.

From the above discussion, we can draw two important conclusions:

(i)   Homo-nuclear diatomic molecules have no dipole moments, as both the nuclei attract electrons equally strongly. Similarly, centrosymmetric polyatomic molecules (such as CO_2) show $\mu = 0$.

(ii) Vibrational spectra are observable only in hetero-nuclear diatomic molecules and non-centrosymmetric polyatomic molecules.

The above discussion points out the change of dipole moment during vibration, but it fails to explain how this phenomenon is taking place.

Let us analyze the problem based on molecular symmetries and look for the selection rule based on symmetry. Therefore, if the infrared absorption is to take place, we have to show that the transition moment given by Eq. 5.2 is nonzero on the basis of symmetries of the wave functions and the components of $\vec{\mu}$. The transition moment given by Eq. 5.2 can be written as

$$M(0, 1) = \int_{-\infty}^{\infty} \psi_0 \left(\hat{\mu}_x + \hat{\mu}_y + \hat{\mu}_z \right) \psi_1 \, d\tau \tag{5.5}$$

According to Eq. 5.5, even if any one of the three components of the dipole moment vector is nonzero, the entire transition moment will be nonzero. To show whether this is possible or not, we have to make use of a general result of quantum mechanics (presented here without proof), according to which an integral of the type

$$\int_{-\infty}^{\infty} \varphi_a \varphi_b \, d\tau$$

will have a nonzero value only if the direct product $\Gamma_a \times \Gamma_b$ contains the totally symmetric irreducible representations of the point group, where the functions φ_a and φ_b form basis for the irreducible representations Γ_a and Γ_b, respectively, for a molecule of the above said point group. Applying this to the present case, if Eq. 5.5 transforms as a totally symmetric representation, the vibrational transition will be infrared (IR) active (allowed), i.e., the absorption will occur. Now in Eq. 5.5, the ground state wave function ψ_0 is totally symmetric for all molecules (except for radicals) and the wave function ψ_1 has the symmetry of the normal mode. This means that the product $\psi_0\psi_1$ has the symmetry of ψ_1. Further, the product $\psi_1\psi_1$ (i.e., the product with itself) is a totally symmetric representation. Thus, if any one of the three components of $\vec{\mu}$ has the same symmetry as ψ_1, the product $\vec{\mu}\,\psi_1$ will be totally symmetric (as $M \neq 0$) and hence the infrared absorption will be allowed.

Since the dipole moment can be represented by vectors, we can see that the symmetry species of $\hat{\mu}_x$, $\hat{\mu}_y$, and $\hat{\mu}_z$ are the same as unit vector transformation x, y, and z, as listed in the third column of the character tables.

5.6 Infrared (IR) Spectrometer

Infrared (IR) spectroscopy is one of the most important and most widely used techniques to study vibrational spectra of a solid, liquid, or gas sample. A polychromatic IR light is allowed to pass through both, the sample and the reference cell, and the transmitted intensities are measured at each frequency (or wavenumber) through a detector/recorder. The result is a characteristic spectrum showing the transmittance (or absorbance) I_s/I_r of the electromagnetic radiation as a function of wavelength (or wavenumber).

Experimental Technique

The schematic diagram of a double beam infrared spectrometer is shown in Fig. 5.2. It consists of the following essential components: IR source, optical system, monochromator, detector with amplifier, and recorder.

The most commonly used sources are the Globar filament (which is basically a silicon rod or helix) and the Nernst rod. A Globar filament can be directly ignited, which has a high emissivity in the short wavelength region and has a burning temperature of 1500 K. On the other hand, a Nernst rod is a non-conductor at room temperature, requires initial heating for ignition, and works at 1900 K. These sources are used

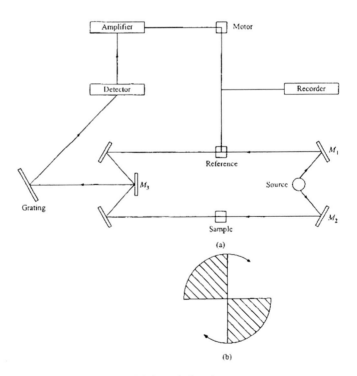

Fig. 5.2 Schematic diagram of a double beam infrared spectrometer

to produce IR radiation in the FIR region ($\sim$400 cm^{-1}) and MIR region ($\sim$2000 cm^{-1}), respectively. In addition, a number of different lasers such as CO_2, CO, NO_2, He–Ne, semiconductor-diode, Ga–As, and Ga–Al–As can also be used as IR sources for different frequency regions. However, their main disadvantage is the low operation temperature (around room temperature or below).

The desired IR beam obtained from the suitable source is divided into two parallel beams of equal intensity with the help of two mirrors M_1 and M_2 (which are silver polished on their faces). The sample is placed in one beam and the other beam is used as reference. The two beams then meet at the rotating sector mirror/chopper M_3 which is also shown separately in Fig. 5.5b. As this mirror rotates, it alternately reflects the reference beam or allows the sample beam through the spaces, into the monochromator slit. Thus, the detector receives the sample beam and the reference beam alternately. Since both the beams travel the same distance through the atmosphere, the extent of extraneous effect due to CO_2 and H_2O in the atmosphere may be taken as the same in both.

As soon as the IR beam from mirror M_3 reaches the monochromator, it splits the polychromatic beam into its component wavelengths (or frequencies). The monochromator consists of a rotatable prism or grating to produce the required frequency, however, grating gives better resolving power. The resolution also depends on the slit width and quality of the mirror. NaCl and KBr prisms are also used as monochromators in some instruments, however in these instruments, the dispersion decreases considerably above 2000 cm^{-1}.

As far as the detectors are concerned, two main types of detectors are in use, one is based on sensing the heating effect of radiation and the other depends on photoconductivity. However, in both cases, the greater the effect (of temperature or conductivity rise) at a given frequency, the greater the transmittance (and less the absorbance) of the sample at that frequency.

Detectors which measure the radiation energy through the heating effect are found in Golay cells. The radiation falls on a very small cell containing air, and the temperature changes are measured in terms of change in pressure within the cell. On the other hand, the detectors are based on photoconductivity, the radiation is allowed to fall on the photoconductor material, and the conductivity of the material is measured continuously by a bridge network. When the energy transmitted by the sample and reference cells are equal, no signal is shown by the detector. However, a signal is produced when the sample absorbs some radiation and the energies of the two beams are unequal. The signal is then fed to the amplifier, which amplifies the low-intensity signal up to the desired level.

The amplified signal received from the detector is used to move an attenuator (comb) which cuts down the radiation coming out of the reference beam until energy balance is restored. This is achieved by a motor which drives the attenuator to the reference beam as soon as an absorption band is encountered and then drives out as soon as it is passed over. A recorder pen is also coupled with this motor so that the attenuator movement is followed by the pen.

Sample Preparation Techniques

The samples to be studied may be in the form of a solid, liquid, or gas. However, they are to be prepared according to the need of the experiment.

Gas samples to be studied at a pressure of 1 atmosphere or greater; they are usually contained in a 5 or 10 cm long sealed glass cell with NaCl windows. Special long path cells may be used for gases at low pressure (less than 100 mm of Hg).

Liquid samples to be studied are held between plates of polished KBr rather than glass. Pure liquids are studied with a thickness of about 0.01 mm, while ordinary solutions are usually 0.1–10 mm thick depending on their dilution.

Solid samples are more difficult to examine because of their high reflectance and scattering, and transmittance is always low. In a solid state, three different techniques are employed to prepare the samples to record their spectra. They are

1.   Mull technique.
2.   Pallet technique.
3.   Thin film technique.

1.   *Mull Technique*

Grind the substance very finely in paraffin oil (nujol) and form its oil suspension, "mull". This can be held like a pure liquid between the salt plates. If the refractive index of the solid or liquid is not very different, scattering will be minimized. The main disadvantage of this technique is that the absorption bends of nujol are found to interfere with certain important group frequencies. Therefore, for better results, the size of the sample particles must be less than the wavelength of the radiation used.

2.   *Pallet Technique*

In this case, the given substance is ground very finely with KBr. At high pressure (0.7–1 Gpa), the mixture is able to cold-flow which can be pressed into transparent pallets with the help of suitable dies. This may then be placed directly in the infrared beam in a suitable holder. This method has a few advantages. They are

(i)   Interfering bends are absent.
(ii)   Examination of samples is easy.
(iii)   There is better control of concentration and homogeneity of the sample.
(iv)   Specimen can be stored for future studies.

However, there are a few disadvantages also. They are

(i)   Anomalous spectra may result.
(ii)   Absorbs water from atmosphere.

3.   *Thin film Technique*

A thin film of the substance can be deposited on a suitable window material as a substrate by using the conventional vapor deposition technique. The thickness of the film can be adjusted by changing the concentration of the solution of the substance. Interference caused by multiple reflections between parallel surfaces affects the accuracy of measurement.

5.7 Fourier Transform Infrared (FTIR) Spectroscopy

FTIR spectroscopy was first of all developed by Astronomers in the early fifties of the nineteenth century to study the infrared spectra of distant stars. However, now it has been developed into a very powerful technique to detect signals not only in the MIR (4000–400 cm^{-1}) region but also in the FIR region (400–10 cm^{-1}) of the electromagnetic spectrum. This technique can provide simultaneous and instantaneous recording of the whole spectrum in all the regions from NIR (12,800–4000 cm^{-1}) to FIR (400–10 cm^{-1}) and hence is preferred over the conventional IR technique.

Principle and Theory

The principle on which it works is based on a mathematical process known as Fourier transform, named after the French mathematician Jean Baptiste Fourier who developed this method in early 1805. The Fourier transform method can resolve a complex wave into its components such as frequencies and intensities. The Fourier transform is also a reciprocal process, i.e., it can also convert a time domain signal into a frequency domain and vice versa.

We know that conventional spectroscopy is basically a frequency domain spectroscopy, where the radiant power G(ω) is recorded as a function of frequency ω. On the other hand, in the time domain spectroscopy, the changes in the radiant power f(t) are recorded as a function of time. In a Fourier transform spectrometer, a time domain function is converted into a frequency domain spectrum. In mathematical form, the Fourier transform of the function f(t) is defined as

$$G(\omega) = \frac{1}{\sqrt{2\pi}} \int_{-\infty}^{\infty} f(t) \exp(i\omega t) \, dt \tag{5.6}$$

and the inverse relation is

$$f(x) = \frac{1}{\sqrt{2\pi}} \int_{-\infty}^{\infty} G(\omega) \exp(-i\omega t) \, dt \tag{5.7}$$

Equations 5.6 and 5.7 represent the Fourier transform of one another.

For example, to illustrate the use of Fourier transform, let us consider the superposition of two sine waves, both have the same amplitude but slightly differ in frequency as shown in Fig. 5.3, where (a) and (b) represent the sine waves of slightly different frequency while (c) represents their sum in time and frequency domains. In a similar manner, even a complex time domain spectra can be transformed into a frequency domain spectra with the help of the Fourier transform, where the actual calculation is done through the use of high-speed computers.

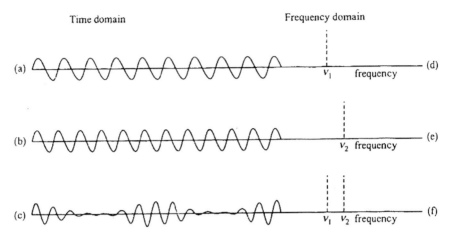

Fig. 5.3 Superposition of two sine waves to study Fourier transform

Experimental Technique

The main part of a Fourier transform (FT) spectrometer is the Michelson interferometer (Fig. 5.4), in which the incoming beam is made to split into two partial beams. They are reflected back to the beam splitter via a fixed and a movable mirror, where they combine and produce interference. If their path lengths are identical or differ by an integral multiple of the wavelength, constructive interference will take place and

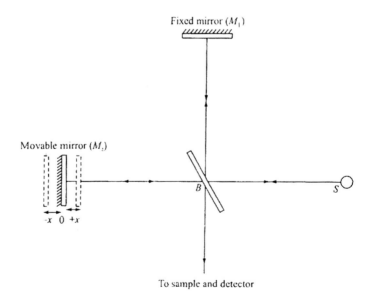

Fig. 5.4 A Michelson interferometer

a bright beam will leave the beam splitter. On the other hand, if their path difference is a half-integral multiple of the wavelength, destructive interference will take place and hence no beam will leave the beam splitter. The two situations will be seen by the sample/detector as an alternation in beam intensity. A Fourier transform of the same will produce the spectrum of appropriate intensity. A similar but more complicated situation will arise if the source contains two close wavelengths (or frequencies) or many wavelengths (white radiation).

If the recombined beam is allowed to pass through a sample (before reaching the detector), any absorption by it will be seen as a gap in the frequency distribution which on Fourier transform will give a normal absorption spectrum. Thus, in the actual experiment, whenever the mirror M_2 is moved the detector receives an intensity signal as a function of optical path length, which is then collected at a multichannel computer for carrying out the Fourier transform and finally the required spectrum is plotted.

There are a number of advantages of FT spectrometers over dispersive ones. They are

1. Fast and simultaneous measurement of wavelengths (frequencies).
2. Shorter measurement time at the same signal-to-noise (S/N) ratio.
3. Higher resolution, 0.001 cm^{-1} against 0.2 cm^{-1} for dispersive case.
4. Fast data processing.
5. Higher sensitivity.

Because of the above advantages, FTIR spectroscopy is preferred over others.

5.8 Role of Functional Groups in Vibrational Spectroscopy

We know that according to the rule 3N − 6 and 3N − 5 (for nonlinear and linear molecules), the number of vibrational modes to be present is governed by the number (N) of atoms in the molecule. We also know that molecules may have simple structures containing linear or branched chains such as

$$-C-O-C-.$$

Or certain associated structural units are

$$-CH_3, \ -NH_2, \ -OH, \ -C \equiv N, \ etc.$$

Accordingly, the vibrational spectrum of a molecule contains two important regions, the group frequency region and the fingerprint (or skeletal) region. Observations suggest that the molecules having similar groups show very similar spectra outside this region, but they show bands typical to the molecule in this region.

The frequencies under fingerprint or skeletal region are due to almost all atoms in the molecule and usually fall in the range 1500–600 cm^{-1}. On the other hand, the group frequencies are usually almost independent of the structure of the molecule as a whole (with a few exceptions) and fall in the ranges, either well above or well below that of the fingerprint region.

The mid-infrared (MIR) frequency region is very important from the point of view to study the nature of group frequencies and can be broadly divided into four regions:

– 4000–2500 cm^{-1}: X-H stretching region.
– 2500–2000 cm^{-1}: Triple bond region.
– 2000–1500 cm^{-1}: Double bond region.
– 1500–600 cm^{-1}: Finger print region.

The fundamental vibrations in the region (4000–2500 cm^{-1}) are generally due to O–H, CH, and N–H stretching. The next region (2500–2000 cm^{-1}) exhibits stretching vibrations of the groups with the triple bonds as well as the asymmetric vibrations of the groups with cumulated double bonds, such as X=Y=Z, where X, Y, and Z represent atoms of typical organic molecules such as C, O, N, S, and halogens. The next region (2000–1500 cm^{-1}) exhibits stretching vibrations of double-bonded X=Y groups such as C=C, C=O, and C=N. The region (1500–600 cm^{-1}) shows stretching vibrations of single-bond atoms of the second period.

Sometimes, a shift in the frequency of vibration is found as a change in the internal or external physical parameters. The two types of physical parameters affecting group frequencies are due to the change of

1. Internal physical parameters such as atomic mass, force constant, resonance field effect, and hydrogen bonding.
2. External physical parameters such as physical state of matter (gas, liquid, solid), solution concentration, and temperature.

In general, the group frequencies are found to approximately obey the expression

$$\bar{v} = \frac{1}{2\pi c}\sqrt{\frac{k}{\mu}}\ cm^{-1}$$

however, for example, an increasing mass tends to decrease the frequency of the series CH, CF, CCl, CBr, or the values of >C=O and >C=S. Similarly, an increasing force constant (the bond length) tends to increase the frequency of the series –C–X, –C=X, –C $\equiv$ X, where X is C, N, or in the first two cases O.

The phenomenon of resonance can also cause a shift in frequency. Let us take the case of a CO_2 molecule; the expected frequencies are v_1 (symmetric stretching) = 1330 cm^{-1}, v_2 (symmetric bending) = 667 cm^{-1}, and v_3 (asymmetric stretching) = 2349 cm^{-1}. However, in the Raman experiment, instead of 1330 cm^{-1}, two strong bands of 1285 cm^{-1} and 1380 cm^{-1} are observed. This is due to strong Fermi resonance between $2v_2 = 1334$ cm^{-1} and $v_1 = 1330$ cm^{-1}, which shifts one line to 1285 cm^{-1} and the other to 1380 cm^{-1}, where v_2 is Raman active.

Similarly, a change in physical state may cause a shift in the frequency of vibration. In general, $v_{gas} > v_{liquid}$ and $v_{liquid} > v_{solid}$. The frequency shift is more prominent in

Table 5.2 Functional groups and their approximate frequencies

Functional group	Approximate frequency (Range in cm^{-1})
– OH	3550–3650
– CH$_3$	2920–2990
	2855–2900
	1445–1475
	1365–1385
– CH$_2$–	2910–2945
	2835–2865
	1455–1485
=CH$_2$	3065–3095
	2960–2990
≡CH	3300–3325
	620–670
C–H (aromatic)	3020–3050
– C≡N	2210–2260
>C=S	1100–1200
– SH	2550–2600
– C≡C–	2190–2250
>C=C	1600–1650
C–F	1050–1150
C–Cl	700–750
C–Br	600–650
C–I	500–550
– NH$_2$	3200–3500
	1580–1675
– NH$_3$	3000–3200
	1550–1650
NH$_3^+$	3050–3150
	2600–3000
	1475–1600
>N–H	3300–3450

polar molecules such as HCl. Some of the important functional groups along with their frequency range are listed in Table 5.2. More extensive correlation charts can be found in the literature.

5.9 Nomenclature of Internal Modes of Vibration

By this time, we are a little bit familiar with the nomenclature of some modes of vibration, such as symmetric stretching, symmetric bending, and asymmetric stretching/bending in simpler molecules such as H_2O and CO_2. However, more

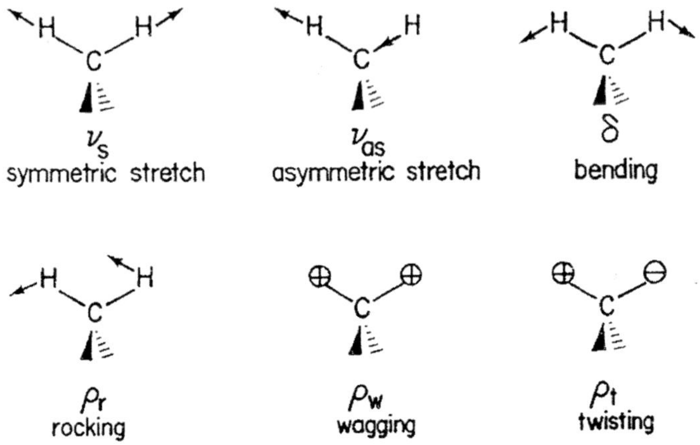

ν_s
symmetric stretch

ν_{as}
asymmetric stretch

δ
bending

ρ_r
rocking

ρ_w
wagging

ρ_t
twisting

Fig. 5.5 Possible modes of vibration

complex molecules require the nomenclature of a larger number of modes of vibration they possess. In describing them, the terms like deformation (δ), wagging (ρ_w), rocking (ρ_r), twisting (ρ_t), etc. are used. Some of them are shown in Fig. 5.5.

5.10 Theory of Raman Scattering

When a monochromatic light of energy $h\nu_0$ strikes a molecular sample (such as benzene), scattering of light takes place either elastically or inelastically. During elastic scattering (also called Rayleigh scattering which has the highest probability), the molecular sample scatters the same quantum of energy $h\nu_0$. On the other hand, during inelastic scattering (also called Raman scattering which has a much lower probability), the molecular sample either gains or loses energy equal to the energy difference between any two of its allowed energy states and gives a Raman spectrum. In the process, the molecule will scatter a quantum of energy, either

$h\nu_R = h\nu_0 - h\nu_s$, when the molecule gains energy and gives a Stokes line
or $h\nu_R = h\nu_0 + h\nu_s$, when the molecule loses energy and gives an anti-Stokes line
where $\nu_s = \frac{\Delta E}{h}$

The scattering processes exhibited by the molecular samples are shown in Fig. 5.6. The theory of Raman scattering can be understood with the help of both classical and quantum considerations.

Classical Theory

Let a molecular crystal as considered above be placed in an electric field E. Polarization of the charges in the sample will take place as shown in Fig. 5.7. Here, the

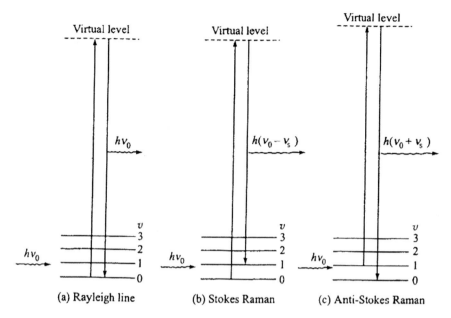

(a) Rayleigh line   (b) Stokes Raman   (c) Anti-Stokes Raman

Fig. 5.6 Different scattering processes exhibited by molecular samples

Fig. 5.7 Polarization of the charges in the samples

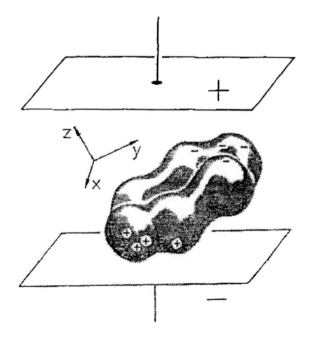

electric field in the x-direction not only induces a dipole in the x-direction but also in the y- and z-directions. The polarization P induced in the sample is proportional to the electric field. That is,

$$P = \alpha E \tag{5.8}$$

where α is a constant and is known as polarizability of the molecule, whose magnitude changes as the molecule oscillates. The polarizability is a tensor and is expressed as

$$\begin{pmatrix} P_x \\ P_y \\ P_z \end{pmatrix} = \begin{pmatrix} \alpha_{xx} & \alpha_{xy} & \alpha_{xz} \\ \alpha_{yx} & \alpha_{yy} & \alpha_{yz} \\ \alpha_{zx} & \alpha_{zy} & \alpha_{zz} \end{pmatrix} \begin{pmatrix} E_x \\ E_y \\ E_z \end{pmatrix} \tag{5.9}$$

where $\alpha_{ij} = \begin{pmatrix} \alpha_{xx} & \alpha_{xy} & \alpha_{xz} \\ \alpha_{yx} & \alpha_{yy} & \alpha_{yz} \\ \alpha_{zx} & \alpha_{zy} & \alpha_{zz} \end{pmatrix}$.

We know that a tensor is always centrosymmetric, therefore,

$$\alpha_{ij} = \alpha_{ji}$$

This gives us,

$$\alpha_{xy} = \alpha_{yx}, \alpha_{xz} = \alpha_{zx}, \; and \; \alpha_{yz} = \alpha_{zy}$$

Therefore, there are only six different components of the polarizability tensor available for any change during the vibration.

Now, suppose an electromagnetic radiation of frequency ν_s is allowed to fall on the molecular sample; each molecule of the sample experiences an oscillating electric field given by

$$E = E_0 \cos(2\pi\nu_0 t) \tag{5.10}$$

Further, let us consider the vibrational motion of the molecule and suppose q_k be the normal coordinate associated with the vibrational mode. In a harmonic approximation, q_k may be expressed as

$$q_k = q_0 \cos(2\pi\nu_s t) \tag{5.11}$$

Like dipole moment, the polarizability α can be expanded in a Taylor series as

$$\alpha = \alpha_0 + \left(\frac{\partial\alpha}{\partial q_k}\right)_0 q_k + \frac{1}{2}\left(\frac{\partial^2\alpha}{\partial q_k^2}\right)_0 q_k^2 + \cdots \tag{5.12}$$

Neglecting higher order terms and substituting Eqs. 5.10, 5.11, and 5.12 in Eq. 5.8, we obtain

$$P = \left[\alpha_0 + \left(\frac{\partial \alpha}{\partial q_k} \right)_0 q_k \cos(2\pi v_s t) \right] E_0 \cos(2\pi v_0 t)$$

$$= \alpha_0 E_0 \cos(2\pi v_0 t) + \left(\frac{\partial \alpha}{\partial q_k} \right)_0 q_k E_0 \cos(2\pi v_0 t) \cdot \cos(2\pi v_s t)$$

Now, using the trigonometric identity

$$2 \cos \theta \cdot \cos \varphi = \cos(\theta + \varphi) + \cos(\theta - \varphi)$$

the above equation can be written as

$$P = \alpha_0 E_0 \cos(2\pi v_0 t) + \frac{1}{2} \left(\frac{\partial \alpha}{\partial q_k} \right)_0 q_k E_0 [\cos 2\pi (v_0 + v_s)t + \cos 2\pi (v_0 - v_s)t]$$

$$(5.13)$$

From Eq. 5.13, it is clear that the induced polarization contains three distinct frequency components.

(i) $v = v_0$ Rayleigh line.
(ii) $v = v_0 - v_s$ Raman Stokes line.
(iii) $v = v_0 + v_s$ Raman anti-Stokes line.

It may further be seen from Eq. 5.13 that when $\left(\frac{\partial \alpha}{\partial q_k} \right) = 0$, no Raman lines will be observed. Therefore, we observe that the classical theory is able to describe three processes of scattering, but it does not say how they happen.

Quantum Theory

Applying quantum mechanics and writing in terms of polarization tensor, the transition moment for Raman activity of a fundamental (i.e., the transition $v = 0 \rightarrow v = 1$) may be expressed as

$$P(0, 1) = \int_{-\infty}^{\infty} \psi_0 \, \alpha E \, \psi_1 \, d\tau$$

$$= E \int_{-\infty}^{\infty} \psi_0 \alpha \, \psi_1 \, d\tau \qquad (5.14)$$

An integral of this type can be written for every component of α_{ij} to obtain the components P_{ij}. Therefore, if there is one out of six components for which $P_{ij} \neq 0$, then the entire moment will be non-vanishing [i.e., $P(0, 1) \neq 0$], and the transition will be Raman active. Similar to the case of infrared (including the symmetry-based selection rule), the integral in Eq. 5.14 will be nonzero only if it is totally symmetric. Similarly, as before ψ_0 is totally symmetric and ψ_1 has the symmetry of normal mode. Therefore, integral in Eq. 5.14 will be nonzero only if ψ_1 and any one component out of six α_{ij} have the same symmetry so that their product will be totally symmetric [as $P(0, 1) \neq 0$] and hence the transition will be Raman active.

Since the polarizability is represented by a tensor, we can see that the symmetry species of its components are the same as that of the binary (direct) products of vectors, such as $x^2, y^2, z^2, x^2 - y^2$, and $x^2 + y^2 + z^2$, whose transformation properties are listed in the fourth column of the character tables.

5.11 Raman Spectrometer

The first Raman spectrum of an organic compound was recorded using the Sun as a source, a telescope as a receiver, and the human eye as a detector. However, since then a lot of advancement in every aspect of the experimentation has taken place. The schematic diagram of the present-day Raman spectrometer is shown in Fig. 5.8. The main components of the experimental setup are source, sample device, monochromator, and detector.

For better detection of Raman scattering, one requires an intense source of small size. Typically UV, visible, and NIR emitting monochromatic laser sources are used. The most commonly used sources are He–Ne laser, Ar$^+$ laser, and diode-pumped solid state lasers or tunable lasers. In certain cases, red radiation is preferred to reduce the fluorescence and decomposition of the sample.

The monochromatic laser beam is made to focus through a lens into a small sample ($\sim 10^{-5}$ cm^3) equivalent to the region in which the beam is most concentrated (called the focal cylinder). This beam is again made to undergo multiple passes with the help of two concave mirrors (Fig. 5.9) to enhance the Raman signal by 8–10 times. Notch filters are used to filter the Raleigh line intensity before the scattered light enters the

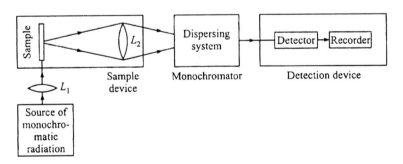

Fig. 5.8 Schematic diagram of the present-day Raman spectrometer

Fig. 5.9 Mirror-lens
assembly to enhance Raman
signal

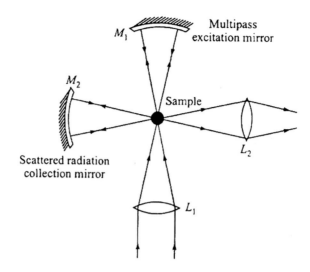

dispersing system. In case a laser source produces fluorescence in the sample, simply
switch to another laser source.

   Radiation scattered from the sample is directed to a monochromator, which
disperses various wavelengths in slightly different directions and only a narrow
band of wavenumbers is allowed to reach the detector at a time. The grating-based
monochromator is often used.

5.12 Fourier Transform (FT) Raman Spectrometer

In order to minimize fluorescence, NIR laser is used which has insufficient energy
to excite electronic transitions. But the In, GaAs, or Ge detectors used in this (NIR)
region have high background noise level and low sensitivity. However, with the application of FT methods (as discussed in the IR case), not only the lost sensitivity is
restored but also the signal-to-noise ratio can be improved. Thus, the FT Raman spectrometer uses NIR lasers for excitation and an interferometer-based system followed
by an FT program to produce the required spectrum. Its added advantage is that,
once the optical path of the spectrometer is properly aligned, it becomes simple to
interchange sources, beam splitters, and detectors so that it can be used as either IR or
Raman spectrometer. A schematic representation of the same is shown in Fig. 5.10.

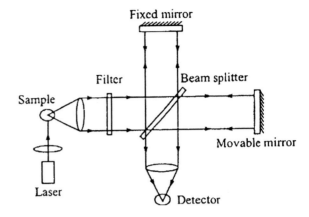

Fig. 5.10 Schematic representation of FT Raman spectrometer

5.13   Symmetry Based on Some Useful General Conclusions

Examining the character tables obtained in Chap. 3 on the basis of transformation properties/basis functions such as symmetries of dipole components (equivalent to unit vector transformations x, y, and z) and the symmetries of polarizability components (equivalent to the binary direct product transformation properties) or alternatively the symmetry properties of atomic orbitals (s, p, d, f), we can draw some important general conclusions related to the spectroscopic activity of the normal modes in IR and Raman spectra.

1. Since the spectroscopic activity is associated with the transformation properties of the three unit vectors x, y, and z, the IR active modes can be distributed at the most three symmetry species. On the other hand, the direct products (the Raman active modes) normally span two to four symmetry species, depending on the point group. For example, the distribution of basis functions in the character table for the point group C_{2v} is shown below.

C_{2v}	Transformation	Properties
	IR active	Raman active
A_1	z	x^2, y^2, z^2
A_2		xy
B_1	x	xz
B_2	y	yz

2. Sometimes, it is found that neither the unit vector nor the direct product transformation spans all species in some point groups. As a result of this, some irreducible representations are neither associated with unit vector nor direct product transformation. Still, the molecule may have a normal mode that transforms due to one of these species. This implies that it is possible to have normal modes that

cannot be observed as a fundamental in either IR or Raman spectra. These spectroscopically inactive modes are known as silent modes. For example, consider XY_4 square planar molecule whose point group is D_{4h}. From the unit vector and the direct product transformations, we obtain the following activities:

A_{1g}	B_{1g}	B_{2g}	A_{2u}	B_{2u}	$2E_u$
ν_1	ν_2	ν_4	ν_3	ν_5	ν_6, ν_7
Raman (Pol)	Raman	Raman	IR	–	IR

It is to be noted that the normal mode B_{2u}, which might be described as deformation of the molecular plane, is not active in either spectrum. This shows that B_{2u} is a silent mode.

3. The totally symmetric representation in every point group is associated with one or more direct product transformations. Therefore, the totally symmetric normal modes will always be Raman active. Further, they may or may not be IR active, depending on whether the point group is non-centrosymmetric or symmetric. For example, consider symmetric stretching mode for H_2O and CO_2 molecules whose point groups, respectively, are non-centrosymmetric C_{2v} and centrosymmetric $D_{\infty h}$. We observe that the symmetric stretching mode is Raman active in both but IR active in H_2O and Raman active in CO_2 (it is being centrosymmetric).

4. In centrosymmetric point groups, the vectors x, y, and z transform as ungerade species but the direct products transform as gerade species. Therefore, Raman active modes will be IR inactive, and vice versa, for centrosymmetric molecules. When a molecule has a center of symmetry, those vibrations which are centrosymmetric w.r.t. inversion symmetry cannot produce a change in the polarizability. For example, let us take the case of a CO_2 molecule to check the behavior of its normal modes.

Mode of vibration	Raman	IR
Symmetric stretching	Active	Inactive
Symmetric bending	Inactive	Active
Asymmetric stretching	Inactive	Active

The information contained in the table suggests that the Raman and IR spectra of centrosymmetric molecules have no fundamental frequencies in common. It shows that Raman and IR measurements complement each other and the complete picture of the problem is clear only when both the techniques are used to study the specimen. This requirement is known as the rule of mutual exclusion.

5.14 Determination of Molecular Structures Using IR and Raman Results

In the previous chapter, we observed that the knowledge of molecular symmetry is helpful in determining the number of normal modes of vibration and the corresponding spectroscopic activities possessed by the molecule. Conversely, it can be said that the IR and Raman results are equally helpful in distinguishing between two or more possible structures that a particular molecule can have. Here, we are going to discuss this aspect by taking different cases of simple molecular systems.

Case I: Diatomic Molecules

Since a diatomic molecule is always linear, the number of fundamentals in such a molecule is given by

$$\Gamma_{vib} = 3N - 5 = 1, \text{ where } N = 2$$

This implies that for a diatomic molecule, there is only one mode of vibration. This is nothing but the symmetric stretching mode which is IR active if the molecule is hetero-nuclear and IR inactive if it is homo-nuclear. On the other hand, if it is allowed as a polarized line, it is Raman active.

Case II: Triatomic Molecules of XY_2 (or AB_2) Type

For triatomic molecules, we know that there are three possibilities. They are (i) Linear symmetric, (ii) Bent symmetric, and (iii) Linear asymmetric.

For linear symmetric and linear asymmetric molecules, the number of fundamentals is given by

$$\Gamma_{vib} = 3N - 5 = 4, \text{ where } N = 3$$

and for bent symmetric molecule, we have

$$\Gamma_{vib} = 3N - 6 = 3, \text{ where } N = 3$$

Let us discuss the three cases separately.

Linear Symmetric Molecule

Because of the presence of center of symmetry, a linear symmetric molecule will obey the rule of mutual exclusion. Let us consider a CO_2 molecule as an example under this category. Only one fundamental is observed in Raman as a polarized band at 1343 cm^{-1}. The IR spectrum shows two fundamentals at 667 and 2349 cm^{-1} (with P, R contour). A CS_2 molecule shows similar behavior as CO_2. In such molecules, it is observed that a Raman active vibration is IR inactive and vice versa.

Bent Symmetric Molecule

In the category of bent symmetric molecules, let us consider the SO_2 molecule as an example. In this category, all the three distinct modes are active in both IR and Raman. However in Raman, the symmetric stretching and bending vibrations are found to be polarized, whereas the asymmetric vibration is depolarized.

Linear Asymmetric Molecule

Like the bent symmetric case, in the linear asymmetric molecules all the three distinct modes are active both in IR and Raman. However, nonlinear asymmetric molecules do not give rise to the IR bands with P, R contours. The number of distinct fundamentals along with their activities in the three categories of molecules is listed in Table 5.3.

Now, let us consider the N_2O molecule as an example under this category. Since N_2O and CO_2 are isoelectronic and CO_2 is linear symmetric, it is expected that N_2O is also a linear symmetric molecule. There are certain experimental observations (according to which the magnetic dipole moment is approximately zero and the two bond lengths are almost equal) which suggest this point of view. Further, the observation of P, R contours in N_2O also indicates its structure is linear but the failure to observe the mutual exclusion principle in it suggests the N–N-O structure. The details of the observed IR and Raman spectra of N_2O molecule are listed in Table 5.4.

Table 5.3 Number of fundamentals and their activities in triatomic molecules

Molecular shape	No. of IR active fundamentals	No. of Raman active fundamentals	No. of IR and Raman coincidences	No. of polarized Raman lines
Linear (sym. CO_2) Y—X—Y	2	1	0	1
Bent (sym. SO_2) X / Y Y	3	3	3	2
Linear (asym. N_2O) Y—Y—X	3	3	3	2

Table 5.4 IR and Raman spectra of N_2O molecule

Frequency $\bar{v}$ (cm^{-1})	Infrared (IR)	Raman	Mode of vibration
1285	Very strong, PR contour	Very strong, polarized	Symmetric stretching
589	Strong, PQR contour	–	Symmetric bending
2224	Very strong, PR contour	Strong, depolarized	Asymmetric stretching

Table 5.5 Some IR and Raman spectral data for planar and pyramidal four-atom molecules

Molecular shape	No. of distinct fundamentals	No. of IR active fundamentals	No. of Raman active fundamentals	No. of IR and Raman coincidences	No. of Polarized Raman lines
Planar (D_{3h})	4	3	3	2	1
Pyramidal (D_{3h})	4	4	4	4	2

Case III: Four Atomic Molecules of XY_3 (or AB_3) Type

Four atomic molecules of XY_3 (or AB_3) type can exist in two simpler forms. They are (i) Planar with point group D_{3h} and (ii) Pyramidal with point group C_{3v}. The number of fundamentals in them can be given by

$$\Gamma_{vib} = 3N - 6 = 6, \text{ where } N = 4$$

However, one stretching mode in symmetric planar and one angle deformation mode in symmetric pyramidal in each is doubly degenerate. Accordingly, the number of distinct fundamentals in both of them is 4, i.e., only four different fundamental frequencies are observed in each case. Table 5.5 provides some IR and Raman spectral data for the two shapes of molecules.

In fact, these two molecular shapes are like symmetrical tops w.r.t. the threefold (principal) axis passing through the atom X (or A) perpendicular to the Y_3 (or B_3) plane. All the vibrations are accordingly described as either $\parallel$ or $\perp$.

The symmetric modes of vibration are parallel and Raman polarized while the asymmetric modes are perpendicular and depolarized. Further, all the vibrations of a pyramidal molecule change both the dipole moment and the polarizability; hence, they are both IR and Raman active. On the other hand, in a planar molecule, the symmetric stretching mode leaves the dipole moment unchanged (it remains zero throughout) and so is IR active, while the symmetric bending modes do not change the polarizability and so is Raman inactive. Therefore, the overall pattern of the spectra is as follows:

Planar AB_3: It contains one vibration of Raman active only, one vibration of IR active only, and two vibrations of both IR and Raman active.

Pyramidal AB_3: All four vibrations are both IR and Raman active.

Non-symmetric AB_3: It possibly contains more than four different fundamental frequencies.

Table 5.6 IR and Raman spectral data of NO_3—(planar) ion and PCl_3 (pyramidal) molecules

NO₃—(planar D₃ₕ)		PCl₃ (pyramidal C₃ᵥ)		Mode of vibration
IR (cm⁻¹)	Raman (cm⁻¹)	IR (cm⁻¹)	Raman (cm⁻¹)	
–	1049 (polarized)	504 (∥)	514 (polarized)	Symmetric stretching
830 (∥)	–	252 (∥)	256 (polarized)	Symmetric deformation
1350 (⊥)	1355 (depolarized)	482 (⊥)	482 (depolarized)	Asymmetric stretching
680 (⊥)	690 (depolarized)	198 (⊥)	184 (depolarized)	Asymmetric deformation

Keeping the above-mentioned overall pattern in mind, let us consider the IR and Raman spectral data of NO_3—(planar) ion and PCl_3 (pyramidal) molecule as summarized in Table 5.6.

Under this category, the ClF_3 molecule is of special interest because its spectral data suggests that this molecule is neither symmetric planar nor pyramidal. Further, the IR and Raman data alone seems to be insufficient to give its complete structure. However, the use of microwave spectroscopy suggests that the molecule is T- shaped with bond angles of nearly 90°.

Case IV: Five Atomic Molecules of XY_4 (or AB_4) Type

Five atomic molecules of XY_4 (or AB_4) type generally exist in two simpler forms. They are (i) Square Planar form with point group D_{4h} and (ii) Tetrahedral form with point group T_d. The number of fundamentals in them can be given by

$$\Gamma_{vib} = 3N - 6 = 9, \text{ where } N = 5$$

However, some of them may become degenerate according to their symmetry. Taking the symmetry into account, the number of distinct fundamentals for the two forms reduces to seven and four, respectively. The analysis of IR and Raman spectral data suggests that many molecules acquire the two forms. The molecules such as $PtCl_4$, $PtBr_4$, $PdCl_4$, and XeF_4 acquire the square planar form, while the molecules such as CH_4, SiH_4, GeH_4, and $GeCl_4$ acquire the tetrahedral form. The vibrational spectroscopy came as a great help in resolving the square planar-tetrahedral anomaly for the XeF_4 molecule. The IR and Raman spectral data related to XY_4 (or AB_4) molecules are summarized in Table 5.7.

Researchers found it difficult to solve the structure of the SF_4 molecule under this category because of some conflicting results obtained from the IR and Raman studies. Ambiguity regarding the structure persisted for quite some time. This molecule actually showed the difficulties faced while using the vibrational spectroscopic method in determining the molecular structure. Fermi resonance, weak intensities, instrumental limitations, presence of overtone bands, and combination bands are some of the factors that create difficulties in fixing the fundamentals. In spite of all these, the procedures based on symmetry, vibrational selection rules, polarization of the observed lines, etc., together with the identification of molecular constituents based

Table 5.7 IR and Raman spectral data related to XY_4 (or AB_4) molecules

Molecular shape	No. of distinct fundamentals	No. of IR active fundamentals	No. of Raman active fundamentals	No. of IR and Raman coincidences	No. of polarized Raman lines
Square planar (D_{4h})	7	3	3	0	1
Tetrahedral (T_d)	4	2	4	2	1

on the group frequency concept (the presence of a functional group in a particular frequency range) make the IR and Raman spectroscopy a powerful tool for the molecular structure determination.

5.15 Correlation Between Super Group-Subgroup Species

Whenever there is a change in the molecular structure/crystal structure, there is a change in the associated symmetries (point group) with them. The change of symmetry has a direct effect on the molecular properties/physical properties of the specimen.

Symmetries may either be added to a group to get its super group (a point group of higher order) or subtracted from the group to obtain its subgroup (a point group of lower order). In either case, there is a direct or indirect super group-subgroup relationship between the old and new point groups. In the first process, there occurs an ascent (increase) in symmetry, while in the second there is a descent (decrease) in symmetry.

Earlier in Sect. 2.2, we studied the general form of the super group-subgroup relationships (Fig. 2.3) while dealing with the crystal systems and their associated symmetries. However, in this section we will relook the super group-subgroup relationships by keeping in view the molecular aspect of a given specimen. Here, our main thrust will be to formulate certain empirical rules which can provide us with the correlation between the species (irreducible representations) of the initial and final point groups in terms of characters of various symmetry operations shared by them.

Based on the character tables obtained in Chap. 3 and related transformation properties, we can formulate the following empirical rules:

1. According to super group-subgroup diagram (also called correlation diagram), it is clear that the order of the super group must be an integral multiple of the order of the subgroup. Conversely, the order of the subgroup must be an integer divisor of the order of the super group.
2. During molecular structural change if the point groups of the initial and final structures maintain a super group-subgroup relationship, then their irreducible representations will also exhibit some definite relationships, and they are said to be correlated.
3. If the final point group is a subgroup (due to descent in symmetry) of the old one, then the degeneracies in certain representations that existed earlier may become distinguishable in the new point group (structure). Conversely, if the final point group is a super group (due to ascent in symmetry) of the old one, then certain distinguishable properties in the old structure may become degenerate in the new configuration.
4. In many cases, degenerate representations of a super group correlate to two or three irreducible representations of the subgroup. A doubly degenerate representation splits into two non-degenerate representations and a triply degenerate representation into either three non-degenerate representations or a combination of one degenerate representation and one non-degenerate representation. However, the degeneracy is completely removed when the subgroup contains rotational symmetry of less than threefold.
5. The physical significance of correlations is a property that transforms as one representation in the super group, and will transform as its correlated representation in the subgroup.
6. Between a super group and any one of its subgroups in the chain, representations arising from the same vectors will have the same character $\chi(R)$ values for all symmetry operations that are shared with the subgroup.
7. Often, two or more bases of separate representations of a subgroup yield the same set of character $\chi(R)$ values for all the symmetry operations that are shared with the subgroup.

For example, let us consider an octahedral molecule MA_6 whose point group is O_h (m3m). When some of the A atoms are replaced by B and C atoms, four different molecular structures with different point groups are obtained as shown in Fig. 5.11. All five point groups and their corresponding symmetry operations are summarized in Table 5.8.

Comparing Fig. 5.11 with Fig. 2.3 (which exhibits the super group-subgroup relationships), we observe that the structures I–IV are related to each other as successive subgroups of one another, and they belong to one out of many possible super group-subgroup hierarchy channels. Structure V is also a subgroup of structure I but belongs to a different channel where the structure (whose point group is D_{3d}) between the structures I and V is missing. Since structure V belongs to a different channel, it

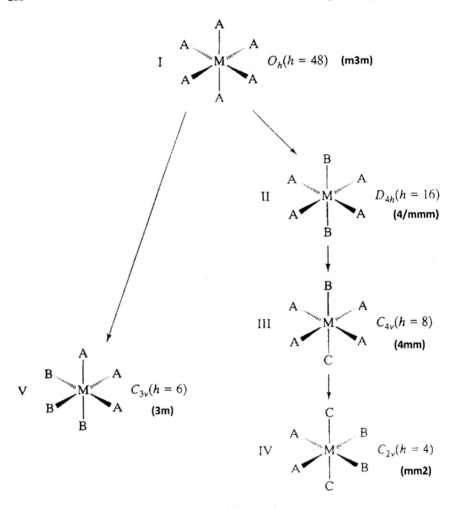

Fig. 5.11 Different molecular structures show different point groups

Table 5.8 Some point groups and their symmetry operations

Structure no.	Point group	Symmetry operations
I	O_h	E, $8C_3$, $6C_2$, $6C_4$, $3C_2$ $\left(=C_4^2\right)$, i, $6S_4$, $8S_6$, $3\sigma_h$, $6\sigma_d$
II	D_{4h}	E, $2C_4$, C_2, $2C_2'$, $2C_2''$, i, $2S_4$, σ_h, $2\sigma_v$, $2\sigma_d$
III	C_{4v}	E, $2C_4$, C_2, $2\sigma_v$, $2\sigma_d$
IV	C_{2v}	E, C_2, $\sigma_v(xz)$, $\sigma_v'(yz)$
V	C_{3v}	E, $2C_3$, $3\sigma_v$

Table 5.9 Character table of D_{4h} showing characters for operations shared with C_{4v}

D_{4h}	E	$2C_4$	C_2	$2C_2'$	$2C_2''$	i	$2S_4$	σ_h	$2\sigma_v$	$2\sigma_d$	C_{4v}
A_{1g}	1	1	1						1	1	A_1
A_{2g}	1	1	1						-1	-1	A_2
B_{1g}	1	-1	1						1	-1	B_1
B_{2g}	1	-1	1						-1	1	B_2
E_g	2	0	-2						0	0	E
A_{1u}	1	1	1						-1	-1	A_2
A_{2u}	1	1	1						1	1	A_1
B_{1u}	1	-1	1						-1	1	B_2
B_{2u}	1	-1	1						1	-1	B_1
E_u	2	0	-2						0	0	E

cannot have a group-subgroup relationship with any point group of the other channel (structures II–IV).

When the point groups have super group-subgroup relationships (as observed among the structures I–IV, and I and V), their irreducible representations are also related. This can be seen in the character table of D_{4h} and its subgroup C_{4v}, where only the characters of shared operations are considered as shown in Table 5.9.

We observe that the characters for the shared operations corresponding to the two sets of irreducible representations are correlated. Accordingly, the 10 irreducible representations of D_{4h}, correlate to 5 irreducible representations of C_{4v}, i.e., two representations of D_{4h} together correlate to one of C_{4v}. The complete correlation process between the point group D_{4h} and its subgroup C_{4v} is shown in Fig. 5.12 (called the correlation diagram). The whole problem can also be seen from the subgroup-super group point of view in an inverse way.

5.16 Summary

1. The infrared (IR) spectral region lies between the visible and the microwave regions of the electromagnetic spectrum. This region of the spectrum is particularly more important from chemists' point of view for their studies on molecular vibrations. Two very important techniques involving IR and Raman spectra are used to characterize the sample materials by all those working in the field of materials science.

2. A polychromatic IR light is allowed to pass through a molecular sample, and the infrared spectrum is usually recorded by measuring the transmittance of light quanta. The frequencies of the absorption bands v_s are found to be proportional to the energy difference ($\Delta E = hv_s$) between vibrational ground and excited states as shown in Fig. 5.1. The absorption bands due to vibrational

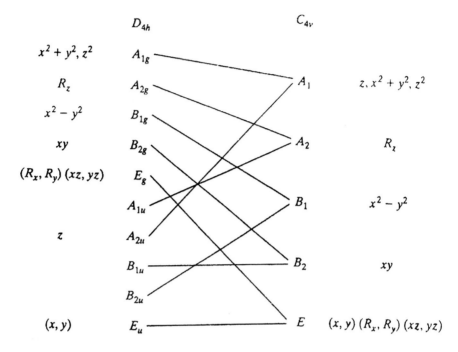

Fig. 5.12 Correlation diagram showing relationships between the species of D_{4h} and C_{4v}

transitions lie in the wavelength region $\lambda = 2.5$ to 1000×10^{-6} m (approximately equivalent to 4000 to 10 cm^{-1} in terms of wavenumbers, where $\bar{\nu} = \frac{1}{\lambda}$). Here, we observe that the sample merely absorbs certain frequencies (energies) of infrared light.

3.   Infrared (IR) spectroscopy is one of the most important and most widely used techniques to study vibrational spectra of a solid, liquid, or gas sample. A polychromatic IR light is allowed to pass through both, the sample and the reference cell, and the transmitted intensities are measured at each frequency (or wavenumber) through a detector/recorder. The result is a characteristic spectrum showing the transmittance (or absorbance) I_s/I_r of the electromagnetic radiation as a function of wavelength (or wavenumber).

4.   FTIR spectroscopy was first of all developed by Astronomers in the early fifties of the nineteenth century to study the infrared spectra of distant stars. However, now it has been developed into a very powerful technique to detect signals not only in the MIR region (4000–400 cm^{-1}) but also in the FIR region (400–10 cm^{-1}) of the electromagnetic spectrum. This technique can provide simultaneous and instantaneous recording of the whole spectrum in all the regions from the NIR region (12,800–4000 cm^{-1}) to the FIR region (400–10 cm^{-1}) and hence is preferred over the conventional IR technique.

5.   The nomenclature of modes of vibration for simple molecules such as H_2O and CO_2 is simple, such as symmetric stretching, symmetric bending, and

asymmetric stretching/bending. However, more complex molecules require the nomenclature of a larger number of modes of vibration they possess. In describing them, the terms like deformation (δ), wagging (ρ_w), rocking (ρ_r), twisting (ρ_t), etc. are used. Some of them are shown in Fig. 5.5.

6. When a monochromatic light of energy $h\nu_0$ strikes a molecular sample, scattering takes place either elastically or inelastically. During elastic scattering (also called Rayleigh scattering), the molecular sample scatters the same quantum of energy $h\nu_0$. On the other hand, during inelastic scattering (also called Raman scattering), the molecular sample either gains or loses energy equal to the energy difference between any two of its allowed energy states and gives a Raman spectrum. In the process, the molecule will scatter a quantum of energy, either

$$h\nu_R = h\nu_0 - h\nu_s,$$ when the molecule gains energy and gives a Stokes line
or $$h\nu_R = h\nu_0 + h\nu_s,$$ when the molecule loses energy and gives an anti-Stokes line
$$\text{where } \nu_s = \frac{\Delta E}{h}$$

The scattering processes exhibited by the molecular sample are shown in Fig. 5.6. The theory of Raman scattering can be understood with the help of both classical and quantum considerations.

7. The first Raman spectrum of an organic compound was recorded using the Sun as a source, a telescope as a receiver, and the human eye as a detector. However, since then a lot of advancement in every aspect of the experimentation has taken place. The schematic diagram of the present-day Raman spectrometer is shown in Fig. 5.8. The main components of the experimental setup are source, sample device, monochromator, and detector.

8. The application of FT methods in the Raman technique not only helps restore the lost sensitivity but also the signal-to-noise ratio can be improved. Thus, the FT Raman spectrometer uses NIR lasers for excitation and an interferometer-based system followed by an FT program to produce the required spectrum. Its added advantage is that, once the optical path of the spectrometer is properly aligned, it becomes simple to interchange sources, beam splitters, and detectors so that it can be used as either IR or Raman spectrometer.

9. Examining the character tables obtained in Chap. 3 on the basis of transformation properties/basis functions such as symmetries of dipole components (equivalent to unit vector transformations x, y, and z) and the symmetries of polarizability components (equivalent to the binary direct product transformation properties) or alternatively the symmetry properties of atomic orbitals (s, p, d, f), we can draw some important general conclusions related to the spectroscopic activity of the normal modes in IR and Raman spectra.

10. In the previous chapter, we observed that the knowledge of molecular symmetry is helpful in determining the number of normal modes of vibration and the corresponding spectroscopic activities possessed by the molecule. Conversely, it can be said that the IR and Raman results are equally helpful

in distinguishing between two or more possible structures that a particular molecule can have.

Bibliography

1. Raman, K.V.: Group Theory and Its Applications to Chemistry. Tata McGraw-Hill Company Limited, New Delhi (1996)
2. Robert, L.: Carter, Molecular Symmetry and Group Theory. Wiley, New York (1988)
3. Harris, D.C., Bertolucci, M.D.: Symmetry and Spectroscopy. Oxford University Press, Inc. (printed in USA) (1978)
4. Turrell, G.: Infrared and Raman spectra of Crystals. Academic Press, London and New York (1972)
5. Chandra, S.: Molecular Spectroscopy. Narosa Publishing House Pvt. Ltd., New Delhi (2011)
6. Banwell, C.N., McCash, E.M.: Fundamentals of Molecular Spectroscopy, 4th edn. Tata McGraw-Hill Company Limited, New Delhi (1997)
7. Wahab, M.A.: Essentials of Crystallography, 2nd edn. Narosa Publishing House Pvt. Ltd., New Delhi (2011)
8. Wahab, M.A., Wahab, K.M.: Resolution of ambiguities and the discovery of two new space lattices. ISST J. Appl. Phys. **6**(1), 1 (2015)
9. Wahab, M.A.: The mirror: mother of all symmetries in crystals. Adv. Sci. Eng. Med. **12**, 289–313 (2020)

M. A. Wahab, *Symmetry Representations of Molecular Vibrations*, Springer Series
in Chemical Physics 126, https://doi.org/10.1007/978-981-19-2802-4

Index

© The Editor(s) (if applicable) and The Author(s), under exclusive license to Springer Nature Singapore Pte Ltd. 2022
M. A. Wahab, *Symmetry Representations of Molecular Vibrations*, Springer Series in Chemical Physics 126, https://doi.org/10.1007/978-981-19-2802-4

Printed in the United States
by Baker & Taylor Publisher Services